Learning in the Anthropocene

To Eve!
In solidarity!
Carl

Environment and Society

Series Editor: Douglas Vakoch

As scholars examine the environmental challenges facing humanity, they increasingly recognize that solutions require a focus on the human causes and consequences of these threats, and not merely a focus on the scientific and technical issues. To meet this need, the Environment and Society series explores a broad range of topics in environmental studies from the perspectives of the social sciences and humanities. Books in this series help the reader understand contemporary environmental concerns, while offering concrete steps to address these problems.

Books in this series include both monographs and edited volumes that are grounded in the realities of ecological issues identified by the natural sciences. Our authors and contributors come from disciplines including but not limited to anthropology, architecture, area studies, communication studies, economics, ethics, gender studies, geography, history, law, pedagogy, philosophy, political science, psychology, religious studies, sociology, and theology. To foster a constructive dialogue between these researchers and environmental scientists, the Environment and Society series publishes work that is relevant to those engaged in environmental studies, while also being of interest to scholars from the author's primary discipline.

Recent Titles in the Series

Learning in the Anthropocene: Reimagining Education in the Twenty-First Century, by Carl A. Maida

The Social Life of Unsustainable Mass Consumption, by Magnus Boström

Everyday Life Ecologies: Sustainability, Crisis, Resistance, by Alice Dal Gobbo

Environmental Legacies of the Copernican Universe, by Jean-Marie Kauth

Anticipatory Environmental (Hi)Stories from Antiquity to the Anthropocene, edited by Christopher Schliephake and Evi Zemanek

Mapping the Environmental Humanities: The Emerging Role of GIS in Ecocriticism, edited by Mark Terry and Michael G. Hewson

The Bangladesh Environmental Humanities Reader: Environmental Justice, Developmental Victimhood, and Resistance, by Samina Luthfa, Mohammad Tanzimuddin Khan, and Munasir Kamal

Loren Eiseley's Writing across the Nature and Culture Divide, by Qianqian Cheng

The Saving Grace of America's Green Jeremiad, by John Gatta

Art and Nuclear Power: The Role of Culture in the Environmental Debate, by Anna Volkmar

Contesting Extinctions: Decolonial and Regenerative Futures, edited by Luis I. Prádanos, Ilaria Tabusso Marcyan, Suzanne McCullagh, and Catherine Wagner

Embodied Memories, Embedded Healing: New Ecological Perspectives from East Asia, edited by Xinmin Liu and Peter I-min Huang

Ecomobilities: Driving the Anthropocene in Popular Cinema, by Michael W. Pesses

Global Capitalism and Climate Change: The Need for an Alternative World System, Second Edition, by Hans A. Baer

Learning in the Anthropocene

Reimagining Education in the Twenty-First Century

Carl A. Maida

LEXINGTON BOOKS
Lanham • Boulder • New York • London

Published by Lexington Books
An imprint of The Rowman & Littlefield Publishing Group, Inc.
4501 Forbes Boulevard, Suite 200, Lanham, Maryland 20706
www.rowman.com

86-90 Paul Street, London EC2A 4NE

British Library Cataloguing in Publication Information Available

Library of Congress Cataloging-in-Publication Data

Names: Maida, Carl A., author.
Title: Learning in the anthropocene : reimagining education in the twenty-first century / Carl A. Maida.
Description: Lanham, Maryland : Lexington Books, [2024] | Series: Environment and society | Includes bibliographical references and index. | Summary: "This book reimagines the education of future generations in our complex society. The author argues that two provinces-the school and society-can join together to afford students greater freedom to produce future knowledge as humanity faces profound challenges to its existence by advancing experiential instructional approaches"— Provided by publisher.
Identifiers: LCCN 2023042008 (print) | LCCN 2023042009 (ebook) | ISBN 9781666924688 (cloth) | ISBN 9781666924695 (epub)
Subjects: LCSH: Environmental education—Research. | Climatic changes. | Environmental policy. | Community and school. | Educational change.
Classification: LCC GE70 .M34 2024 (print) | LCC GE70 (ebook) | DDC 333.7071—dc23/eng20230927
LC record available at https://lccn.loc.gov/2023042008
LC ebook record available at https://lccn.loc.gov/2023042009

∞™ The paper used in this publication meets the minimum requirements of American National Standard for Information Sciences—Permanence of Paper for Printed Library Materials, ANSI/NISO Z39.48-1992.

To the Memory of
Paul Heckman

Contents

Acknowledgments

The content and organization of the book, as with my two previous books, were inspired by the curricular frameworks of the Association of American Colleges & Universities, the Association for the Advancement of Sustainability in Higher Education, and the *New Liberal Arts Program* of the Alfred P. Sloan Foundation, with the shared goal of integrating knowledge and skills across disciplinary areas. I have used their models of liberal education to develop interdisciplinary curricula for undergraduate and graduate students in the liberal arts, and also the environmental, health, and life sciences, with core concepts derived from the humanities, the social sciences, the arts, and science, technology and society studies.

Grants and fellowships to support the work discussed in this book have come from Ford Foundation, Robert Wood Johnson Foundation, Howard Hughes Medical Institute, The California Endowment, Anthony and Jeanne Pritzker Family Foundation, Sally and Dick Roberts Coyote Foundation, University of California Office of the President, National Institute of Mental Health, National Science Foundation, and United States Environmental Protection Agency.

I am grateful to the UCLA Institute of the Environment and Sustainability for the opportunity to develop and teach courses in action research on sustainability for students in the College of Letters and Science.

I have benefited from the resources of the Louise M. Darling Biomedical Library at the University of California, Los Angeles, and the University Library at California State University, Northridge.

The historical and ethnographic research studies described in the book were previously reported in technical sessions at annual meetings of the American Anthropological Association and the American Association for the Advancement of Science.

The shaping of this book has benefited from the support and comments of many people. I have appreciated the critical advice of Richard Adler, Jorge Alvarez, Roberto Belloso, Bonny Bentzin, John Briere, Gayle Byock, Robert Chianese, Joe Cicero, Tim Dagodag, Lou Di Giacomo, Robert Emerson, Tom Fehrenbacher, Linda Fishman, Michael Frondelli, Krista Harper, Javier Iribarren, Nurit Katz, Mary Keipp, Llon King, Cheryl Lanktree, Nan Levine-Mann, Brian McKenna, Marvin Marcus, Ken Mazey, Cully Nordby, Erik Osugi, Dianne Philibosian, Rob Riordan, Dick Roberts, Bill Roller, Liseth Romero-Martinez, Donald Quintana, Larry Rosenstock, Johnnie Savoy, Mark Stringer, Steven Stumpf, Sharon Szmolyan, Chuck Thegze, John Tipre, and Yong Zhao.

The late Paul E. Heckman, to whom the book is dedicated, brought warmth, wisdom, and inspiration as a colleague and mentor in educational research about how youth learn, school culture, and the community. Sam Beck has been a source of inspiration and critical insight for over two decades, when I engaged in the fieldwork and historical study leading to the writing of this book. Marlene Grossman introduced me to advocacy on behalf of preserving the aesthetic qualities of the city and its diverse landscapes. Bruce Woych guided my thinking on the book's argument and organization, read countless drafts, and provided extensive revisions for each chapter. I am grateful to Courtney Morales at Lexington Books for her encouragement and support of this book project. I trust that my daughters, Vanessa and Alexandra, and their peers in the millennial generation, may appreciate how the forces and ideas discussed in the book have shaped their lives and times. I am again especially grateful to Christian Kilpatrick for her steadfast efforts to provide a warm and facilitating home environment to support both the writer and the task of writing.

All previously published materials have been substantially rewritten and reorganized for the purposes of this book.

Earlier versions of some sections of chapter 2 were published in *Common Worlds: Paths Toward Urban Sustainability* (Lanham, MD: Rowman and Littlefield, 2019).

Earlier versions of some sections of chapter 3 were published in *Common Worlds: Paths Toward Urban Sustainability* (Lanham, MD: Rowman and Littlefield, 2019); and under the titles, "Introduction," *Public Anthropology in a Borderless World*, edited by Sam Beck and Carl A. Maida (New York and Oxford: Berghahn Books, 2015); "Project-Based Learning: A Critical Pedagogy for the 21st Century," *Policy Futures in Education* 9:6, 2011.

Earlier versions of some sections of chapter 4 were published under the title, "Project-Based Learning: A Critical Pedagogy for the 21st Century," *Policy Futures in Education* 9:6, 2011.

Earlier versions of some sections of chapter 5 were published under the title, "Knowing Sustainability: Building Communities of Practice in Urban Ecology at High Tech High," in *Global Sustainability and Communities of Practice*, edited by Carl A. Maida and Sam Beck (New York and Oxford: Berghahn Books, 2018).

Earlier versions of some sections of chapter 6 were published in *Pathways Through Crisis: Urban Risk and Public Culture* (Lanham, MD: Rowman and Littlefield, 2008); *Common Worlds: Paths Toward Urban Sustainability* (Lanham, MD: Rowman and Littlefield, 2019).

Earlier versions of some sections of chapter 7 were published under the title, "Expert and Lay Knowledge in Pacoima: Public Anthropology and an Essential Tension in Community-Based Participatory Action Research," *Anthropology in Action* 16:2, 14–26, 2009.

Earlier versions of some sections of chapter 8 were published in *Common Worlds: Paths Toward Urban Sustainability* (Lanham, MD: Rowman and Littlefield, 2019).

Earlier versions of some sections of chapter 9 were published in *Pathways Through Crisis: Urban Risk and Public Culture* (Lanham, MD: Rowman and Littlefield, 2008); *Common Worlds: Paths Toward Urban Sustainability* (Lanham, MD: Rowman and Littlefield, 2019).

Earlier versions of some sections of chapter 10 were published in *Common Worlds: Paths Toward Urban Sustainability* (Lanham, MD: Rowman and Littlefield, 2019).

Earlier versions of some sections of chapter 11 were published in *Pathways Through Crisis: Urban Risk and Public Culture* (Lanham, MD: Rowman and Littlefield, 2008); *Common Worlds: Paths Toward Urban Sustainability* (Lanham, MD: Rowman and Littlefield, 2019).

Earlier versions of some sections of chapter 12 were published under the title, "Challenged Academy: Engagement after Neoliberalism," *ANUAC La rivista dell'Associazione Nazionale Universitaria Antropologi Culturali*, 6:1, 41–45, 2017.

Part I

MILLENNIAL ADOLESCENCE

Chapter 1

Reimagining Education

This is the moment to reimagine the education of future generations in our complex society. We are living in an era of transition that has been repeatedly called an age of acceleration. It may be more accurate to call this an era of escalation defined by pathways forming discrete transitions in a time of crisis. Climate change and pandemic disease are direct experiences of this crisis. Social media has transcended the boundaries of previous communication modes, expanding reach yet isolating individuals in technological silos. Smart phones answer information, connect youth to each other at a touch, even as unseen market-based attention economies capture unwitting participants. Young people are caught up in a social media environment where they are compelled to select their beliefs to adjust to a particular situation, often based upon inconsistent information and inauthentic online interpersonal perceptions. Individuals are encouraged to engage with a cyber-social identity that renders interpersonal communication and even learning as surface commodities. All-consuming minute-by-minute text exchanges superficially bind the young in the immediate moment while the historical moment alters their present and future directions.

Through this transition, many seek an education model beyond institutional acceleration, which can enhance school cultures to emphasize deeper learning. This involves creating flexible school environments where, as Rick Mintrop and his colleagues state: "students are explorers of the world and in charge of constructing knowledge."[1] Yet, such reorientation of learning takes place within educational settings connected by cybertechnology to communication systems of unprecedented sophistication. An impetus to create all-encompassing computerized educational platforms has led to inequalities and asymmetrical class interests. The current educational environment deriving from these divisions is by no means inclusive. Technology-embedded classrooms and electronic learning (e-learning) systems can hinder interpersonal communication and, for many students, the retention of streaming content. Still, when used in support of democratic schooling, such platforms may hold the promise of inclusiveness and connectedness that is sensitive to a range of educational experiences.

Over a century ago, forces and ideas from any number of disciplines guided educational processes and inspired the redesign of spaces and activities for innovative approaches, such as project-based learning. Technological and organizational innovations were rapidly transforming the United States from a largely agrarian world to a modern industrial society. Institutional modernization likewise shaped the lives of youth. Core institutions established at this time, notably the public high school and apprenticeships in the new industrial and service workplaces, all "top-down processes," reshaped the learning experience. Paul Willis sees a dialectical relationship between institutional forces of technical modernization ("from above") and cultural modernization ("from below")[2] in part attributable to young people's responses to disruptive changes:

> Youth are always among the first to experience the problems and possibilities of the successive waves of technical and economic modernization that sweep through capitalist societies. Young people respond in disorganized and chaotic ways, but to the best of their abilities and with relevance to the actual possibilities of their lives as they see, live, and embody them. These responses are actually embedded in the flows of cultural modernization, but to adult eyes they may seem to be mysterious, troubling, and even shocking and antisocial.[3]

As their life trajectories were reshaped from above in the technical and economic spheres, the emergence of project-based learning in schools coupled with young people's involvements in tightly knit peer cultures enhanced their cognitive and interpersonal skills. John Dewey articulated the primacy of lived experience and reflection as the basis of learning and effective self-direction at this historical moment. He described how the flow between informal and formal educational practices in school and beyond had the power to expand learning and personal development. Critical reflective activities associated with this learning model underlie current experiential pedagogical approaches and occupational techniques. These methods along with efforts to enculturate contemporary worldviews can again move youth toward greater understanding of themselves and their places in the world. When project-based learning works in practice, students' images of themselves as learners change. Students become more consistent with Paulo Freire's understanding of people "as beings of praxis, [who] in accepting our concrete situation as a challenging condition [. . .] are able to change its meaning by our action."[4]

The cognitive, learning and education sciences and developmental psychology converged in their findings on how youth learn best. Their studies confirm what many have observed: that good learning involves direct experience in an activity, one that is "situated in and supported by a more or less enabling cultural setting,"[5] according to Jerome Bruner. Learning works best when young people can focus in depth on a few things at a time and see a clear purpose in learning activities. Such learning comes about when youth take an active role, notably co-constructing, interpreting, applying, making sense of, and making connections. Realizing the potential of good learning requires an understanding of adolescents' developmental capacities and their need to be responsibly engaged through personal connections to a community of learners, as Robert Halpern, Paul Heckman, and Reed Larson state:

> To engage adolescents in learning, society needs to take them seriously—as sentient and purposeful human beings. We must respect, nurture, provide opportunities, and challenge young people to develop their skills, including their new potentials for self-direction and executive thinking. To see how this can be done, we need go no further than the many successful non-school settings that embody the principles of good learning. These include organized programs and apprenticeships in technology, science, arts, community leadership, etc. Youth in these settings take on roles (such as camera-person, editor, investigator, machine operator, committee chair) and responsibilities (e.g., for getting a task or project done, for mentoring others) that give them agency over their work and learning, but within a structure of norms, goals, and high expectations. Youth experience agency within a context of input from and accountability to peers, senior collaborators, and often members of the community.[6]

This approach to learning involves, beyond mastering core academic content, the ability to think critically and solve complex problems, to work collaboratively, to communicate effectively, and to learn how to learn. Good learning supports adolescents' appreciation of differences across race and class in an aspiring democracy and borderless world. Erik Erikson sees the developmental transition in a broader context, namely, as an opportunity to both support and identify with a sustainable future:

> Adolescence has always been seen as a stage of transition from an alternately invigorating and enslaving sense of an over-defined past to a future as yet to be identified—and to be identified with. It seems to serve the function of committing the growing person to the possible achievements and comprehensive ideals of a viable or developing civilization.[7]

Erikson addressed adolescent challenges in his stage theory of psychosocial development. In *Childhood and Society*, Erikson explained how individuals negotiate life through developmental stages each with a defining crisis that of identity versus role confusion, for adolescents, requiring resolution to achieve succeeding levels of maturity.[8] Since mid-century, American youth have been categorized, generationally, as the Silent Generation, Baby Boomers, Generation X, Millennials, and Generation Z. Each cohort faced challenges navigating personal developmental crises in a precarious world, requiring them to deal with unpredictable risks and uncertainties appearing at every turn. The composition of schools has shifted significantly over recent decades, toward greater ethnic and racial diversity, growing income inequality, and with many students who do not speak English as their first language. The politics and economics of public education narrowed opportunities for sustained involvement in high-quality learning experiences. Knowledge technology modified independent learning and social communication in a cyber-media environment. Erikson also understood that life history, or psychogenic development, requires an understanding of the historical development of the times. Experiential instructional approaches likewise occurred within a historical and a cultural moment. Knowing the historical conditions that required the redesign of education can assure sustaining the influence of these approaches beyond the present moment.

VARIETIES OF PRAXIS

Learning in the Anthropocene provides a synthesis of historical, comparative, and field-based social dynamics of project-based learning and knowledge-based social movements. Three themes are explored in the book. The first surrounds active learning in adolescence and dialogue on its behalf. A second considers social formations in the school and society during three periods of technological innovation, creativity, and social transformation. The Machine Age, Postwar, and the Millennium decades are illustrative of changes in knowledge, skilled work, and engagement that affect learning and human development. Across these pivotal decades, major organizational shifts occurred in societal structures and processes, such as education, with consequences for our social experience today. According to Charles Tilly,

> We must look at them comparatively over substantial blocks of space and time, in order to see whence we have come, where we are going, and what real alternatives to our present condition exist. Systematic comparison of structures and processes will not only place our own situation in perspective, but also help in the identification of causes and effects.[9]

Through historical analysis, case studies of the three eras examine societal and cultural influences shaping youth. The interplay of historical and contemporary case narratives is meant to convey multiple perspectives on acquisition and transmission of cultural knowledge and on innovative instructional approaches. The third theme winding throughout is the impact on various aspects of life of the most recent period of earth history, known as the Anthropocene, when experimentation and active thinking are fundamental to locating humanity's place in this transition.

The first section introduces project-based learning in adolescence from the transition to technological modernity through the present era. Our time of accelerated energy use, greenhouse gas emissions, and population growth created increased economic activity, a higher standard of living, but also ecological disruption.

A second section discusses project-based learning, critical and public pedagogies as innovative responses to challenges posed by national educational policies. The youth I am addressing are sixteen-to-eighteen year olds or late adolescents on the verge of emerging adulthood. Three extended case studies address project-based learning in practice in Southern California locales, focusing on projects carried out by students and their mentors during the last two years of high school. The first takes place in Point Loma, a seaside community in the city of San Diego where a high school was established based upon design principles that promoted project-based learning, in this case using nearby San Diego Bay as the locale for junior-year environmental science projects. The second case takes place in Pacoima, a residential community in the northeast San Fernando Valley of Los Angeles where intergenerational mentoring spurred youth-oriented design projects to transform the landscape and a month-long academy where high-school juniors pursued environmental action research projects. The third case takes place at a public research university on Los Angeles' Westside where a two-year life sciences apprenticeship, including an intensive summer residential laboratory-based program, was collaboratively designed and realized for high-school juniors and seniors. The case studies help elucidate the pathways that mediate acquisition of social and craft knowledge for youth seeking a place in the world.

The third section employs historical analysis to understand the "consequences of modernity," in Anthony Giddens' words by examining collective efforts to mitigate "discontinuities of modernity" in the three historical eras under study.[10] Each case depicts the conditions that gave rise to social movements and how these movements addressed structural changes and their consequences, notably how these promote learning through participation and inform public pedagogies. Historical cases also address experiential and cognitive processes in social movements and the ongoing influence of pragmatic thought on knowledge-based movements.

A final chapter returns to the concerns discussed in the first section, namely, the interplay of sociocultural and environmental factors that influence education and the life chances of youth. Social ecological conditions related to digital technologies and energy resources are examined as these affect education, work, and community life. External challenges to the human prospect are then considered, including the existential risks related to technological advances in cybertechnology and artificial intelligence and the threats to the earth, from climate change and environmental degradation to the fate of the commons and accelerated loss of biodiversity. Woven throughout is the need for critical pedagogies to support reflective inquiry and democratic schooling amid unprecedented global change.

The various perspectives thus reconfigure experiential learning and public pedagogies across time and within contemporary spaces. Educational practices may be reconciled with anthropology's praxis orientation, as each recognizes how culture shapes consciousness, action and interaction in society and history. Recent practice models within anthropology hold that people make their own history through their actions and interactions. At pivotal historical moments, collaborative efforts within social movements then influenced education among other knowledge-based practices. Anthropologists have employed collaborative methods to delineate the gaps between knowledge and action in these practices. Kathryn Kozaitis frames the foundation of recent anthropological praxis as "intellectually mediated, ethically sound, and socially responsible work."[11] Taking an engaged and liberatory approach, Donald Nonini understands that such praxis can be "undertaken in anticipation of future collaborations that have not yet emerged or are in process of emergence."[12]

Liberation underlies Paulo Freire's contribution to education. What Freire called critical consciousness (*conscientizacao*),[13] implies raising the consciousness of the educator and the public, as their relationship must engage each in dialogue, co-participation, and co-construction. Brian McKenna sees Freire's work overlapping with anthropological praxis:

> Freire was, in fact, an anthropological educator. He founded an educational movement based, in part, on conducting an ethnographic evaluation of a community to identify the generative themes (or "dangerous words") which matter profoundly to people and which, for just this reason, contain their own catalytic power.[14]

Denis Goulet uncovers a shared vulnerability in Freire's efforts and those of the many researchers, teachers, and community organizers that work among diverse groups:

> . . . research, education, and mobilization for change are seen as parts of a single strategy best carried out when change agents share structural vulnerability with the populace. New kinds of human relationships are thereby instituted, based on

vulnerability and leading to reciprocity. For without reciprocity, neither change agents nor the people are free to transcend the value crises generated by technological expansion.[15]

Rob Riordan finds a widening sense of shared vulnerability in our current global crisis stating "we are *all* vulnerable to climate change, disease, and the vagaries of social media. Shared vulnerability promises to become an increasingly salient part of the human condition."[16] Friere's pedagogy, based upon critical dialogues around such existential concerns, can be a powerful tool to make the invisible visible, surface assumptions, identify contradictions, and improve possibilities for critical practice.

Paul Rabinow rethinks the various styles of inquiry, accepted practices, and scholarly products of the interpretive sciences, beginning with experimentation or "simply trying out different configurations of inquiry and critique."[17] Such inquiry comes about through assemblages of people and projects within common, or shared, venues for experimentation, namely "collaborative zones of inquiry."[18] Collaborative work essentially redesigns anthropology to better address the contemporary situation. Following Dewey's assertion that objects are "both the product of previous inquiry and the objectives of a new state of inquiry,"[19] Rabinow creates a new "problem-space"[20] for collaborative work or assemblages based upon object-centered work. Object One focuses on formulating a common problem. Object Two redefines the first object by delineating between projects and problems. Object Three involves recognizing that the project and the people are part of a "constituted tradition,"[21] with an authority that transcends the group, freeing participants to see new possibilities. Object Four, central to collaborative work, is to develop a conceptual toolkit that can be adapted for various uses and to solve ongoing problems. Object Five forwards the conceptual process following the precept, "increase capacities while not increasing relations of domination."[22] The very nature of collaborative work can lead to methodological concerns involving power.

Michael Burawoy also understands that collaborative work only succeeds by addressing the multiple dimensions of power and their hierarchies. The Manchester School's extended case method for Burawoy "elevates dialogue as its defining principle and intersubjectivity between participant and observer as a premise"[23]—an element essential to reflexive praxis. This method has used ethnography to analyze conflicts in diverse arenas. Within them, experts and laypersons try to resolve disputes over competing claims about how an issue is defined. The various participants bring competing professional and lay perspectives to discussions. These disparate viewpoints can result in disputes, and more often than not, bottlenecks. Extended cases depict such ensuing clashes in perspective as social processes, taking into consideration participants' multiple experiences in their lives and cross-cutting ties within their communities. This approach discloses unexplored strategies and solutions to existing problems, thus allowing stakeholders to collaborate toward the production of new knowledge.

Visual anthropologists engage people being observed and enable their co-participation in the graphic productions that express their culture from their point of view.[24] These projects include video ethnography, photography, and computer-based multimedia, but also film and videography, where cameras are placed in the hands of the people whose culture anthropologists want to translate to the screen. Visual ethnographic techniques document public culture and public art in traditional fieldwork settings and in community-based participatory action research.

The products of such work have often ended up in public events, such as displays, festivals, exhibits in museums, and in other venues. Participation in production of public media has entered the global health communication arena through lay-constructed *photonovelas* and *video voice* approaches, in many cases to reach populations with low literacy skills. The public production of art as an expression of culture, whether either aesthetic or political, is a thematic also explored by visual anthropologists. Participatory approaches to the study of visual and public culture thus informed the ongoing conversation on anthropological praxis.

Anthropology's "crisis of representation" involved acknowledging the limitations of rhetorical strategies ethnographers employed in their narrative accounts of cultural reality.[25] To resolve the crisis, anthropologists experimented with new methods, including dialogue and memoir. The intent was to gain a deeper understanding of how people construct knowledge, together with the relationship between a particular group's knowledge production and its behavior and lived experience. A reflexive turn in the human sciences moved anthropologists to reconsider the biases that shape their observations and the questions they ask of participants. A fully reflexive interpretive anthropology then studies knowledge concerns in broader historical and political economic frames, and also critically evaluates the ethnographer's stance in the field encounter. George Marcus points out that the edges of ethnography's contemporary application are where anthropologists are redefining practical boundaries of their projects in multiple theaters of reception.[26] This involves asking basic questions of scale, function, purpose, and ethics as ethnography shifts away from the study of culture and toward the process of knowledge production.

Following George Marcus' idea that academic disciplines build upon "habits of thought and work," Julie Pearson Stewart and Barbara Brizuela map various disciplines' conventional approaches to qualitative methodologies informing research in education. Disciplines employing these methodologies, notably anthropology, linguistics, psychology and sociology, have entered into mainstream educational research.[27] Questions of disciplinary authority arise, as a result, along with challenges in defining how teachers and students can participate in such research to represent their lived experiences and collaborate in co-constructing knowledge. Reba Page looks at this "turn inward in qualitative research,"[28] starting with the problems that Arno Bellack surfaced in a critique of the traditional process-product model of research on teaching, characterizing this approach as a "mainstream scientific ideology."[29] For decades, educational researchers used methods, such as randomized experiments derived from the natural sciences, to answer the question: "What teaching practices most effectively produce student learning?"[30] Page understands that a similar crisis of representation is implied in Bellack's critique and sees analogies with anthropology's aesthetic and political challenges to conventional qualitative research that Marcus addressed. Finding similarities in carrying out this style of research in educational settings, she restates the critique that ethnography's qualitative accounts, rather than objectively representing social reality, actually create the worlds studied in the field. Page addresses how anthropologists then moved toward collaborative authorship and autobiographical accounts to expand the scope of rhetorical practices, in part to convey the impact of ideologies and power structures on the locales represented in their ethnographies. Calling for a reconstruction of qualitative methodology in education research, Page states:

> Dewey proposed that academic disciplines have to be continuously reconstructed if the knowledge they store and organize is to be useful in conditions that differ from the conditions in which the knowledge was originally produced. The conversation about reconstruction is again before the educational community. . .[31]

To fully represent knowledge reconstruction, qualitative research can provide a comparable collaboratively informed framework emphasizing the ideologies and systems of power that influence contemporary pedagogies. Michael Apple views critical education as a relational practice that is "intimately connected to the inequalities that structure our society and to the movements that seek to interrupt such inequalities."[32] He suggests "repositioning" critical education practice, including the educational researcher, and the research, itself, to address these structural conditions. Gary Anderson also sees educational critical ethnographers moving in this direction, stating that "the tendency toward collaborative action research and the negotiation of research outcomes with informants indicate a growing willingness among researchers to truly ground their critical analyses in the 'trenches' of educational practice."[33] Grounded approaches then move researchers beyond simply documenting the sources of domination toward a critical reflexivity and a theory of action that teachers may employ in their classrooms.

Project-based learning shares a collaborative approach to problem solving for the production of new knowledge, through several steps, including "identifying the problem; agreeing on or devising a solution and potential solution path to the problem; designing and developing a prototype of the solution; and refining the solution based on feedback from experts, instructors, and/or peers."[34] Implementing project-based learning typically involves various processes that lead to co-creating the end product or artifact. Initially, the tasks are "that of defining problems in terms of given constraints or challenges (and) generating multiple ideas to solve a given problem."[35] Then there are "prototyping—often in rapid iteration—potential solutions to a problem (and) testing the developed solution products or services in a 'live' or authentic setting."[36] Projects are usually displayed in public venues through multimedia technologies. Students document projects through participatory visual and digital methods and share them through all-school exhibitions, presentations of learning, and digital portfolios. From the varied styles of creativity in project design to the cross-disciplinary alliances in implementation, students can make connections across diverse practices and their conventions.

Riordan and Stacey Caillier view project-based learning as a hopeful pedagogy toward sustainability by which schools become equitable centers of community inquiry and action:

> We define equity in schools as a condition where everyone exercises voice and choice, engages in work that is accessible and challenging, and connects with the world beyond school—in short, where all have access to deeper learning experiences that prepare them to lead a purposeful life once they graduate. We link equity to sustainability because the world in every corner is becoming more diverse, a phenomenon that problematizes the issue of equity as a subject for inquiry, action and reflection.[37]

To reorient schools as "reflective communities of inquiry," Riordan and Caillier propose three ways to rethink the curriculum. The first foundation, following Freire's framing, is the platform of "experience as text,"[38] where students' observations and questions are valued and critically analyzed as "mediating objects," just as with other cultural artifacts, such as books and films. Project-based learning thus affords the opportunity to view student experience as text to explore, articulate, and share with the world. Such a change necessitates constructing classrooms and learning groups where everyone belongs, since everyone brings experience to the table. Second is "collegial pedagogy"[39] that changes authority relations in the teacher–student relationship. This requires engaging teachers and students together in pursuit of questions to which none of them knows the answer. The pursuit of new knowledge is therefore co-designed and implemented through a relational approach with a broader range of interactions beyond the classroom, assuring agency for both students and teachers.

Project-based learning allows educators to bridge the gap between institutional (top-down) priorities and individual (bottom-up) perceptions, concerns, and goals. The approach creates new, more authentic teacher–student authority relationships, based in shared purpose rather than predetermined hierarchies. The third, "assessment as dialogue," provides a more integrated approach that is dialogical in both design and execution. Such assessment techniques support an understanding that every episode of learning occurs in a context, and so the context must be assessed along with the learning. Dialogical assessment thus makes project-based learning reflexive, grounded in reflection and action. The prospects of future generations remain in projects that promote collaborative inquiry as the new digital economy moves forward and renders both skilled and middle-skilled work more efficient. These pedagogies offer hope in an ongoing climate of despair and uncertainty amid ecological, epidemiological, and economic crises in the transition to the Anthropocene—a time for planetary social thought and an appreciation of the considerable challenges facing humanity toward creating a sustainable future.

NOTES

1. Rick Mintrop, Elizabeth Zumpe, Kara Jackson, Drew Nucci, and Jon Norman, *Designing for Deeper Learning as an Equity Approach: Schools and School Districts Serving Communities Disadvantaged by the Educational System Benefiting from Deeper Learning* (Stanford, CA: The Carnegie Foundation for the Advancement of Teaching, 2022), 7.
2. Paul Willis, "Foot Soldiers of Modernity: The Dialectics of Cultural Consumption and the 21st Century School," *Harvard Educational Review* 73:3 (2003): 391.
3. Willis, "Foot Soldiers of Modernity," 391.
4. Paulo Freire, *The Politics of Education: Culture, Power and Liberation* (South Hadley, MA: Bergin and Garvey, 1985), 155.
5. Jerome Bruner, *The Culture of Education* (Cambridge, MA: Harvard University Press, 1996), 68.
6. Robert Halpern, Paul E. Heckman, and Reed W. Larson, *Realizing the Potential of Learning in Middle Adolescence* (West Hills, CA: Sally and Dick Roberts Coyote Foundation, 2013), 17.
7. Erik H. Erikson, *Life History and the Historical Moment* (New York: Norton, 1975), 195.
8. Erik H. Erikson, *Childhood and Society* (New York: Norton, 1963).
9. Charles Tilly, *Big Structures Large Processes Huge Comparisons* (New York: Russell Sage Foundation, 1984), 11.

10. Anthony Giddens, *The Consequences of Modernity* (Stanford, CA: Stanford University Press, 1990), 4.

11. Kathryn A. Kozaitis, "The Rise of Anthropological Praxis," in *The Unity of Theory and Practice in Anthropology: Rebuilding a Fractured Synthesis*, edited by Carol E. Hill and Marietta L. Baba (Washington: American Anthropological Association, 1997), 46.

12. Donald Nonini, "Praxis," *Dialectical Anthropology* 40:3 (2016): 248.

13. Paulo Freire, *Education for Critical Consciousness* (New York: Seabury Press, 1974), 4.

14. Brian McKenna, "Paulo Freire's Blunt Challenge to Anthropology: Create a Pedagogy of the Oppressed for Your Times," *Critique of Anthropology* 33:4 (2013): 448–449.

15. Denis Goulet, "An Ethical Model for the Study of Values," *Harvard Educational Review* 41:2 (1971): 226.

16. Robert C. Riordan, personal communication.

17. Paul Rabinow, *The Accompaniment: Assembling the Contemporary* (Chicago: University of Chicago Press, 2011), 114.

18. Anthony Stavrianakis, Paul Rabinow, and Trine Mygind Korsby, "In the Workshop: Anthropology in a Collaborative Zone of Inquiry," in *The Composition of Anthropology: How Anthropological Texts are Written*, edited by Morten Nielsen and Nigel Rapport (London: Routledge, 2018), 169–192.

19. Rabinow, *The Accompaniment*, 125.

20. Rabinow, *The Accompaniment*, 125.

21. Rabinow, *The Accompaniment*, 125.

22. Rabinow, *The Accompaniment*, 126.

23. Michael Burawoy, *The Extended Case Method: Four Countries, Four Decades, Four Great Transformations, and One Theoretical Tradition* (Berkeley: University of California Press, 2009), 39.

24. Aline Gubrium and Krista Harper, *Participatory Visual and Digital Methods* (Walnut Creek, CA: Left Coast Press, 2013); Aline Gubrium, Krista Harper, and Marty Otanez, editors, *Participatory Visual and Digital Research in Action* (Walnut Creek, CA: Left Coast Press, 2015).

25. George E. Marcus and Michael M. J. Fischer, *Anthropology as Cultural Critique: An Experimental Moment in the Human Sciences*, second edition (Chicago: University of Chicago Press, 1999), 7.

26. George E. Marcus, "Collaborative Options and Pedagogical Experiment in Anthropological Research on Experts and Policy Processes," *Anthropology in Action* 15:2 (2008): 47–57.

27. Julie Pearson Stewart and Bárbara M. Brizuela, "Habits of Thought and Work—The Disciplines and Qualitative Research: An Introduction," *Harvard Educational Review* 70:1 (2000): 22.

28. Reba Page, "The Turn Inward in Qualitative Research," *Harvard Educational Review* 70:1 (2000): 23.

29. Arno A. Bellack, *Competing Ideologies in Research on Teaching*. Uppsala Reports on Education 1 (Uppsala: Uppsala University Department of Education, 1978).

30. Page, "The Turn Inward in Qualitative Research," 25.

31. Page, "The Turn Inward in Qualitative Research," 29.

32. Michael W. Apple, "On Doing Critical Policy Analysis," *Educational Policy* 33:1 (2019): 279–280.

33. Gary L. Anderson, "Critical Ethnography in Education: Origins, Current Status, and New Directions," *Review of Educational Research* 59:1 (1989): 262.

34. Boston University Center for Teaching & Learning, *Project-Based Learning Teaching Guide* (Boston, MA: Boston University Center for Teaching & Learning), https://www.bu.edu/ctl/guides/project-based-learning/ Accessed February 1, 2023.

35. Boston University Center for Teaching & Learning, *Project-Based Learning Teaching Guide.*

36. Boston University Center for Teaching & Learning, *Project-Based Learning Teaching Guide.*

37. Robert Riordan and Stacey Callier, "Schools as Equitable Communities of Inquiry," in *Sustainability, Human Well-Being, and the Future of Education*, edited by Justin W. Cook (Cham: Palgrave Macmillan, 2019), 122.

38. Riordan and Callier, "Schools as Equitable Communities of Inquiry," 131.

39. Riordan and Callier, "Schools as Equitable Communities of Inquiry," 135.

Chapter 2

Learning from the Anthropocene

ENCLOSING THE COMMONS

Contemporary students are entering a world of near-chaotic perplexity, complex histories, world crises, and a future they are inheriting to shape themselves. The Anthropocene, in this respect, is arguably the period of time beyond the Holocene characterized by the human domination of nature as a result of scientific and technological progress. In his geographical treatise, *Man and Nature*, published in 1864, George Perkins Marsh predicted that unless humans altered their current practices, the earth would be reduced "to such a condition of impoverished productiveness, of shattered surface, of climatic excess, as to threaten the depravation, barbarism, and perhaps even the extinction of the species."[1] Echoing Marsh, Italian geologist Antonio Stoppani in 1873 acknowledged the impact of the profoundly altered extraction and production conditions, referring to the time when the earth was fully transformed by human action as the "Anthropozoic era."[2]

The human triumph over the earth system, according to many scientists, took root with the capture of matter and energy and their industrial transformations, as suggested by Paul Crutzen and Eugene Stoermer, who, together, in 2000, proposed a term for a new geological era:

> Considering these and many other major and still growing impacts of human activities on Earth and atmosphere, and at all, including global, scales, it seems to us more than appropriate to emphasise the central role of mankind in geology and ecology by proposing to use the term "anthropocene" for the current geological epoch. The impacts of current human activities will continue over long periods.[3]

For many other scientists, the Anthropocene designates a geological age that dawned when humanity experienced "the great acceleration,"[4] a term used to denote fast-moving global industrialization after the Second World War. Some scientists even date the new epoch with the first atomic bomb test in Alamogordo, New Mexico, at 5:30 a.m. on July 16, 1945. Recalling the moments after the explosion, J. Robert Oppenheimer, scientific director of the Manhattan Project at the Los Alamos Laboratory where the bomb was produced, said:

> We knew the world would not be the same. A few people laughed, a few people cried, most people were silent. I remembered the line from the Hindu scripture, the Bhagavad-Gita; Vishnu is trying to persuade the Prince that he should do his duty and, to impress him, takes on his multi-armed form and says, "Now I am become Death, the destroyer of worlds." I suppose we all thought that one way or another.[5]

Since that time, geologists and physical geographers argue that anthropogenic activities became the major driver impacting the earth system. Increased wildfire activity, together with the steady depletion of global energy resources and climate change, contribute to precariousness of the conditions of life with the human transformation of the terrestrial biosphere.

Robert Kates, B. L. Turner, and William Clark view the present as a "special period in history"[5] because of the magnitude of human-induced transformation of the biosphere on a global scale driven by population growth. Writing in the 1990s, they indicate that such transformation is no longer confined to landscape changes, stating:

> From the middle of this century until the present, human population, water withdrawal, and cumulative releases of sulfur, phosphorous, nitrogen, lead, and a variety of organic chemicals have doubled and more, although several of these have slowed in recent years. The growth in population and the even larger growth in material and energy consumption mirrored in these indicators have been accompanied by unprecedented movements of material and concentrations of people.[6]

Every conceivable commons, including water and minerals deep within the earth, were expropriated by corporate and state capture, enclosure, and accumulation during this period of explosive economic growth and societal transformation that began in the nineteenth century and accelerated after the Second World War with extensive industrialization. Both human and nonhuman inhabitants of these ecosystems inherited the consequences of the various enclosures. All were captured through the extraction of raw materials and their commodification within a capitalist world economy. These include primary consumers, monetized labor in both state and industrial enterprises, and the bodies and minds, conscious and subconscious, of individual end users of the manufactured commodities of captured ecosystems.

In the Anthropocene, humanity experiences rampant urbanization, mechanized monocropping, ubiquitous pollution, warming trends, deforestation, and species loss, as a consequence of global industrial civilization. People also suffered horrifying advanced industrialized warfare during this epoch, from chemicals and atomic bombs, to drones and cyberwar. At the same time, military logistics planners promote war as a solution to disruptive conditions, often protecting resource-rich regions from state and paramilitary violence surrounding extractive industries. Beyond ensuring the flow of oil, militaries support the supply chain for strategic minerals, such as chromium, cobalt, manganese, and the platinum group metals critical to industrialized weapons production. Hardened attitudes hold that warfare remains an effective way to resolve human conflicts; consequently, global military operations leave in their wake a toxic legacy of pollution, despoiled landscapes, and war-ravaged ecosystems.

Equally distressing is that global elites perpetuate a carbon economy despite adverse anthropogenic effects in the current geological epoch, by its very name indicative of human domination over the earth system. These include crossing the nine planetary boundaries, the very processes that regulate the stability and resilience of the earth system, which would lead to irreversible environmental change. Will Steffen, Katherine Richardson and their colleagues caution, that crossing two core boundaries, climate change and biosphere integrity, may have the potential "to push the Earth system out of the Holocene state,"[7] a 11,700 year-long epoch that has been shown to support complex human societies like our own.

Even in the face of scientific evidence, many still view negative consequences, such as extinction events and habitat destruction, as simply synergistic effects of contemporary life. Some find solace in a protective political realm, seeking state-based solutions to the consequences of human-induced terrestrial change. However, many elites will eschew a policy agenda on behalf of "a safe operating space for humanity on Earth"[8] that many scientists advocate, based upon tighter constraints on land and resources. Still others hold to an ideological certainty that technological ingenuity will fix any contingency. This, as humanity confronts a world on fire from burning fossil fuels, detonating ballistics and explosives, and ferocious wildfires burning across the globe. There is much, then, to be learned from the diverse approaches to the *Anthropocene* and the urgent environmental issues confronting humanity in our time, notably biodiversity decline, climate disruption, population growth, and consumption. In light of these challenges, socioenvironmental research has explored the various problematics of human–nature relationships, providing "structured inquiry about the reciprocal relationships between society and the environment."[9] One strong research paradigm, that of the systems approach, which introduced the idea of resilience, seeks a pragmatic holistic understanding of these relationships and the prospects of adaptive change.[10]

LEARNING FROM THE ANTHROPOCENE

Ours is a time of extreme weather events, violent hurricanes and tornadoes, firestorms, melting glaciers, rising sea levels, in part a consequence of anthropogenic climate change. Together with pandemic disease, economic recessions, large-scale population movements, and centuries-old social injustices within and between countries, climate change poses existential threats. While many in the media and professional networks, and in educational settings, discuss these crises associated with the Anthropocene, there is limited consensus regarding solutions, though ecological theories of adaptive change may suggest a way forward. Theoretical ecologist C. S. Holling introduced resilience ecology as a way to frame nonlinear dynamics in ecological systems. Holling defined "ecological resilience," according to his colleague Lance H. Gusterson, "as the amount of disturbance that an ecosystem could withstand without changing self-organized processes and structures (defined as alternative stable states)."[11] Holling and Gunderson hold that all systems go through an adaptive cycle they metaphorically refer to as "panarchy."[12] In Holling's words, "the 'front loop' of that cycle is the loop of growth. The 'back loop' is the loop of reorganization."[13] While the front loop assures order and stability, for a time, amid chaos and disruption, the back loop provides opportunities for creative transformation.

Stephanie Wakefield explains that the front loop is a progression from an initial phase toward conservation and stability, to a back loop characterized by the coming apart of these structures. This latter phase, in Wakefield's words, consists of "destabilization, fragmentation, confusion, and release but also great potential for experimentation, reorganization, and transformation."[14] The Anthropocene front loop involved "stitching the whole earth together,"[15] through historical transformations that include colonization, slavery, resource extraction, industrialization, fossil fuel combustion, and the transformation of land, labor, and capital as resources. The Anthropocene back loop, or "splintering whole earth,"[16] is marked by increased concentration of carbon dioxide, ice melts, rising sea levels, mass extinction, biodiversity loss, deforestation, ocean acidification, fisheries collapse, and a dismantling of Fordist production regimes. At the back loop, according to Wakefield, is a "vast proliferation of experimentation in redefining what human life will be."[17] She goes on to state:

> We are amidst a wave of experimentation with new ways of transforming bodies, minds, lives, and the world around them: from hacking, making, modding,

> prepping, and weight lifting to citizen science, eco-design, solar energy grids, and wireless mesh networks. People everywhere are searching their souls, scouring the Earth for tools, and trying in a million ways to reinvent what it means to be human and to dwell on Earth.[18]

The back loop, then, represents a phase where individuals and small groups relate across what were previously viewed as "unbridgeable divides," in Wakefield's words, "and in so doing create something fundamentally original."[19] Wakefield proposes the idea of a repurposed ecological back loop, based upon creative experimentation across disparate groups holding incongruent ideas and practices. This approach aligns with project-based learning pedagogies, although fundamental questions remain about the direction and consequences of adaptive change at the back loop.

Already occurring are widespread global climate migration and resettlement of climate refugees. There are concerns as to whether governmental agencies and corporate fire and flood insurers will provide requisite funds to rebuild communities in disaster-prone regions that predictably experience future effects of anthropogenic climate change. In North America, these include the Pacific Northwest, California, and lowland areas of Atlantic Coast, Gulf Coast, and Mississippi Valley states. Perennial anxieties over personal and planetary health continue to surround climate change, epidemic disease, and economic insecurity. New technologies have as yet to appear that permit a path forward in resolving the climate crisis, as with the microchip during the third industrial revolution and the digital transition. Still, governmental and citizen efforts toward assuring a just transition to a post-carbon economy can be advanced to counter contemporary nihilism and despair. The present moment continues to be characterized by steady depletion of global energy resources and climate change that, together, contribute to precariousness of the conditions of life with the human transformation of the terrestrial biosphere in the Anthropocene. Amid these concerns, there remains the perennial search for happiness in the face of one's own and the planet's fate.

The human-induced environmental transformation of Southern California, arguably a region where many millions have sought happiness and "the good life" over the last century, is instructive for an understanding of front and back loop dynamics advanced by resilience ecologists. Mike Davis explains how the region's rise to urban technological and social modernity at the front loop of the Anthropocene was undeterred by cataclysmic disasters during its boom times in the 1920s and again in the postwar decades through 1968, when developers, power brokers, and filmmakers, alike, both promoted and depicted the region as a "Land of Sunshine" in the American imagination:

> The urbanization of the Los Angeles area has, it seems, taken place during one of the most unusual episodes of climatic and seismic benignity since the inception of the Holocene; or put another way, twentieth century Los Angeles has been capitalized on sheer gambler's luck.[20]

Millennial Southern California at the back loop, by contrast, provokes an imaginary "apocalyptic temper"[21] according to Davis, as the region's mix of natural disasters and both social and environmental contradictions synergistically produced an "ecology of fear"[22]within a landscape of risk with attendant economic and emotional trauma.

SOUTHERN CALIFORNIA: INNOVATION IN A LANDSCAPE OF RISK

There is a tragic aspect to any wildland fire, and especially to the 2020 Bobcat Fire in the Angeles National Forest where Mt. Wilson Observatory is located, a place where beginning in the 1920s, astronomer Edwin Hubble revolutionized our understanding of the universe. Though, the story of innovation amid a landscape of risk begins much earlier in the social, cultural, and environmental transformation of Southern California at the turn of the twentieth century.

Southern California developed as a modern region through settler colonialism, a settlement pattern supported by a racialist ideology of "civilizing the land."[23] Settlers' beliefs of their right to homestead on land occupied by Native Americans and Mexicans, who together with their distinctive procedural knowledge bases, were adaptive to the needs of the moment. This was a time when the nation's elites were engaged in an immense nation-building project, often embedded within a mythic narrative of Manifest Destiny. Less entitled European and American settler farmers, miners, and homesteaders arriving from the Midwestern United States, and also from England, Germany, Scotland, and Ireland, expropriated indigenous lands and their soils, remaking these into orchards, ranches, mines, and ultimately agroregions. Settlers harnessed the water resources of the Colorado River and the Sierra Nevada range to build a hydraulic civilization in a dryland region. Settler efforts focused on gaining control over regional space and natural resources and were accompanied by the eventual concentration of corporate power and disenfranchisement of indigenous populations through federal governmental land allocation policies. Following the imposition of a modernist property regime were coercive patterns of discrimination, containment, and inequality of *indigenes*.[24]

Amid privatization and other technological modernization processes, as Kevin Starr indicated, Southern California realized the "material dreams" of regional elites.[25] Beyond the various legal and financial instruments to assure dominant Anglo-American property arrangements supported by the military and federal agencies, such as the Bureau of Indian Affairs and the Bureau of Land Management, only limited abstract thought took root in these early years of settlement. Frontier conditions instead provided an open-ended and actively experiential, and at times an experimental, way of life that William James would characterize in his late ideas on radical empiricism.[26] By the turn of the century, James understood the creative energies of an experiential self as a source for pragmatic thought, one with more latitude than that of a bourgeois self, cultivated among elites in nineteenth-century Victorian America. For James, pragmatism elevated individual experience as a transaction with nature, where the truth of a concept is tested through scientific experimentation and examination of its practical outcome.

Following James' pragmatic outlook, a modern "culture of experience" grew in fin de siècle America, according to John McDermott.[27] Within this new cultural ethos, members of a restless, educated younger generation sought the promise of a better life in the "Land of Sunshine," as promoted by booster campaign advertisements. During the Gilded Age and Progressive Era, when America was arguably a developing country, a nascent Southern California culture and society advanced at the interface of the Intermountain West and the Pacific. The region's environmental conditions, notably a Mediterranean climate, and innovative social formations, beginning with an emerging film industry, fostered experiential synergies during this time. Together, these conditions opened up alternative ways of living for those seeking an escape from technological rationality and its institutionalization amid the rapid urbanization of the Northeast and industrial Midwest, even providing a temporary landing place before traveling into Mexico or across the Pacific.

Jackson Lears writes about how overly socialized, repressed, often Calvinist, youth raised in the Northeast felt the need to escape from that region and, for many, from familial expectations, ostensibly as a potential cure for their neurasthenia.[28] In their search for personal fulfillment, the young among certain elites sought out authentic experiences and through them discovered the means to liberate themselves of an earlier social conditioning. Embracing a "therapeutic ethos,"[29] they engaged in such intense experiences to unburden themselves of modernist anxieties and to revive their youthful bodies and minds that had become physically and psychically depleted, exhausted from intellectual overwork during a time of nation building. Many of the young that were discontented with Victorian mores moved into the Southwest, including Santa Fe, New Mexico, to explore Native American arts and culture. Others arrived in Southern California before traveling on to the Pacific Islands to immerse themselves in a Polynesian culture and its ethos and to Asia to study the contemplative humanities and martial disciplines. Some eschewed an actual "Passage to India," seeking spiritual experiences among theosophists and later within the community following Jiddu Krisnamurti's teaching in the Ojai Valley, eighty miles northwest of Los Angeles. Still others remained in Southern California to participate in the Arts and Crafts movement, a search for cultural self-sufficiency amid industrial design, which also thrived in this era when disenchanted elites sought the "good life" in authentic materials, including pottery, ceramic tiles, metalwork, leather, stained glass, and furniture.[30] The movement spurred innovations in garden and interior designs and in residential architecture, notably the craftsman houses and bungalows designed by the Greene and Greene firm, established in 1894 in Pasadena. Growing side by side with these craft traditions and their arguably charmed lifestyles was a dynamic regional economy and society based upon mass production technologies, including the refining of oil and the manufacture of steel, aircraft, and automobiles.

By the 1920s, California, itself, underwent fast-paced demographic and technological change and built up formidable research institutions. Seven decades after the Gold Rush, the state's universities had invested in engineering, meteorology and geology, crop and range science, chemistry, and applied physics. Hubble used photographic plates from the 100-inch Hooker telescope at the Mr. Wilson Observatory,[31] on a mountaintop at an altitude of 5,000 feet, to measure galactic distances and to prove that Andromeda is a galaxy. Hubble's inquiries also proved that the universe was expanding, which supported the Big Bang theory. Albert Michelson would also use the observatory in an experiment to accurately measure the speed of light. Breakthroughs across the sciences provided the varied conceptual frameworks that guided hydrologists to build dams and aqueducts, horticulturalists and foresters to establish nurseries and manage forested areas for commercial uses, and manufacturing engineers to innovate technologies based on discoveries in atomic and nuclear physics, and the space sciences. The humanities were not especially privileged in this earlier era of rapid economic growth; though the state's research universities recruited faculty trained in the East and Midwest, many of whom produced sound critical and analytical studies.

The interwar years were relatively stable times in Southern California: the land, ocean beaches, forest areas, and mountains were bathed in sunlight and clean air; fresh water flowed from the Owens Valley and the dams along the Colorado River. It was a wonderful hydraulic civilization for Northern European-Americans to enjoy a quality of urban life far different from that of industrial cities east of the Mississippi. Red Car lines wove in and out of the region, connecting the hillside communities to the center of town and then out to the oceanside communities through an interurban network. The movie industry filmed throughout the region's diverse landscapes, many of which became iconic in the American imagination. CalTech, USC, and UCLA, the region's research universities, were located, respectively, in Pasadena, Downtown Los Angeles, and the Westside, all predominately white enclaves maintained by tight-knit networks of power and influence before the Second World War and for some time after. City Hall, the Chandler family and their *Los Angeles Times*, and the Catholic Archdiocese were influential power brokers in the region. The movie colony maintained its own tightly controlled studio-based enterprises and dwelled lavishly in Hollywood Hills, Beverly Hills, and Pacific Palisades residences. The film community welcomed the many émigré artists, musicians, and writers fleeing Europe. Central and Eastern European artists and intellectuals formed a liminal "cultural third space" creating projects reflecting a hybrid of Continental and North American aesthetic attitudes and affinities.[32] The Hollywood Bowl amphitheater provided Pasadena and Los Angeles elites with orchestral music, opera, and ballet under the stars on crystal-clear summer evenings.

All of this was in place to assure that "the dream endures"[33] in Kevin Starr's words—a system designed to provide affluent white residents a sense of freedom to move about the region wherever and whenever they wished and to purchase land and then build homes wherever they could afford the asking price. African Americans, Latinos, and Asians were at the same time constrained by restrictive covenants in housing and brutal policing of their neighborhoods, sustaining a long-standing urban ecology of fear that protected elites and exposed others to worlds of pain. Before the war, there were approximately 36,000 Japanese Americans living in Los Angeles. In February 1942, President Roosevelt issued Executive Order 9066 that established military zones on the West Coast authorizing military commanders to exclude civilians from military areas.[34] As a result, there was forced removal and incarceration of individuals of Japanese ancestry, many of whom were American citizens, to temporary detention centers, or "assembly centers," and later to "relocation camps," also known as "internment camps." Violent clashes known as the "Zoot Suit Riots" occurred in summer 1943 between hundreds of white soldiers, sailors, and Marines stationed in and around Los Angeles and Mexican American youth.[35] The *pachucos*, as they were known, dressed in oversized, flamboyant zoot-suits, and were vilified by the servicemen who viewed them as "unpatriotic" draft dodgers, though many were too young to serve in the military. These polarizing conditions, a mix of affluence and the search for the good life, together with indifference and systemic discrimination, were foundational to "front loop" processes in Anthropocene Southern California.

Also at the front loop, the Second World War transformed the place into a defense region, with a proliferation of aerospace and defense enterprises supported by federal governmental largesse, from San Diego at the United

States–Mexico Border to the San Fernando Valley. With the onset of the Cold War, the Rand Corporation was established in 1948 by the Pentagon as a nonprofit research and development military and national security think tank across from the beach in Santa Monica. The social sciences grew at this time in the region's universities, especially their application to medicine, health care, science and technology, urbanism, and modernization, through social research that employed chiefly quantitative methods. In their empirical work, social scientists applied theories practiced at universities in the Northeast and Midwest, where an intellectual "third space" grew with the arrival of European émigré scholars on these campuses.

Auto assembly plants that had been converted to military production were retooled to build multiple car lines and hired workers at every skill level into union-organized workplaces, which would initiate the fabled automobile culture of the 1950s. Just after the war, the Regional Planning Commission adopted the 1947 Master Plan for Metropolitan Los Angeles Freeways.[36] Freeway construction began in the early 1950s, the end of an era for the Pacific Electric Railway Red Cars, an interurban transit system that criss-crossed the Los Angeles area for half a century. A personal transportation boom helped the real estate industry, together with the savings and loan banks, to sell people suburban lots where there once were orange groves. Others built homes in canyons, on coastal cliffs, and against the forested lands essentially establishing a wild land–urban interface. The automobile and fossil fuel-dependent urban civilization in Southern California added new conditions to the ongoing ecology of fear-a mix of threats from ambient air pollution and smog, traffic congestion, residential fires in the area's canyons, and renewed racism, and in some cases, nativism, with the in-migration of diverse strangers to work in the industrial plants.

Major League baseball arrived from the East Coast as Dodgers' owner Walter O'Malley moved the team from Brooklyn to Los Angeles in 1957, and a new 56,000-seat stadium was built in 1962 on 300 acres of land in Chavez Ravine, which uprooted a long-standing Latino community, displacing 1,000 families.[37] At the same time, Dorothy Buffum Chandler, the wife of *Los Angeles Times'* publisher Norman Chandler, spearheaded efforts to build the Music Center, a three-theatre complex on Bunker Hill, near the Civic Center, the largest concentration of public sector buildings and workers outside of Washington, D.C. The Music Center anchored the City of Los Angeles' Bunker Hill Urban Renewal Project that uprooted 6,000 residents in the ethnically diverse working-class neighborhood.[38] These included many Native Americans who migrated to Los Angeles after the war as part of the federal government's forced termination policies that dissolved treaties, eliminated tribal governments, and closed reservations. Completed in 1964, the performing arts complex supported the Los Angeles Philharmonic's rise to global prominence, with internationally renowned conductors Zubin Mehta and Carlo Maria Giulini serving as music directors during this florescent era.

With the postwar population growth and the federal government's extensive support to Southern California industries and public initiatives, the region attracted creative and entrepreneurial types, and also offered a social landscape for both spiritual seeking and secular lifestyle experimentation. For many new elites, the emphasis was on personal fulfillment within an affluent postwar culture of experience, which privileged lived experience over a rational bureaucratic ethos. Though, by this time, the state and its municipalities by necessity had enacted for decades the Progressive ideas of efficiency and bureaucracy to manage the massive demographic shifts and burgeoning rural and urban economies emergent during the first half of the twentieth century.

In a broader perspective of adaptive cycles, the places, processes, and creative ideas that converged in Southern California at the front loop of the Anthropocene—when engineering, chemistry, physics, the earth sciences, and astronomy at CalTech, together with the arts and design worlds, and modern Southern California itself, were ascendant—began experiencing back loop conditions or the coming apart of those structures. In stark contrast to the elite technological society of modern Southern California affluent development, new ethnic, artistic, and lifestyle scenes were emerging across Los Angeles that did not share the wealth. During the 1960s, there were poor living conditions and negligent schooling and high rates of unemployment among minorities. Conditions erupted in what historically and notoriously became known as the Watts Riots over the conditions that were growing, in fact, intolerable. Then members of black and Latino communities demonstrated for social justice and improved health care. In that next decade, Martin Luther King Jr. Hospital opened in Watts, and county-operated public health clinics were established in South-Central and East LA. At the same time, wild youth lived and caroused high on weed and acid in Laurel Canyon, Topanga Canyon, and on Southland beaches. The local music club scene, a vibrant art world, together with the surfer, beach and custom car cultures, all grew to become commercial spectacles. Organized opposition to the Vietnam War in Southern California, including the "Chicano Moratorium" that mobilized a broad-based coalition of Mexican American antiwar activities across the Southwest,[39] became pivotal to the transition from front loop to back loop. The decades-long pattern of planned regional development and residential segregation took a turn in the late 1960s as a consequence of these divergent processes.

Two events signaled troubling consequences for the postwar optimism that Southern California represented to many Americans. Robert Kennedy was assassinated in 1968 at the Ambassador Hotel, home to the Cocoanut Grove nightclub and the site of countless Oscar ceremonies, and after, in 1969, there was the Santa Barbara oil spill. A bright shining moment amid societal discord and ecological destructiveness was the moon landing on July 20, 1969. Thousands of the region's aerospace workers at the North American Aviation plant in Downey produced the engine, the lunar module, and the lunar lander for the Apollo mission. These events were followed by the post-Vietnam War economic downturn, and the closing of Southern California's auto, steel and rubber plants, and later the region's aerospace and defense plants as the Cold War receded.

Deindustrialization of older industries was accompanied by advanced industrial transformation, including the physics and engineering worlds that Cal Tech helped to grow, incubated as applied technologies by the NASA-Jet Propulsion Laboratory in space exploration, robotics, and advanced imaging systems. Global design studios at Pasadena's Art Center College of Design shaped future transportation design and related technologies. Lockheed-Martin's "skunk works" engineers developed experimental aircraft and military surveillance and reconnaissance technologies. San Diego's high-technology hub on Torrey Pines Mesa, encompassing among others the University of California, San Diego, Salk Institute for Biological Studies, and Scripps Research Institute, brought together academic and industrial research to inform commercial applications by the region's 1,500 information technology, telecommunications, and biotechnology companies. As the principal international trade gateway to the Asian-Pacific region, electronic goods, aircraft products, and computers flowed through harbors in the San Pedro Bay port complex on the Pacific Ocean, the nation's largest container-shipping facility. Southern California also became the principal distribution hub for the country's own export goods, including the commodities and services produced by the region's high-technology, multimedia, and entertainment industries.

Amid a recovery based on economic globalization, the region was also being transformed by early phase back loop processes. Southern California experienced a decade of disasters in the 1990s, including earthquakes, wildfires, and urban riots, following on the previous decade's crises of homelessness, gang violence, transnational immigration, and refugee resettlement. These conditions continued into the millennium and led to a tightening up of community life through surveillance technologies and the training and mobilizing of large cadres of first responders to manage and control urban policing, urban and wild land fires, and acute, chronic, and pandemic disease in the population. This growing service sector was reinvented through new response models to handle the emergencies and other contingencies of back loop conditions, including the weeks-long threat to the very existence of the Mt. Wilson observatory in 2020, an iconic location. This event may be symbolic of a generalized existential insecurity, and for some even precarity, that would define Southern California as place of innovation amid a landscape of risk and for over a century, of "the new life out there,"[40] as Tom Wolfe observed decades ago.

A KNOWLEDGE ECONOMY

The ethnographic case studies take place during two pivotal decades in millennial Southern California, a core region for the production and distribution of innovations in microchip-based technologies and biotechnology, and also of health care, financial, and telecommunications services. These industries have come to define the regional economy, the national economy, and an emerging pattern of global economic relations. The region underwent postindustrial transformation following the closing of the auto, rubber, and steel plants, then the reduction of jobs in the aerospace and manufacturing sectors. This industrial transition was accompanied by degraded environmental conditions, as many plants left communities facing toxins from their heavy metals seeping deep into the soil and the aquifer. Change within the region's manufacturing areas brought about a new set of labor demands. Workers displaced by large producers were reemployed in medium-size manufacturing operations, trucking, and warehousing. Younger women with at least high school and some postsecondary training found essentially screen-based office work. Younger men followed their fathers and uncles into the midsize manufacturing operations and trucking fields and various semiskilled trades. Recent immigrants found work in auto dismantling, granite and marble shops, and the landscaping, home-building, and renovating areas, including painting, roofing, flooring, and stone masonry.

During this time when national elites abandoned a militarized domestic economy in postwar defense production areas, places like Los Angeles and San Diego moved toward "swords to plowshares" economies, embracing high-technology and biotechnology industries. These cities, the sites of knowledge-based industries, underwent shifts in local governmental structures and fiscal strategies. Similar to other urban regions caught up in the postindustrial transition, Southern California experienced significant institutional changes, including rearranged class relations, retooled educational systems, and a local-level politics, all in line with the new economy.

Knowledge work requires training in engineering or bioscience at the professional level, or a mix of technology, software application, and social skills for technoservice occupations. Public charter schools, magnet schools, and public school academies made curricular changes to prepare students for postsecondary education in the science, technology, engineering and mathematics (STEM) fields. These served as pathways for youth to develop contemporary skill sets while producing new knowledge, along with an appreciation of the changed ways of working. The schools adopted project-based instructional approaches as a consequence of the new crafts and their industries that put down roots in former defense production areas. Synergies between newer technology and cognitive-cultural industries, schools, and local governmental and nonprofit-sector organizations ensured acquisition of the new skill sets through project-based and work-based learning.

As the knowledge economy advanced, the challenge of sustainability endured. Computerized methods rendered both skilled and semiskilled work more efficient. The shift toward computer-assisted office work, light manufacturing, building trades, and regional transportation, each primarily dependent on nonrenewable resources, brought divergent impacts on the natural and built environments. Social-class disparities persisted in these urban regions that opted for knowledge work to replace older technologies. Suburbanization dynamics continued, as more affluent families moved away from the cities to seek a better life and the educational opportunities to gain their children a competitive edge in the new economy. Beyond the boundaries of the retooled workplaces and affluent community infrastructures are the youth within the unestablished, immigrant generation, marginalized, and frequently challenged by economic precarity, along with social risks and environmental hazards in their older neighborhoods.

Social suffering consequent to systemic change thus continued during the transition, compounded by increased street crime, environmental and mental illnesses, family dissolution, and homelessness. The 2008 Great Recession contributed to a growing unemployment-driven loss of shelter and homelessness in millennial Southern California. The recession brought job destruction, mass layoffs, and unemployment, placing the young, twenty-four years of age and under, those with less than a four-year college degree or without a high-school diploma, at the highest risk. Policy actions to stem downward mobility of younger workers included employing existing trade apprenticeship programs as models of skill-based retraining.

A historically informed narrative of events, issues, controversies, and pivotal decades may help to construct the scaffolding needed to critically view educational processes. If we are to seize the historical moment, along with its potential momentum for democratic education, we must define it accurately. However, there remain many divisive issues that challenge education today. Identity politics influence discourse in various political arenas and are a compelling feature in educational dialogues to increase equity; although, frequently acrimonious conflict of clashing perspectives threatens to drive a wedge within education and the public sphere. Divisions between states and localities, within and across regions, and even within sectors of metropolitan areas that form the hubs of innovation, lead to contentious debates over public and private educational priorities owing to class segmentation. Amid fragmenting conditions, school systems adopted electronic media and computational models as a future foundation for instruction and administrative control. Throughout these incongruities, public pedagogies hold the power to unify purpose and standards in classrooms, by emphasizing the necessity of schooling for democracy as a preparation for life.

NOTES

1. George Perkins Marsh, *Man and Nature*, edited by David Lowenthal (Cambridge, MA: Harvard University Press, 1965), 43.
2. Eugenio Luciano and Elena Zanoni, "Antonio Stoppani's 'Anthropozoic' in the Context of the Anthropocene," *The British Journal for the History of Science* 56:1 (2023): 103–114.
3. Paul J. Crutzen and Eugene F. Stoermer, "The Anthropocene," *Global Change Newsletter* 41, The Royal Swedish Academy of Sciences, Stockholm, Sweden, 17.
4. John R. McNeill and Peter Engelke, *The Great Acceleration: An Environmental History of the Anthropocene since 1945* (Cambridge, MA: Harvard University Press, 2014).
5. Kai Bird and Martin J. Sherwin, *American Prometheus: The Triumph and Tragedy of J. Robert Oppenheimer* (New York: Knopf, 2005), 309.
6. Robert W. Kates, Billie Lee Turner II, and William C. Clark, "The Great Transformation," in *The Earth As Transformed by Human Action*, edited by Billie Lee Turner II, William C. Clark, Robert W. Kates, John F. Richards, Jessica T. Mathews, and William B. Meyer (Cambridge: Cambridge University Press, 1990), 12.
7. Will Steffen, Katherine Richardson, Johan Rockström, Sarah E. Cornell, Ingo Fetzer, Elena M. Bennett, Reinette Biggs, Stephen R. Carpenter, Wim de Vries, Cynthia A. de Wit, Carl Folke, Dieter Gerten, Jens Heinke, Georgina M. Mace, Linn M. Persson, Veerabhadran Ramanathan, Belinda Reyers, and Sverker Sörlin, "Planetary Boundaries: Guiding Human Development on a Changing Planet," *Science* 347:6223 (2015): 744.
8. Steffen et al., "Planetary Boundaries," 736.
9. Simone Pulver, William R. Burnside, Kathryn J. Fiorella, Meghan L. Avolio, and Steven M. Alexander, "Introduction: Foundations of Socio-Environmental Research," in *Foundations of Socio-Environmental Research*, edited by William R. Burnside, Simone Pulver, Kathryn J. Fiorella, Meghan L. Avolio, and Steven M. Alexander (Cambridge: Cambridge University Press, 2023), 2.
10. Sophie Adams, "The Pragmatic Holism of Social-Ecological Systems Theory: Explaining Adaptive Capacity in a Changing Climate," *Progress in Human Geography* 45:6 (2021): 1580–1600.
11. Lance H. Gunderson, "Ecological Resilience—in Theory and Application," *Annual Review of Ecology and Systematics* 31:1 (2000): 425.
12. Lance H. Gunderson and C. S. Holling, editors, *Panarchy: Understanding Transformations in Human and Natural Systems* (Washington, DC: Island Press, 2002).

13. C. S. Holling, "Foreword: The Backloop to Sustainability," in *Navigating Social-Ecological Systems*, edited by Fikret Berkes, Johan Colding, and Carl Folke (Cambridge: Cambridge University Press, 2003), xv.

14. Stephanie Wakefield, *Anthropocene Back Loop: Experimentation in Unsafe Operating Space* (London: Open Humanities Press, 2020), 22.

15. Wakefield, *Anthropocene Back Loop*, 20.

16. Wakefield, *Anthropocene Back Loop*, 29.

17. Wakefield, *Anthropocene Back Loop*, 31.

18. Wakefield, *Anthropocene Back Loop*, 31.

19. Wakefield, *Anthropocene Back Loop*, 33.

20. Mike Davis, *Ecology of Fear: Los Angeles and the Imagination of Disaster* (New York: Metropolitan Books/Henry Holt and Company, 1998), 37–38.

21. Mike Davis, "Golden Ruins/Dark Ruptures: The Literary Destruction of Los Angeles," in *Critical Views: Essays on the Humanities and the Arts*, edited by Teresa Stojkov (Berkeley: University of California Press, 2011), 85.

22. Davis, *Ecology of Fear: Los Angeles and the Imagination of Disaster*, 365.

23. Benjamin Shultz, "Land of Abundance: A History of Settler Colonialism in Southern California," Electronic Theses, Projects, and Dissertations (2020), 985, https://scholarworks.lib.csusb.edu/etd/985.

24. Shultz, *Land of Abundance*, 20.

25. Kevin Starr, *Material Dreams: Southern California through the 1920s* (New York: Oxford University Press, 1990).

26. William James, *Essays in Radical Empiricism*, edited by Ralph Barton Perry (New York: Dutton, 1971).

27. John J. McDermott, *The Culture of Experience: Philosophical Essays in the American Grain* (New York: New York University Press, 1976).

28. Jackson Lears, *No Place of Grace: Antimodernism and the Transformation of American Culture, 1880–1920* (New York: Pantheon, 1981), 47.

29. Lears, *No Place of Grace*, 195.

30. Leslie Greene Bowman, "The Arts and Crafts Movement in the Southland," in *The Arts and Crafts Movement in California: Living the Good Life*, edited by Kenneth R. Trapp (New York: Abbeville Press, 1992), 163–201.

31. Daniel J. Kevles, *The Physicists: The History of a Scientific Community in Modern America* (Cambridge, MA: Harvard University Press, 1987), 83.

32. Jeffrey Fear and Paul Lerner, "Behind the Screens: Immigrants, Émigrés and Exiles in Mid-Twentieth-Century Los Angeles," *Jewish Culture and History* 17:1–2 (2016): 1–21; Jarrell C. Jackman, "German Émigrés in Southern California," in *The Muses Flee Hitler: Cultural Transfer and Adaptation*, edited by Jerrell C. Jackman and Carla M. Borden (Washington, DC: Smithsonian Institution Press, 1983), 95–110; Anthony Heilbut, *Exiled in Paradise: German Refugee Artists and Intellectuals in America from the 1930s to the Present* (Boston: Beacon Press, 1984).

33. Kevin Starr, *The Dream Endures: California Enters the 1940s* (New York: Oxford University Press, 2002).

34. Executive Order 9066, February 19, 1942; *General Records of the United States Government*, Record Group 11, National Archives.

35. Richard Griswold del Castillo, "The Los Angeles 'Zoot Suit Riots' Revisited: Mexican and Latin American Perspectives," *Mexican Studies/Estudios Mexicanos* 16:2 (2000): 367–391.

36. Eric Avila, "All Freeways Lead to East Los Angeles: Rethinking the LA Freeway and its Meaning," in *Overdrive: L.A. Constructs the Future, 1940–1990*, edited by Wim de Wit and Christopher James Alexander (Los Angeles: Getty Research Institute, 2013), 35–47.

37. Linda Christensen, "Stealing Home: Eminent Domain, Urban Renewal and the Loss of Community," *Rethinking Schools* 27:4 (2013): 34–41.

38. Eric Avila, *Popular Culture in the Age of White Flight: Fear and Fantasy in Suburban Los Angeles* (Berkeley: University of California Press, 2006), 59–61.

39. George Mariscal, editor, *Aztlán and Viet Nam: Chicano and Chicana Experiences of the War* (Berkeley: University of California Press, 1999).

40. Tom Wolfe, "The New Life Out There," *New York Herald Tribune*, November 21, 1965, reprinted in *McLuhan: Hot and Cool*, edited by Gerald Emanuel Stearn (New York: The New American Library, 1969), 30–48.

Chapter 3

Preparing for Life

MILLENNIAL ADOLESCENCE

With the transition to the Anthropocene and the more contingent modern history of Southern California, it may be useful to consider the cultural significance of the millennium itself. The new epoch brought fundamental societal transformations, characterized by globalization and digital interconnectedness, with the prospect of innovative ecologies of learning, working, and community life. Marshall McLuhan's idea of a "global village"[1] interconnected by media technologies bodes well for youth seeking to expand their horizons through Internet and electronic communications. Such exploration frequently occurs in familiar locales, notably the school and the community, where creative efforts can provide the young with both competence and confidence in the tools required to understand and perhaps embody globalizing changes. An earlier period of innovation, in the years around 1900, or the fin de siècle, evoked comparable expectations of a civilizational transition.[2] An accompanying generational crisis took place then, characterized by a cultural modernist revolt against tradition and a turn away from the certainties provided by classical science.

During this time, the very idea of modern adolescence emerged, together with more active pedagogical approaches and a nascent high-school movement. There was a growing understanding that adolescents' personal agency comes about through a reciprocal relationship between the social environment and an emerging sense of self. In 1904, G. Stanley Hall conceptualized adolescence as a universal experience, one denoting the confluence of biological and social forces that produced emotional and behavioral upheaval at this stage of life.[3] Hall's ideas provided a rationale for high school as an institution that permitted adolescents the freedom to mature before carrying out adult tasks.[4] A transition where biological and psychological capacities and social context intersect, adolescence is a life stage when personal identity concerns are central, prompting critical decisions about the future. The modern idea of adolescence thus appeared when machine technologies freed youth from factory and farm work to become educated during their teenage years. This was also a time when child labor laws were enacted, and compulsory education was being taken seriously, as was the place of the high school in modern American society. Industrial transition and educational reform, together, reshaped the minds and the bodies of modern youth and brought new insights on adolescents' development and their struggles toward maturity. Youth require new ways of learning that provide creative opportunities for them to explore and experiment. A reoriented educational landscape thus opened spaces where youth could engage in more embodied learning activities. Experiential pedagogies were therefore designed to motivate youth to think critically and solve complex problems, to work collaboratively, to communicate effectively, and to learn how to learn through self-reflection.

Adolescent identity formation also changed as youth flocked to movies in 1920s America, where they saw images of themselves on the big screen. Jazz Age musical rhythms and social dancing freed their bodies to experience the new physical demands that accompanied frequently complex dance steps and accompanying positions. Gaining the right to vote, women adopted fashions that provided more freedom as well. Like womens' wear, men's clothing moved toward comfort and simplicity of style. These cultural trends along with changing sexual norms influenced a vibrant youth culture that emerged along with modern high school in a time when adolescents embraced cultural modernist ideas, trends, and ways of living. High school would come to define adolescence, as it became a central place that sustained intense peer group interactions. School life spent among peers in a separate world away from home helped form teenage identities. The historical moment of urban life in the second industrial revolution that reinvented high school joined with the cultural moment, which brought forth a generational cohort of culturally modernist youth deeply involved in adolescent peer culture.[55] Paula Fass argues that youth embodied the dynamic changes that characterized the new era; however, more than a microcosm of the time, youth

> was a social product made possible, even necessary, by those changes which various historians called industrialization, modernization, or rationalization of American life. What the society experienced in larger terms as a maturing economy, an urbanized geography and a nationalized culture—all bewildering and threatening to those who remembered when things were otherwise—individuals born in the new century experienced on a more intimate (but no less significant) level in the changing experience of growing up.[6]

Core institutions shaped by industrial transition, notably the public high school and the workplace, influenced cultural transmission of cognitive, technical and interpersonal skills, and in turn increased adolescents' life chances and future opportunities. Whether students followed an academic or manual training program in the comprehensive high school, new occupational techniques and habits of mind reoriented teaching and learning, and the course of young lives. Project-based learning, one such instructional approach, thus arrived during a historical and a cultural moment, that of the Machine Age in the early twentieth century.

PREPARATION FOR LIFE

Around the turn of the century, improved machine tools and Frederick Taylor's "scientific management" principles had revolutionized production techniques.[7] In Taylor's system, scientific managers and engineers controlled production techniques from above, while workers carried out tasks in compliance with the predetermined work process. The workers' technical understanding of the work process was limited, as their sole function was to carry out certain specialized task in the overall production routine. Prior to these changes in production relations, the "working craftsman was tied to the technical and scientific knowledge of his time in the daily practice of his craft," as "apprenticeships commonly included training in mathematics, including algebra, geometry, and trigonometry, in the properties and provenance of the materials common to the craft, in the physical sciences, and in mechanical drawing."[8] With "the specialized and rationalized control of large scale production"[9] and the intensification of the labor process came the destruction of craftsmanship and, as a result of the separation of craft knowledge from craft skill, the "deskilling" of the worker. At this time, public school reformers and policy elites also adopted the language of scientific management on behalf of reorganizing the educational system toward both centralization and control from above by experts, and the adoption of practices, such as "the age-grading of students, the division of knowledge into separate subjects, and the self-contained classroom with one teacher."[10]

In Gilded Age America, the public school was the domain where children and their parents were socialized to comply with the rules and demands of the new industrializing society and its governing elites. The Progressive culture that was taking root in the American 1890s held that "children were a precious gift to be protected and nurtured."[11] To this end, the public school was a place that protected the young from being exploited as child workers in industrial mills and mines, and the reformers' call for compulsory education was a clear attack against child labor and the reliance on children's wages for the family economies of poor urban immigrant homes and farm households in the agricultural countryside.

Pragmatic philosopher John Dewey founded an experimental school in Chicago in 1894 and opened its doors to twelve children two years later. He envisioned the school as having features similar to "the workshop, the laboratory, [with] the materials, the tools with which the child my construct, create, and actively enquire."[12] For Dewey, the classroom was a "miniature community" where play was integral to learning social roles and actively engaging with the physical environment.[13] He viewed the school as providing students "with the instruments of effective self-direction,"[14] tools that would help them to gain greater control over their cognition and social behavior, as well as over their social and physical environments. In *Democracy and Education*, Dewey defined education as "that reconstruction or reorganization of experience which adds to the meaning of experience, and which increases ability to direct the course of subsequent experience."[15] According to Lawrence Cremin, Dewey's conception of growth was central to his view of the aim of education as a directive structure that would "expand the range of social situations in which individuals perceived issues and made and acted upon choices."[16]

Dewey thus understood the school as a place where students develop the habits of mind that would enable them "to control their surroundings rather than merely adapt to them."[17] For Dewey, social reconstruction could only occur after individuals used scientific inquiry to reflect upon their experience and to understand the social consequences of their behavior. Hence, it is only in and through a "community of inquirers"—an idea that Dewey received from philosopher Charles Sanders Peirce[18]—such as the progressive school, that cognitive processes for regulating human behavior could be developed and tested. Dewey's progressive school was a setting where rules, based on such experientially-derived knowledge, were socialized and used to guide further inquiry, presumably for community betterment.

Dewey's educational ideas, especially the notion of the school as a "social laboratory," inspired teachers and influenced educational policy, so that by the 1920s, public schools were charged with preparing youth "for their dual responsibilities to the social order: citizenship, and more important, learning to labor."[19] As David Nasaw states, "the school's goal was to take the children from the streets and the streets from the children,"[20] and city kids experienced a profound discontinuity between the "as if" world of the classroom and the "real world" of the street. The street presented a separate world, with a set of rules and social codes vastly different from those encountered in the school. As "a larger play community and a smaller subgroup defined by age and gender,"[21] the block brought together children from any number of ethnic, language, and religious groups who would manage the space and the activities within it. The lessons learned on the home block included loyalty to peers and family members, rather than to "outside" adult authorities such as policemen and teachers, but also to the community as a place to play, to socialize, and to feel safe within.

The rise in public education in the Gilded Age, between 1870 and 1900, brought about higher levels of basic literacy, but also helped to shape the class structure of the emerging modern industrial nation. During this period, the percentage of American adults completing high school rose from 2 to 6.4 percent, with the percentage for native-born Anglo youth rising considerably faster than for immigrants and African Americans, leading to social and economic disparities that would continue for generations.[22] Youth completing high school and going on to college in these years, albeit without standardized testing and College Boards, increased their chances of entering the professions or well-paying jobs in finance, journalism, and advertising—careers that characterized the expanding urban middle classes. The expansion of secondary schooling also increased the formal education of young people entering into various trades and in the skilled manufacturing sector, as machinists, electricians, and technicians. Students now learned in high schools the cognitive skills that were previously learned on-the-job, including mechanical drawing, algebra, geometry, and trigonometry.[23]

As the national population increased dramatically between 1900 and 1930, from 76,094,000 to 123,076,741,[24] so too did the secondary school population—from 630,000 in 1900 to 4,740,580 in 1930. The expansion of educational attainment in the twentieth century—"by 5.27 years for those born in the United States from 1895 to 1975"[25]—is attributable to the spread of mass secondary schooling, which accounts for about 50 percent of that gain. According to Claudia Goldin, the United States underwent a "great transformation" in its educational history,[26] resulting partly from the "high school movement" on behalf of "the publicly funded academic secondary school for the masses,"[27] where the focus was "schooling for life" rather than for entering college. As a result of this change toward mass secondary schooling, and the shift from training "for college" to training "for life," in 1910, 49 percent of high-school graduates continued to some form of higher education; by 1933, only 25 percent did.[28]

Early adherents of the high-school movement were the Prairie states, such as Iowa, North Dakota, and Nebraska, where progressive farmers understood the value of the natural and applied sciences to operating a modern farming enterprise. Moreover, their state and local governments were clearly able to assume the cost of secondary schooling because "the prairie states were wealthy farming areas during the first three decades of the twentieth century and wealth in land is more easily taxed than its portable forms."[29] Small towns and villages in the West and Midwest, especially embraced the high school movement as these places offered young people little opportunity to do much during most of the year, relative to the work available to youth in the larger cities and in the open country; moreover, "small-town America was a locus of associations (religious, fraternal/sororal, business, and political organizations) that could have played an important role in galvanizing support for the provision of local publicly provided goods, including high schools."[30] Along with grassroots efforts to expand public secondary schooling were state financial incentives toward the creation of large school districts through consolidation and uniform graduation standards.[31] By 1938, the regions that led the high-school movement—New England, West North Central, and Pacific—attained graduation rates around or exceeding 60 percent and enrollment rates greater than 80 percent.[32]

The high-school curriculum changed dramatically to accommodate the occupational diversification that took place during this transition, from nine subjects in 1890 to forty-seven in 1928.[33] High-school graduates, as educated labor, brought a valued cognitive skill set to the twentieth-century shop floor; "they could read manuals and blueprints, knew about chemistry and electricity, could do algebra and solve formulas, and . . . could more effectively converse with nonproduction workers in high-technology industries."[34] With the phenomenal growth of schooling for vocational achievement came the expansion and professionalization of teaching, including an upgrading of teacher preparation along progressive lines.[35] Amid the whirlwind of initiatives and interests that fostered changes in those years, the earlier foundational work of Dewey and other pragmatists grounded progressive instructional approaches.[36]

In decades following the Second World War, the American educational system underwent a process of "pluralist accommodation" that required a reorientation of educational perspectives to a new set of economic forces, such as "the 'free market' choices of students, the school bond-issue referenda, the deliberations of elected school boards and the like."[37] Educational reform was focused on integrating into the wage-labor system "uprooted southern Blacks, women, and the once-respectable, 'solid' members of the pre-corporate capitalist community—the small business people, independent professionals, and other white collar workers."[38] Through the recent cycle of pluralist accommodation, school systems have taken on the social relations of production characteristic of corporate enterprises by embracing the "free marketplace in ideas," decentralized decision-making, and top-down control of information, programs, and outcomes. This was an effort to align educational processes toward life and work within the new economy of the late 1970s and early 1980s, marked by a slowdown in productivity, rising oil prices, inflation shocks, deindustrialization, and the abandonment of city living by white working-class families.[39]

The advent of the globalization of the neoliberal 1980s brought on a "fragmented centralization" to local school districts. Under the differential interest and economic incentives the local districts increased their centralized access to power, influencing infrastructure, purchasing, and technology decisions; meanwhile "social control and punishment for failure is decentralized" thereby holding individual schools responsible for poor student and/or teacher performance.[40] Educational attainment slowed considerably during these years for native-born Americans, so that "the educational attainment of a child born in 1975 was just 0.50 years more than that of his or her parents born in 1951, but the educational attainment of a child born in 1945 was 2.18 years more than that of his or her parents born in 1921."[41] With growing bureaucratization, union demands, and school board pressures for better student outcomes, K-12 educational institutions were drawn into adopting more rigid pedagogies. The schools were subjected to corporate-style performance measures, forced into using both standards-based textbooks from a handful of publishers and mandated standardized tests originally devised by experts to rank students but now used to measure the quality of teaching and learning. Research has shown that non-instructional factors in the home, such as parents' educational levels, and in the community, such as poverty and neighborhood disorganization, accounted for most of the variance among test scores.[42] Despite these data, the top-down model of educational reform, with its standards-based approach to educational achievement and a reliance on testing for graduation and promotion, would carry into college and even graduate coursework.

Under neoliberal reforms, conventional schooling is a consumer good, with high school becoming more mechanistic. Schools are driven by accountability through testing, data collection, ratings, and satisfaction scores, all measures designed to boost competitiveness within market forces. The reforms have neither reduced inequality of opportunity nor improved college-readiness in low-performing schools where poorer students are more likely to attend. Students thus face the dual challenges of mechanized instruction and unequal opportunities. Still, experience-based pedagogies, notably project-based learning, mentored internships, and apprenticeships, are designed and carried out at certain progressive high schools. As nonmechanistic approaches to instruct and inspire youth, these pedagogies provide a balance between experiential and didactic learning. These modalities also forward occupational techniques and habits of mind that move youth toward greater understanding of themselves and their world. As developmentally sensitive approaches, they promote a learning that supports adolescents' need to lead purposeful, self-reliant, and socially engaged lives.

COMMUNITIES OF PRACTICE

The many critical responses to neoliberal schooling are instructive for current efforts to reinvent high school. Robert Halpern advocates for high-school apprenticeship as a style of learning that supports adolescent development and affords youth optimal chances for productive living. Halpern's approach requires buy-in outside the school, notably industries and practice communities in the new digital economy. Within these worlds, adult mentors must be willing to persevere with the youth in their steep learning curves, incongruous family worlds, and the false starts and setbacks in their education to expose the young to a culture of artisanship. Work-based learning provides "the means to grow up,"[43] in Halpern's words, as youth explore pathways for career and personal success, easing their sense of ontological precariousness. To create the conditions where a shift toward project-based learning can occur, progressive educators a century ago went beyond their classrooms to explore work within reinvented Machine Age organizations. This entailed the design of inspiring classroom-based projects that, through their active engagement with materials and processes, opened pathways for youth that connect the classroom to the world. The same is required today, as teachers seek cooperation beyond the school from skilled professionals in new economy workplaces and advocates in youth-serving organizations to establish communities of practice.

Efforts on behalf of children and youth increased significantly in the early twentieth century, a period of civic inventiveness when informal networks evolved into formal organizations to mediate the painful effects of industrialization, urbanization, and immigration on youth and their families. Networks of civic engagement defined styles of growing up through public and lay support of youth-serving organizations and activities. Civic networks addressed youth and their needs, typically forming around better high schools, parks, playgrounds, camping, and team sports. Each programmatic effort supported a dual ethos of voluntary social action and belonging, ideological components of American national identity development since frontier times. Since that time, experts and lay people together became actively engaged in the public sphere on behalf of schools and the life chances of children and youth. Direct experiences in these same public arenas are necessary for youth as they encounter situations beyond home and neighborhood in their struggle for social identity and a place in the world. For such a space to open within an enlarged public sphere requires intergenerational communities of practice to promote collaborative learning opportunities for youth.

Jean Lave and Etienne Wenger hold that situated learning activities require a deepening commitment to engage in a community of practice.[44] The community of practice is an organizational form that complements the current economy focused on the "production and services based upon knowledge-intensive activities,"[45] as defined by Walter W. Powell and Kaisa Snellman. This learning model has accelerated with advances in information production and dissemination. Promising to "radically galvanize knowledge sharing, learning, and change,"[46] as Wenger and William M. Snyder assert, a community of practice provides a framework for understanding social learning in complex organizations, specifically the notion of knowing. Knowing within a community of practice is based upon socially defined competence, the ability to act and be viewed as a competent member, and ongoing experience within a learning community.

This style of learning involves collaborative practices, historically situated within the everyday, local knowledge of a specific community that is also interconnected with the wider society. Belonging to a particular community is based upon engagement, imagination, and alignment within learning ecologies supportive of group members and the community itself. Communities of practice provide a framework for social learning when members share a sense of joint enterprise as they commit significant energy to the task. Members interact on the basis of mutuality and through participating in these encounters develop an "enactive intersubjectivity,"[47] a relational knowledge of how to deal with others. They also share a repertoire of resources based upon levels of participant self-awareness and reflective capacities. This framework—of knowing, belonging, and social learning through more informal styles—provides members with the skills to meaningfully engage in knowledge production, exchange, and transformation in complex organizations.

Julian Orr observed how a culture of artisanship exists among technicians through an informal community of practice based upon "becoming and remaining a competent practitioner,"[48] within the community. Seasoned practitioners employ stories to celebrate their own competency and to instruct novices about the nuances of their craft. An occupational community is maintained through technicians' war stories about confronting and overcoming problematic situations that arise in technical work. These stories "preserve and circulate hard-won information and are used to make claims of membership and seniority within the community,"[49] in Orr's words. Practicing artisanship requires facility in occupational techniques and holding a set of perspectives that, together, permit strategic mastery of a workplace culture. These habits of mind and specialized attitudes are then re-enacted and improved in professional or occupational practices.

Enculturation of craft knowledge and occupational perspectives and their acquisition occurs through situated learning within a community of practice where experienced mentors guide novices toward a path to mastery. A novice's craft consciousness, then, emerges through intimate familiarity with materials and processes, embodiment of craft techniques, and proficiency in social relations within these "separate worlds of knowledge,"[50] following Joseph Bensman and Robert Lilienfeld. As youth become immersed in learning a skilled craft, through the sum of work-based learning experiences and attendant embodied knowledge they subsequently symbolically create and re-create these separate worlds through their occupational practices and social interactions. Craft work thus socializes a skill-oriented occupational identity that privileges individual and interpersonal competence and a broader social consciousness.

CRAFT AND CONSCIOUSNESS

Cognitive psychologists and anthropologists recently discovered what artisans have always known—the value of project-based learning activities, notably work-based learning through mentored internships and apprenticeships, in articulating craft knowledge. Craft refers to any skilled work music, carpentry, surgery, handicraft, writing, drafting, experimental science, cartography, animation, photography, filmmaking, sound engineering are older crafts made more efficient through digital computer technology. The professional and service sectors use the digital screen as a supportive device to map information on behalf of their practices, including information, data, record keeping, billing, appointments, and the maintenance required to keep hardware running.

Through ethnographies of apprenticeship, Jean Lave identified a process that connects students' attainment of classroom knowledge to skill acquisition in the world beyond the school, namely, "learning in practice."[51] Exploring learning as participation in both apprenticeship and in American high schools, Lave found that learning in school is not fundamentally different from learning anywhere else. She forwards a contextualized approach to learning, arguing for analyses of "learners learning,"[52] rather than "learners being taught."[53] Lave suggests that to be effective, teachers will need to acknowledge, relationally, the collective identity making of their students "in their practice as learners learning."[54] She maintains that learning viewed as "situated activity" has as its central characteristic a process called "legitimate peripheral participation" (LPP).[55] Learning is therefore a function of the activity, context, and culture in which it occurs (i.e., is situated). Learners participate in communities of practice, moving toward full participation in the sociocultural practices of a community. LPP, as a process of "co-participation,"[56] provides a way to speak about crucial relations between newcomers and established members, and also about their activities, identities, knowledge, and practice. Lave embraces the various strategies of inquiry that situate learning within more interactive frames, such as craft apprenticeships, that offer more counterintuitive approaches to skill mastery.[57]

Enculturating craft knowledge requires progressive academic teachers, vocational-technical educators, and school-to-work program administrators to advocate on behalf of work-based learning across academic and vocational education. A first step is to determine demographic factors that condition this knowledge, and then ask: who works where, uses what, lives near and succeeds or fails within the new economy based upon the digital technologies, notably in biotechnology, information technology, allied health, and design industries. According to Robert Halpern, high-school apprenticeships will remain "an ad hoc, decentralized, idiosyncratic phenomenon,"[58] one that will rely upon students and mentors, together, to articulate next steps once the apprenticeship ends. A major task is that of redefining pathways toward greater inclusiveness in acquiring craft techniques for entry in the new economy's workplaces. This will entail taking actions to assure wider paths to youth apprenticeship opportunities, broadly understood to include the scientific and technology sectors. Practitioners with mastery maintain the occupational boundaries of their craft, controlling both entrance and training opportunities. Expert practitioners influence the shape of their craft within a regional economic landscape, as they articulate the selection of novices and the modes of early training. Residential patterns and economic prospects also follow the path of craft knowledge within a region. Assuring students entry into regional career pipelines requires high schools that offer such apprenticeships to sustain collaborations with expert practitioners within the new technological workplaces. Beyond the classroom, pedagogies grounded in the design of quality projects are typically continued as work-based learning in extended and summer internships.

By articulating informal and in-school learning, the new knowledge created in each mutually influence efforts to sustain high-quality student work. Ron Berger holds that by encouraging such work, teachers build a "culture of craftsmanship" with their students, which may then translate to a schoolwide "ethic of excellence."[59] Craft-based instruction provides a foundation that can lead to developing routine expertise or core competencies in well-practiced routines. Early exposure to craftsmanship at school may motivate a later adaptive expertise that involves both innovation and efficiency. Adaptive expertise, according to John Bransford and his associates, may be driven by: "a metacognitive awareness of the distinctive roles and tradeoffs of the innovation and efficiency dimensions of expertise, and the active design and creative structuring of one's learning environment in order to support their dual utilities."[60] Designing spaces for craft-based learning in high school can then provide paths toward routine expertise and for attaining adaptive expertise later in their careers.

Craft-focused pedagogy resurfaced in certain schools today, as a century ago, to expose youth to a knowledge economy based upon innovative technologies and their products, and also new ways to deliver services. Then as now, these approaches fit within Hannah Arendt's view of education as a way of providing future generations with tools for "setting right" a world that is "out of joint" and perhaps "wearing out" because of prior generations' demands and uses.[61] David Orr asks educators to consider including ecological design arts into the curriculum to promote "a symbiotic relation between learning and locality."[62] Following Dewey's critique of conventional schools as being "so isolated from the ordinary conditions and motives of life,"[63] Orr proposes that campuses become laboratories for the study of ecological design, focusing on food, energy, water, materials, and waste with the following objectives so as "to equip young people with a basic understanding of systems and to develop habits of mind that seek out 'patterns that connect' human and natural systems."[64] With these objectives in mind, accessible spaces can be created to promote learning beyond conventional schooling where individuals are exposed to what Thomas Bender refers to as "urban knowledges."[65] These professional, creative, and social forms of knowledge define a "new metropolitanism"[66] that is both cosmopolitan and socially inclusive.

Craft may also be an effective way of integrating traditionally accepted and more innovative views of education and vocation. Project-based learning articulates how experiential pedagogies can embody core elements of a culture of artisanship. In communities with a mix of immigrant and second-generation families, many young people often are the first to complete an academic program in high school and go on to college. For these adolescents, success in navigating school and family worlds in the critical last two years of high school often requires a demonstrated connection to the artisanal tradition, broadly understood. In high schools committed to a culture of craftsmanship, where teachers collaboratively design curriculum and projects for their students, a widening space is being created for the broad dissemination of project outcomes. End-of-year large-group exhibition night events highlight students' work within the classroom and beyond, providing opportunities to discuss project achievements with other members of the school community. Family members can observe their students' work, meet their teachers and mentors, and listen as youth discuss their projects with other attendees, thereby inspiring them toward greater confidence in their children's prospects.

A sense of "knowing-in-action"[67] derives from participation in mentored classroom projects, work-based internships and apprenticeships during later adolescence, a time for exposure to and appreciation of occupational knowledge and values. The intent is that the youth will carry with them the tools and experiences gained through project-based learning beyond high school. By creating the conditions for a reoriented identity associated with academic interest and success, these pedagogies hold the promise of more fulfilling life trajectories. Communities of practice inspired by the craft tradition afford youth opportunities for expansive learning activities to gain technical skills along with social and emotional competencies. There, they can discuss academic and career aspirations, and also interpersonal challenges in the school, among their peers, and within their families. Mentored learning experiences encourage youth to move beyond outmoded class reproduction practices and embrace a style of conscious praxis where they can express their educational and occupational talents and speak on behalf of their projects and their futures.

EMBODIED COGNITION

Adolescence is marked by higher neuroplasticity and is a time when the brain is especially sensitive to social environmental influences.[68] The adolescent brain thus responds to new information and experiences, forming neural connections that integrate and consolidate different regions of the developing brain.[69] These are years of increasing exploration and experimentation, and certain types of learning experiences can give shape to behavior and future direction.[70] Mentoring opportunities take on greater importance during this time of complex neural reorganization when adolescents are especially sensitive to interpersonal influences.[71] Cognitive psychologists recently found social and cultural practices that contribute to adolescent cognitive development by increasing their capacity to engage in systematic thinking.[72]

An embodied cognition approach helps frame cultural processes underlying social cognitive development and has influenced contemporary field-based research in thinking, learning, and action in adolescence. Francisco Varela, Evan Thompson, and Eleanor Rosch introduce human experience and enactive cognition into cognitive science, based upon the phenomenological perspectives of Maurice Merleau-Ponty grounded in perceptual experience, bodily expression, and institutional life.[73] The body is a community of senses that organizes an individual's experience and gives meaning to the world. The concept of "enaction" emphasizes how the world of lived experience can be understood as mutual interactions between bodily activities of the organism, its sensorimotor circuit, and the environment. Cognition is essentially the coupling of brain, body, and world through activity, to derive meaning through an engagement with the environment.

Thompson advances this initial argument, adopting a biological construct, "autopoiesis," introduced by Humberto Maturana and Varela to denote the self-maintaining chemistry of living cells and the organization of living systems.[74] Living systems maintain and reproduce themselves through self-regulatory strategies. Thompson embraces autopoiesis to explain cognition as a sensorimotor activity, viewing the world of lived experience as emerging through bodily activities of the organism. For Thompson, "cognition is behavior or conduct in relation to meaning or norms that the system itself enacts or brings forth on the basis of its autonomy."[75] The trend toward autonomy is encapsulated in adolescence. How self-regulation will develop is largely dependent upon the varieties of opportunities for autonomous learning in the social environment.

Albert Bandura's social cognitive theory understands human learning as occurring in the social environment. Enactive learning, or learning from the consequences of one's actions, involves self-regulation and the encoding and storage of action-based information in the memory.[76] Although generally associated with earlier stages of cognitive growth, enactive learning together with discovery learning, conceptualized as "all forms of obtaining knowledge for oneself by the use of one's own mind,"[77] as Jerome Bruner states, informs a hypothetical mode and necessitates a cooperative position of teacher, or mentor, and learner. Social neuroscience views cooperative approaches as ways to enhance the social interaction essential to learning. Learning encounters are supported by neural circuits linking perception and action for close coupling and attunement between youth and mentor and for synaptic plasticity, following Andrew Meltzoff.[78] These interactions then bring about powerful cognitive effects and neuronal connections. Experiential learning occurs between youth and their mentors as dialogues through which the mentoring relationship is constructed and negotiated. Interactions derived from such cooperative engagement support learners' conceptual judgment and reflective insight, thus bridging enactive and discovery learning. Youth also learn and comprehend the worlds around them through a shared intentionality from their everyday interactions. In more directed learning with their peers, they not only modify routine practices but also create new knowledge through collaborative activity. Such processes are critical in an ecological transition.

An autonomous unit denoting self-organization, a "holon" is an apt model for the restructured habits of mind essential for engaging the next culture's technologies, societal arrangements, and emergent mindsets.[79] Within education, the term may represent individual integration and personal agency attuned with comparable processes taking place in the larger world. It may also be analogous to the connectivity that unifies the learning experience, both cognitively and behaviorally. Active learning, then, may be adaptive in this regard, providing the ability to effectively navigate rapidly changing sociocultural and environmental conditions.

Neural networks are attuned to such learning, as the requisite engagement and reflection require flexibility. Likewise, these networks are themselves undergoing shaping and reconfiguration, internally constructed from external situations, notably interactions with peers and mentors that provide meaning and purpose within learning encounters. The emergent feedback loops sustain ongoing developmental capacities based upon experiential practices within a socially mediated learning environment. Such processes compel a path toward sustaining both resilience and creativity by establishing new learning potentials for educational success. Flexibility may even permit the young from becoming entangled in societal arrangements adaptive for previous generations, but less so for rising ones. By efficiently processing new information, while letting go of earlier, outmoded conscious and subconscious patterns of thought, the young can then move more flexibly within the next economy and society. Current accelerating conditions of life require the young to embrace active and flexible learning styles into their late 20s after which neuroplasticity, characterized by rapid synaptic pruning and growth, tends to diminish.

CULTURE AND LIFE SPACE

Michael Cole's cultural-historical approach "assumes that mind emerges in joint mediated activity of people."[80] Mind is thus "co-constructed" through a unity of material and symbolic elements within a culturally informed context. Developmental change takes place through intergenerational culturally mediated situations that weave together various practices to socialize the young. Using this approach, Barbara Rogoff defines human development and learning as processes of participation in cultural communities, holding that "development can be understood only in light of the cultural practices and circumstances of their communities—which also change."[81] Children's observational learning is then apprenticeship learning through "guided participation" that "resembles the structure of learning and assisting of mastery in apprenticeships."[82] Cole and Rogoff provide a social developmental framework that includes culturally contextualized practices; one that resonates with experience-based instructional approaches consciously designed by teachers and mentors to produce deeper learning outcomes.

Antecedent to this line of thought is the cultural-historical psychology of Alexander Luria and Lev Vygotsky. Alexander Luria, studying developmental neuropsychology, conceptualized the "working brain,"[83] with co-active attention, programming, and energetic processes. For Luria, the brain is organized, hierarchically. At the primary level are the sensory functions; at the secondary level, sensory information is associated with prior learning; at the tertiary level, the brain integrates this information for decision-making and problem solving. This last level can be viewed as the attention association area. Children's cognitive development occurs through a "process of objective activity and communication with adults,"[84] as Luria, states. The child's attention and problem solving are mediated by mastery of language and the symbolic function, and later internalized, giving rise to higher mental functions that are, per Luria, "complex and self-regulated and are social in origin, mediated in their structure, and conscious and voluntary in their mode of functioning."[85]

Lev Vygotsky, Luria's colleague, understood that higher mental functions develop by a process of mediation, primarily through language and other symbolic tools, including art, writing, and mathematics.[86] Human mental functioning develops in sociocultural settings through mastery of complex systems socialized through historically constructed institutions. Culture plays a role in shaping the mind through a set of complex social relationships. A child internalizes knowledge through everyday experiences within a social world constructed through language and tools. These secondary experiences give rise to new cognitive groups or psychological systems within the developing child that are both mental and physiological.

Following Luria, Vygotsky understood that social mediation helped form the new system, wherein "two points in the brain which are excited from outside have the tendency to work in a unified system and turn into an intracortical point."[87] This mediation takes place within a "zone of proximal development," a spatial construct indicating the distance between individual performance and assisted performance, through which a mentor establishes activities, or experiential and cultural tools, for a child's social and cognitive development. Conceptual thinking in adolescence occurs through specific social interactions, presumably through authentic instructional activities with both teachers and peers to facilitate autonomous problem solving and decision-making.

For Vygotsky, analytical skills and hypothetical reasoning are "connected with the process of growing into the cultural, professional, and social life of adults."[88] Ron Ritchhart sees Vygotsky's position as informing a "culture of thinking"[89] in school and elsewhere, one where "learning to learn is an apprenticeship in which we don't so much learn from others as we learn with others in the midst of authentic activities."[90] Along these lines, Vygotsky's ideas also inform the "Third Space" construct in education.[91] Denoting a hybrid formation, a Third Space emerges through dialogues that bridge dominant discourses and knowledges enculturated through conventional instructional approaches with diverse epistemologies derived from home and community discourses and students' historical experiences.

Yrjö Engeström builds upon the Third Space for his concept of "expansive learning," which creates "a radically wider horizon of possibilities than in the previous mode of activity"[92] through enaction of two interactive activity systems. Kris Gutiérrez found that expansive learning can include a "collective Third Space"[93] comprised of multiple spaces derived from formal learning in the classroom and what students learn outside school. Diverse "repertoires of practice" form a particular learning ecology mediated by "'sociocritical literacy,' a historicizing literacy that privileges and is contingent upon students' sociohistorical lives, both proximally and distally."[94] A mix of formal learning and cultural knowledge derived from everyday life, these repertoires mutually influence the individual and the practices themselves. Through their movements within particular learning ecologies, youth "appropriate cultural concepts as they move in and across a range of social practices—attending to the ways boundary crossing reinforces, extends, and conflicts with individuals' dispositions and repertoires of practices."[95] Vygotsky's cultural-historical activity theory thus supports dynamic enculturation of intellectual, or thinking, dispositions as a mediated activity in the classroom and beyond in a widening social world, both expanding the horizons of the young while also reshaping learning.

The later work of the Russian cultural-historical school was influenced by German social psychologist Kurt Lewin. The interplay between psychological and nonpsychological factors exemplifies Kurt Lewin's theories of human behavior as a function of the interplay between the person and the environment.[96] For Lewin, the crucial factor in understanding an individual's psychological field is that person's own interpretation of it. Lewin's phenomenological conception of the environment focuses on the way that an individual perceives personal "life space."[97] Lewin established a topological psychology of how regions, boundaries, pathways, and barriers in this space direct human action. A person's behavior is not merely steered from within but is motivated by the environment. Objects, activities, and other people in the psychological field also direct a person's behavior. Lewin views the relation between psychological and nonpsychological factors as central to understanding the boundary conditions of the individual or group. Culture and group life, for example, are quasi-stationary processes that serve to maintain the present situation; therefore, the starting point for any form of change will need to begin with cultural lifeways, or habits, as these result from "forces in the organism and its life space, in the group and its setting,"[98] in Lewin's words.

Influenced by Vygotsky and Lewin, Urie Bronfenbrenner conceives the ecological environment as a nested arrangement of four concentric structures, each contained within the next: the micro-, meso-, exo-, and macrosystems.[99] The microsystem comprises the activities, roles, and interactions of the young person in a specific setting where people engage in face-to-face behavior. The mesosystem comprises the interrelations among two or more settings in which the youth actively participate. The young person's development is affected by the interconnections among home, school, neighborhood, work, and volunteer service venues. The exosystem refers to those settings beyond the youth's immediate experience that nevertheless affect him or her, such as the parent's place of work. The macrosystem does not refer to a specific setting, but to the consistencies across the micro-, meso, and exosystems, and to the values and beliefs of the culture or society in which the young person lives. Bronfenbrenner's ecological systems theory frames human development as a dynamic, mutually reciprocal process. This model may serve to understand the boundary situations that confront youth. Within this framework, the young person not only influences the variety of settings encountered, such as the home, school, park, and the workplace, but is simultaneously influenced by these environments.

PLURAL LIFEWORLDS

Contemporary youth dwell within many small social worlds amid the fluid dynamics of late modernity. Their "plural lifeworlds,"[100] or worlds of everyday life shared with others, are fragmented across various domains in private life and public institutions, often without having any relation to one another. Francis Müller understands how each small lifeworld has its own "cultural grammar,"[101] which requires the young person to draw upon stored social knowledge and then take a specific role:

> A student, for example, operates in her parents' house, in her shared flat with her roommates, at a university, at a Kung Fu school, in a relationship, at a bar, where she works on weekends, etc. These "places" constitute specific social identities: She behaves differently at her parents' house, in her flat, and at the university.[102]

Within multiple social realities, youth are exposed to a multiplicity of information that compels their understanding of specific scripts and interacting accordingly. They call upon certain situation-dependent conceptual and communicative styles, which may be grounded in place, or more recently, are entirely placeless. The latter situations capture more of their lives, in part, through encompassing digital technologies that can both enhance learning, but also disrupt the tenor of personal and community life. Varying technologies tend to both integrate and fragment human experience, separating time and space, so that social relations are articulated beyond face-to-face encounters. Along with micro-encounters in cyberspace, youth confront abstract systems, such as science, technology, finance, and the law, which produce disembedding mechanisms that separate social relations from familiar locales. Once recognizable institutions, including schools, hospitals, courts, and workplaces, that originally operated through sustained face-to-face encounters, now routinely use digitally based information to manage operations within their settings. Altered structural conditions within these domains influence interpersonal relations making adolescence especially difficult, indicating the need for increased intergenerational developmental support of tasks associated with this life stage.

Adolescents are faced with two main developmental tasks: to integrate and adapt to the physiological revolution within themselves and to prepare themselves for the tangible adult tasks ahead of them. They are involved with how to connect their roles and skills with the occupational prototypes of their culture and are hampered by concern over how they appear to be in the eyes of others. Conformity is the outstanding characteristic of adolescents, but it is essentially conformity with their peer group and its standards. Adolescents have a great need to appear competent to the world around them, especially to their family and friends. They are struggling to achieve independence from the family and are torn between the desire for increasing responsibility and the ambivalent wish to maintain the more dependent role of childhood. Conflicts may occur with the family because the peer group seems to have become more important than the parental world. In the usual course of events, the struggle between adolescents and family plays itself out and, depending on the basic relationship between the adolescents and the family, the ordeals and problems are eventually resolved.

Resilience is a key factor in the development of coping patterns that play a role in effective adaptation to changing circumstances. These include the dual challenges of the transition from adolescence to emerging adulthood, namely, emancipation from parental control, concurrent with the transition to college and career paths. Resilience generally refers to manifested competence in the context of significant challenges to adaptation or development. Resilient youth have the capacity to spring back, or to recover rapidly, from the disruptions that they experience. Resilience is, in part, related to the adolescent's disposition, familial strengths and resources, quality of parenting, and stability of the home environment. Three protective factors found in resilient adolescents include positive dispositional features, such as mindfulness and gratitude, family cohesion, and the availability of external support systems.

Project-based learning supports and enriches the position of youth in the world beyond the family. Here, they can develop meaningful intergenerational bonds through mentoring and exposure to work practice activities in the community. Frequently, their communities are experiencing disruptions to their physical and social environments following any number of structural crises that can marginalize adolescent lives. Adolescents can regain their broken trust in the community as a safe place through helping to restore the locale. Intergenerational efforts on behalf of planning, design, mitigation, and even governance provide the sense of belonging that is a condition of place-based learning. Through engagement in such work-based activities, youth are exposed to urban knowledge opportunities in science, technology, and information, which may include the health, design, media, and environmental professions. Beyond such exposure to knowledge-based careers, mentoring experiences in advocacy, community organization, and communication skills create opportunities for engagement in the public sphere. As they acquire work practice skills, youth may take part in collaborative practices for any number of public concerns that surface in a community undergoing structural change. These activities can then lead to intergenerational partnerships for care and concern for community locales.

Projects exposing youth to urban professional knowledge thus serve as catalysts for community building and sustained civic responsibility. Through them, youth can better understand the broad structural contexts that affect their horizons. Such activities also strengthen civic networks that include nonprofit organizations, which support voluntarism, social trust, and reciprocity. These networks produce social capital and can frequently generate a civic arena around a common concern. The localized network may then facilitate intergenerational trust, cooperation, and communication, all qualities necessary to resolve adolescent life crises in a precarious world. Rationality and negotiation are central to action within the various networks that coalesce into civic arenas. These are the larger cultural settings crucial to maintenance of a public sphere where participants freely engage in political discourse and debate over common concerns. The styles of action within these arenas surround questions of identity, rights, and strategic behavior that are debated, negotiated, manipulated, and even coerced by their members. Typically, these arenas tend to only include the participation of adults on behalf of their own interests. Through their experiential learning activities, youth may also become active in civic arenas where they can give voice to making their future. Enlarging arenas to include youth also expands social capital within a locale. Youth participation contributes to character development beyond the school and may even sustain generational continuity of these arenas over time.

For well over a century, a number of arenas have emerged in North American communities as experts and laypersons, together, sought a path toward modernity. John Dewey and his fellow pragmatists considered such interaction within arenas as central to the construction and transformation of social worlds. The pragmatic theory of action encompasses the need for the interpretation of meaning as a process for overcoming roadblocks, whether environmental or situational, that impede the flow of routine activity within these worlds. Built into their scheme is the primacy of reflection upon and discussion of alternative paths of action as interactive processes necessary for transforming the course of everyday life.

ETHNOGRAPHIC MODERNISTS

At the dawn of the twentieth century, American pragmatic thinkers and progressive educators sought to make sense of the diverse national and subnational groups that moved into modern industrial cities. In-migrating families brought cultural styles and cognitive frames divergent from dominant worldviews. Ethnographic studies of their cultural ethos and lifeways informed progressive educators' efforts to design more authentic learning approaches. Ethnographers and educators at the time were cognizant that divergent local knowledges within the school and repertoires of practice in the family and home community brought distinct challenges. Then as now, they understood how such conditions can lead to tensions across school-based peer cultures, where conflicts between youth holding differing cultural styles and world images play out in distressing interpersonal encounters. These concerns are a part of the larger "project of modernity,"[103] as defined by Jürgen Habermas, where a systemic sphere of scientific and technical problems constituted by experts is distinguished from a lifeworld, a sphere of varying communicative practices within everyday social experiences of people interacting within their communities. Societal modernization depends upon rationality, specifically technological reason, to resolve the various problems and pathologies of humanity, often "colonizing" the lifeworlds of cultural "others" in a search for expert solutions. Anthropologists as ethnographic modernists focus on the lifeworld and processes of cultural modernization where differentiation remains a norm within local knowledges and practices.

Anthropologists, together with sociologists and journalists, began to document the assimilation of diverse cultural groups, their worldviews and divergent repertoires of practice within complex urban environments. They observed how liberal cities encountered challenges as they struggled to integrate people with differing cultural norms and ways of living into their civil societies. Before the First World War, Vienna, Berlin, London, Paris, and New York engaged with cosmopolitan modernity. All were liberal urban societies in the nineteenth-century sense, which grappled with essential questions of cultural differences. Many questioned how to assimilate "cultures into civilization" as defined by the elites at the time, including academics, clergy, politicians, entrepreneurs, and corporate leaders. Certain members of these groups blamed "cultural others," including immigrants, typically Catholics, "Negroes," and "the Jews," for the ills of modernity. Culture then became designated as "other" relative to the "civilizing" properties of urban cosmopolitan modernity, as Terry Eagleton asserts:

> Civilization is precious but fragile; culture is raw but potent. Civilizations kill to protect their material interests, whereas cultures kill to defend their identity. These are seeming opposites; yet the pressing reality of our age is that civilization can neither dispense with culture nor easily coexist with it. The more pragmatic and materialistic civilization becomes, the more culture is summoned to fulfill the emotional and psychological needs that it cannot handle—and the more, therefore, the two fall into mutual antagonism.[104]

Distinguishing various urban "others," categorically, arose with elite appropriation of technology, new forms of managerial organization, and social knowledge. Progressive-era religious leaders practiced social gospel to abate poverty in liberal cities and adopted these same categories, notably to promote alcohol Prohibition for "Negroes" and "foreigners," and Temperance for native-born Americans. Many late Victorians viewed immigrants, particularly the women, as morally compromised for adopting new urban lifeways. As for the men, labor leaders of the time shared similar Victorian backgrounds and viewed the workers coming into their mills and factories from the various cultures as clearly other, relative to "Anglo-Saxon civilization." These attitudes coalesced immediately after the First World War during the New Era's culture wars, from the Red Scare and the Scopes Monkey Trial to the sexual fundamentalism of those outside the big cities, following Ann Douglas.[105] After the 1929 Crash, the presidency of Franklin D. Roosevelt, the Repeal of Prohibition, and mobilization of diverse recruits during the Second World War, this late Victorian assault on alterity seemingly declined. Contentious debates surrounding difference and otherness then re-emerged with the ending of the war. The core issues revolved around integration (Negroes as culture; whites as civilization) and immigration (immigrants as culture; native-born whites as civilization). Amid a half-century of debates about human differences, American anthropologists set out through field encounters to understand alterity, inclusion, and consequent conflicts between dominant conceptions of culture and civilization that inspired these disputes in liberal societies.[106]

Within the discipline, cultural anthropology emerged during the 1920s and engaged in contentious debates over eugenics and scientific racism. After twenty-five years of intensive field studies, the subdiscipline refined an empirical method that would define and understand cultural differences in the postwar decades. North American anthropologists reinvented a professional discipline established decades before by Franz Boas, the founder of American anthropology.[107] Boas originally conceptualized a disciplinary matrix built upon four fields, namely, physical anthropology, archaeology, linguistic and cultural anthropology, to enable holistic understandings of "the other" in relativist terms. From its beginnings, cultural anthropology's empirical method, namely ethnography, used participant observation and face-to-face interviewing to provide carefully detailed descriptions of a particular human society, documenting cultural similarities and differences. Trained ethnographers set out to the field to attain an intimate familiarity with non-Western traditional societies worldwide, referred to at the time as "primitive" cultures. Intensive fieldwork permitted anthropologists to examine Western notions of alterity, while reflecting on diverse human ways of being, defined by geography, material life, kinship, ethnicity, cognition, and cultural ethos. Using ethnographic data, cultural anthropologists provided scientific generalizations about behavior in human communities across time and space, or ethnology, arguing for the psychic and cognitive unity of humanity.

Cultural anthropology repositioned itself professionally after the Second World War with younger practitioners setting out to establish an entirely new framework for their subdiscipline. Ethnographers sought ways to integrate a humanistic tradition with emerging evolutionary and ecological models to better depict traditional cultures. In the volume, *Anthropology*,[108] Eric Wolf introduced a more encompassing theory of culture that privileged the impacts of external forces on less developed societies, notably a world system dominated by technologically advanced capitalist countries. Wolf and others in the postwar generation "reinvented anthropology"[109] toward an understanding of the state as the most recent phase of political evolution. Their intent was to document how states reshape local societies and their cultures and thereby engender disruptive crises. Wolf locates dual epistemological crises in our civilization and our definition of human nature: "The crisis in the Western world and its imperial hinterland, which is also the crisis of humanity, cannot be confined to social, economic or technological 'problems': it inheres in our definition, our very understanding of man."[110] Using abstract models to study humanity as a problem propagates an "official definition of reality" and thereby denies the "integral relation of theory and praxis," Wolf argues, calling for an anthropology that "can also posit new possibilities for us" and "political action consistent with our insights."[111]

Along these lines, Wolf's project entailed documenting and theorizing what took place under societal and economic transformation since the late nineteenth century, notably global capitalism and the conquest and capture of the global South's resources by the global North. He considered how global Northern socioeconomic systems impacted traditional economies and local societies in the global South. Wolf also discussed how communities worldwide faced changing lifeways with the imposition of Euro-American capitalist socioeconomic relations accompanying the second industrial revolution. These include production, movement, and distribution of goods, by way of commodity chains that cut across vast swaths of the earth. For Wolf, such forms of control over people, property, and production, or power relations were embedded within social-class relations and in the culture itself.[112] Such delocalization results when people become less affected with local concerns, especially in decisions about the management of common resources, and in their stance toward their neighbors who have been marginalized by consequences of global change.

Through encounters with people undergoing these displacements, ecologically informed anthropologists came to view the community as embedded within larger systems at the regional, national, and international levels and to study the impact of a multitiered and globalizing world on the locality. This recent paradigm recognizes the importance of the state and cultural mediations in ecological processes at a time when local ethno-ecologies are being altered by development, biodiversity conservation, environmentalism, and the influence of nongovernmental organizations. Global Southern peoples derived local responses to live within or resist hegemonic domination of their lands and ecologies of knowledge in the face of structural changes wrought by global transformation that reshaped the human condition in their societies.

Immanuel Wallerstein's world systems perspective likewise explores sources of social and economic discontinuities within global Southern societies.[113] Following his conceptualization, core countries with capital-intensive production systems appropriated the resources of labor-intensive countries at the periphery. Following Wallerstein, Boaventura de Sousa Santos argues that "global social justice is linked to global cognitive injustice."[114] Santos forwards an "ecology of knowledges" that seeks an intersection and a "co-presence" between nonscientific and scientific knowledges, through local-global linkages, where one form of knowledge does not preclude the other.[115] Global Northern corporate, governmental, and nongovernmental organizations may hold limited interest in preserving local and indigenous knowledges. Rather, these entities invest in technology transfer of Green Revolution initiatives to increase crop yield through chemical fertilizers and pesticides; experimental biotechnology; and exploration, extraction, and patenting of minerals and biologicals in Southern hemisphere rainforests, mountains, rivers, virgin soils, bays, and oceans.

Throughout, global Northern elites continue to excuse the impact of such initiatives on diverse lifeworlds within global Southern regions. David Barkin[116] describes a counter-hegemonic movement for sustainability and self-sufficiency in the global South, articulated in the exchange of indigenous knowledges across national and global alliances. An emerging communitarian ethos underlies this initiative, including indigenous epistemologies forwarding "the common good," reciprocity in social relations, and an "eco-territorial feminism" privileging the perspectives of women in defense of both geographical and personal spaces. Eschewing these plural and hybrid social and cognitive paths, elites persist in encouraging global Southern societies to embrace Western technological modernity and its epistemologies, even though features of this worldview provoked centuries of social suffering in their populations.

The global Northern worldview was formed through changes in material culture and technologies since the Renaissance, including scientific discovery, worldwide expansion of capital, and state power. Defining features of this worldview have a long history, with their roots in late medieval and early modern Europe. Western modernity encompasses an ethos that developed within certain European national states. European voyages of discovery and systematic colonization of pre-Columbian America, together with the brutality of the Atlantic slave trade, framed modes of thought and social relations that were redefined and tempered in the encounter with traditional societies of sub-Saharan Africa, and later across the Pacific. Critical tensions endure within the disparate world system developed over five centuries and surface amid the disruptive tendencies in culture, science, and politics that frame global North–South relations. The civilization-culture dichotomy and the tension between adherents of steadfast tradition and a more fluid social modernity repeat themselves in our time. For over a century, advanced capitalism and its technologies contributed to conditions leading to modernization of traditional social and cultural arrangements in localities, globally.

MOVEMENTS AND MODERNITY

Cultural and technological modernization imposed striking changes in fin de siècle America. A restless tenor of life was felt within the crowded metropolises and the nation's isolated, relatively autonomous "island communities" outside of them.[117] New structural conditions, such as accumulation, dispossession, and exclusion, surfaced feelings of alienation, anger, and even despair among youth and their families in both city and small town. Sustained efforts ensued to improve conditions in traditional neighborhoods and communities caught up in the complexity of modern life. Youth-focused institutions, notably parks, playgrounds, trade apprenticeships, and high schools, were established or reinvented at a time of technological modernity. Various social movements supported changing dynamics in these vital spaces for human development, even enlarging policy arenas to challenge managerial impositions. Then as now, learning diverse social knowledges is central to participation in movement cultures constituted as responses to modernity.[118]

Tricia Niesz and her co-authors conducted a review of educational research that addressed learning in, and from, contemporary social movements.[119] They found five types of knowledge in these studies. The first involved participants' learning of expert knowledge to advance movement goals. Second was their learning organizing skills on behalf of collective action. A third comprised participants' learning about a movement's vision and value set. Fourth was knowledge supporting individual and collective identity formation through movement participation. Fifth was critical or emancipatory learning so that participants acquire a sense of personal agency and engage in activities that stimulate social critique to bring about, following Freire, *conscientization*, or critical consciousness. A focus on the "forms of consciousness" articulated in social movements holds the prospect for change in educational policy and practice. This is the case with critical and experience-based pedagogies that are linked to contemporary movements for social and environmental justice. Furthering these pedagogies as foundational for future knowledge construction can increase critical consciousness in students and their teachers.

Education thus resembles a social movement, as both "develop worldviews that restructure cognition, that re-cognize reality itself,"[120] as Ron Eyerman and Andrew Jamison state. Educational practices are also essential to social movement knowledge production and its promotion, as Tricia Niesz indicates:

> Movements both require and promote learning and education (of varied types, informal, nonformal, and formal) at every stage in their lifecycle, from the articulation and framing of the movement's vision, to organizing, to engaging in collective action, to influencing policy, law, institutions, and social life.[121]

Modern movements have held diverse aesthetic, existential, and material goals. Some championed rights and social justice had mass appeal, and were contentious in character, often carried out in public spaces. Others sought aesthetic release from older forms, practices, and structures and were subdued and limited to fewer participants often in salons, studios, and workshops, though they were no less politically engaged. Still others combined their skill sets in material activities to design and "make" new objects in the physical world. Through this mix of craft and design consciousness participants remade their vocational worlds, often reinventing schools and apprenticeships along the way. Previous social movement knowledge scaffolds current educational, artistic, and activist efforts to enlarge "spaces of affirmative resistance," that, argue Gary L. Anderson, Dipti Desai, Ana Inés Heras, and Carol Anne Spreen, "not only work to collectively counter neoliberal ideology, but also work to create networks of third spaces that prefigure the alternative world we envision."[122] Successful movements thus combine the skills of experts and lay people to deliver hybrid social knowledge for both individuals and communities caught up in tumultuous change. Social movements occur within a historical moment and are influenced by larger events. Historical and cultural moments, together, advanced the ways adolescents and youth entering adulthood have learned to actively shape their social environments and form their personal identities.

NOTES

1. Marshall McLuhan and Quentin Fiore, *War and Peace in the Global Village* (New York: Bantam Books, 1968).

2. Mikulás Teich and Roy Porter, editors, *Fin de Siècle and Its Legacy* (Cambridge: Cambridge University Press, 1990).

3. John Demos and Virginia Demos, "Adolescence in Historical Perspective," *Journal of Marriage and Family* 31:4 (1969): 635–636.

4. Joseph F. Kett, "Reflections on the History of Adolescence in America," *The History of the Family* 8:3 (2003): 355–373.

5. Paul Howe, *Teen Spirit: How Adolescence Transformed the Adult World* (Ithaca, NY: Cornell University Press, 2020).

6. Paula Fass, *The Damned and the Beautiful: American Youth in the 1920s* (New York: Oxford University Press, 1977), 120.

7. Samuel Haber, *Efficiency and Uplift: Scientific Management in the Progressive Era, 1890–1920* (Chicago: University of Chicago Press, 1964).

8. Harry Braverman, *Labor and Monopoly Capital: The Degradation of Work in the Twentieth Century* (New York: Monthly Review Press, 1974), 133.

9. Paul Willis, *Learning to Labor: How Working Class Kids Get Working Class Jobs* (New York: Columbia University Press, 1977), 180.

10. David Tyack and Larry Cuban, *Tinkering Toward Utopia: A Century of Public School Reform* (Cambridge, MA: Harvard University Press, 1995), 8–9.

11. Brian Gratton and Jon Moen, "Immigration, Culture, and Child Labor in the United States, 1880–1920," *Journal of Interdisciplinary History* 34:3 (2004): 356.

12. John Dewey, *The School and Society* (Chicago: University of Chicago Press, 1900), 173–174, quoted in James Bowen, *A History of Western Education, Volume 3: The Modern West, Europe and the New World* (New York: St. Martin's Press, 1981), 219.

13. Dominick Cavallo, *Muscles and Morals: Organized Playgrounds and Urban Reform, 1880–1920* (Philadelphia: University of Pennsylvania Press, 1981), 67.

14. Dewey, *The School and Society*, 27.

15. John Dewey, *Democracy and Education* (New York: Macmillan, 1916), 89–90.

16. Lawrence A. Cremin, *The Transformation of the School: Progressivism in American Education, 1876–1957* (New York: Random House, 1961), 123.

17. Cremin, *The Transformation of the School*, 123.

18. Charles Sanders Peirce, *Collected Papers of Charles Sanders Peirce, Volumes V and VI, Pragmatism and Pragmaticism and Scientific Metaphysics*, edited by Charles Hartshorne and Paul Weiss (Cambridge, MA: Harvard University Press, 1935); Richard J. Bernstein, *The Pragmatic Turn* (Cambridge: Polity, 2010), 36.

19. Stanley Aronowitz, "Against Schooling: Education and Social Class," *Social Text* 22:2 (2004): 13.

20. David Nasaw, *Children of the City: At Work and Play* (Garden City, NY: Doubleday, 1985), 26.

21. Nasaw, *Children of the City*, 28.

22. Rebecca Edwards, *New Spirits: Americans in the Gilded Age, 1865–1905* (New York: Oxford University Press, 2006), 116–117.

23. Claudia Goldin and Lawrence F. Katz, "The Origins of Technology-Skill Complementarity," *Quarterly Journal of Economics* 113:3 (1998): 714.

24. US Census Bureau, *Statistical Abstract of the United States: 2000* (Washington, DC: US Census Bureau, 2000).

25. Claudia Goldin and Lawrence F. Katz, *The Race between Education and Technology* (Cambridge, MA: Harvard University Press, 2008), 22.

26. Claudia Goldin, "Egalitarianism and the Returns to Education during the Great Transformation of American Education," *Journal of Political Economy* 107:6 (1999): Pt. 2, S65–S66.

27. Claudia Goldin, "The Human-Capital Century and American Leadership: Virtues of the Past," *Journal of Economic History* 61 (2001): 263–292.

28. Claudia Goldin and Lawrence F. Katz, "Human Capital and Social Capital: The Rise of Secondary Schooling in America, 1910–1940," *Journal of Interdisciplinary History* 29:4 (1999): 689.

29. Claudia Goldin and Lawrence F. Katz, "Education and Income in the Early Twentieth Century: Evidence from the Prairies," *Journal of Economic History* 60:3 (2000): 784.

30. Goldin and Katz, "Human Capital and Social Capital," 707.

31. Goldin, "The Human-Capital Century and American Leadership," 281.

32. Goldin, "The Human-Capital Century and American Leadership," 279.

33. C. Wright Mills, *Sociology and Pragmatism: The Higher Learning in America*, edited by Irving Louis Horowitz (New York: Oxford University Press, 1964), 335.

34. Claudia Goldin and Lawrence F. Katz, "The Origins of Technology-Skill Complementarity," *Quarterly Journal of Economics* 113:3 (1998): 718.

35. William Heard Kilpatrick, *Foundations of Method: Informal Talks on Teaching* (New York: Macmillan, 1925).

36. Phyllis Chiasson, "Peirce's Design for Thinking: An Embedded Philosophy of Education," *Educational Philosophy and Theory* 37:2 (2005): 207–226; Christine L. McCarthy, "Knowing Truth: Peirce's Epistemology in an Educational Context," *Educational Philosophy and Theory* 37:2 (2005): 157–176.

37. Samuel Bowles and Herbert Gintis, *Schooling in Capitalist America: Educational Reform and the Contradictions of Economic Life* (New York: Basic Books, 1976), 237.

38. Bowles and Gintis, *Schooling in Capitalist America*, 235.

39. Cynthia Goldin and Lawrence F. Katz, *The Race between Education and Technology: The Evolution of US Educational Wage Differentials, 1890–2005*, Working Paper 12984 (Cambridge, MA: National Bureau of Economic Research, 2007), 11.

40. Torin Monahan, *Globalization, Technological Change, and Public Education* (New York: Routledge, 2005), 10.

41. Goldin and Katz, *The Race between Education and Technology*, 19.

42. Alfie Kohn, *The Case against Standardized Testing: Raising the Scores, Ruining the Schools* (Portsmouth, NH: Heinemann, 2000).

43. Robert Halpern, *The Means to Grow Up: Reinventing Apprenticeship as a Developmental Support in Adolescence* (New York: Routledge, 2009).

44. Jean Lave and Etienne Wenger, *Situated Learning: Legitimate Peripheral Participation* (New York: Cambridge University Press, 1991).

45. Walter W. Powell and Kaisa Snellman, "The Knowledge Economy," *Annual Review of Sociology* 30 (2004): 199.

46. Etienne Wenger and William M. Snyder, "Communities of Practice: The Organizational Frontier," *Harvard Business Review* 78:1 (2000): 139.

47. Thomas Fuchs and Hanne De Jaegher, "Enactive Intersubjectivity: Participatory Sense-Making and Mutual Incorporation," *Phenomenology and the Cognitive Sciences* 8 (2009): 465–486.

48. Julian E. Orr, *Talking about Machines: Ethnography of a Modern Job* (Ithaca, NY: Cornell University Press, 1996), 147.

49. Orr, *Talking about Machines*, 126.

50. Joseph Bensman and Robert Lilienfeld, *Craft and Consciousness: Occupational Technique and the Development of World Images*, second edition (New York: Aldine de Gruyter, 1991), xxii.

51. Jean Lave, *Learning and Everyday Life: Access, Participation, and Changing Practice* (Cambridge: Cambridge University Press, 2019), 1–9.

52. Lave, *Learning and Everyday Life*, 2.

53. Lave, *Learning and Everyday Life*, 5.

54. Lave, *Learning and Everyday Life*, 84.

55. Lave, *Learning and Everyday Life*, 135.

56. Lave, *Learning and Everyday Life*, 141.

57. Jean Lave, *Apprenticeship in Critical Ethnographic Practice* (Chicago: University of Chicago Press, 2011).

58. Halpern, *The Means to Grow Up*, 14.

59. Ron Berger, *An Ethic of Excellence: Building a Culture of Craftsmanship with Students* (Portsmouth, NH: Heinemann, 2003).

60. John C. Bransford, Brigid Barron, Roy D. Pea, Andrew Meltzoff, Pat Kuhl, Phil Bell, Reed Stevens, Daniel L. Schwartz, Nancy Vye, Byron Reeves, Jeremy Roschelle, and Nora H. Sabelli, "Foundations and Opportunities for an Interdisciplinary Science of Learning," in *The Cambridge Handbook of the Learning Sciences*, edited by R. Keith Sawyer (New York: Cambridge University Press, 2006), 26.

61. Hannah Arendt, "The Crisis in Education," in *Between Past and Future: Eight Exercises in Political Thought* (New York: Viking, 1968), 192–193.

62. David W. Orr, *Earth in Mind: On Education, Environment, and the Human Prospect*, second edition (Washington: Island Press, 2004), 109.

63. Orr, *Earth in Mind*, 109.

64. Orr, *Earth in Mind*, 110.

65. Thomas H. Bender, *The Unfinished City: New York and the Metropolitan Idea* (New York: The New Press, 2007), 234.

66. Bender, *The Unfinished City*, 219–238.

67. Donald A. *Schön, The Reflective Practitioner: How Professionals Think in Action* (New York: Basic Books, 1981), 49–54.

68. Delia Fuhrmann, Lisa J. Knoll, and Sarah-Jayne Blakemore, "Adolescence as a Sensitive Period of Brain Development," *Trends in Cognitive Sciences* 19:10 (2015): 558–566.

69. William T. Greenough, James E. Black, and Christopher S. Wallace, "Experience and Brain Development," *Child Development* 58:3 (1987): 539–559.

70. Linda P. Spear, *The Behavioral Neuroscience of Adolescence* (New York: Norton, 2010), 279–282.

71. Sandra Johnson, "The Neuroscience of the Mentor-Learner Relationship," *New Directions in Adult and Continuing Education* 110 (2006): 63–69.

72. Robert A. Zucker, Raul Gonzalez, Sarah W. Feldstein Ewing, Martin P. Paulus, Judith Arroyo, Andrew Fuligni, Amanda Sheffield Morris, Mariana Sanchez, and Thomas Wills, "Assessment of Culture and Environment in the Adolescent Brain and Cognitive Development Study: Rationale, Description of Measures, and Early Data," *Developmental Cognitive Neuroscience* 32 (2018): 107–120.

73. Francisco J. Varela, Evan Thompson, and Eleanor Rosch, *The Embodied Mind: Cognitive Science and Human Experience* (Cambridge, MA: MIT Press, 2016); Maurice Merleau-Ponty, *The Phenomenology of Perception*, translated by Donald A. Landes (New York: Routledge, 2012).

74. Evan Thompson, *Mind in Life: Biology, Phenomenology, and the Sciences of Mind* (Cambridge, MA: Harvard University Press, 2007), 91–127.

75. Thompson, *Mind in Life*, 159.

76. Albert Bandura, *The Social Foundations of Thought and Action: A Social Cognitive Theory* (Englewood Cliffs, NJ: Prentice-Hall, 1986), 106–107.

77. Jerome S. Bruner, "The Act of Discovery," *Harvard Educational Review* 31 (1961): 22.

78. Andrew N. Meltzoff, Patricia K. Kuhl, Javier Movellan, and Terrence J. Sejnowsky, "Foundations for a New Science of Learning," *Science* 325:5938 (2009): 284–288.

79. Arthur Koestler, *The Ghost in the Machine* (New York: Macmillan, 1968), 48.

80. Michael Cole, *Cultural Psychology: A Once and Future Discipline* (Cambridge, MA: Harvard University Press, 1996), 104.

81. Barbara Rogoff, *The Cultural Nature of Human Development* (New York: Oxford University Press, 2002), 4.

82. Rogoff, *The Cultural Nature of Human Development*, 323.

83. Alexander R. Luria, *The Working Brain: An Introduction to Neuropsychology*, translated by Basil Haigh (New York: Basic Books, 1976).

84. Alexander R. Luria, "L.S. Vygotsky and the Problem of Functional Localization," *Journal of Russian and East European Psychology* 40:1 (2002): 21.

85. Alexander R. Luria, *Higher Cortical Functions in Man*, translated by Basil Haigh (New York: Basic Books, 1980), 31.

86. Jerome Bruner, "Vygotsky: A Historical and Conceptual Perspective," in *Culture, Communication, and Cognition: Vygotskian Perspectives*, edited by James V. Wertsch (Cambridge: Cambridge University Press, 1985), 21–34.

87. Lev S. Vygotsky, *The History of the Development of Higher Mental Functions, The Collected Works of L. S. Vygotsky: Vol. 4*, translated by Marie J. Hall, edited by Robert W. Rieber and Aaron S. Carton (New York: Plenum Press, 1997), 106.

88. Lev S. Vygotsky, "The Development of Concept Formation in Adolescence," in *The Vygotsky Reader*, edited by René Van der Veer and Jaan Valsiner (Oxford: Blackwell, 1994), 213.

89. Ron Ritchhart, *Intellectual Character: What It Is, Why It Matters, and How to Get It* (San Francisco: Jossey-Bass, 2002), 44.

90. Ron Ritchhart, *Creating Cultures of Thinking: The 8 Forces We Must Master to Truly Transform Our Schools* (San Francisco: Jossey-Bass, 2015), 20.

91. Kris D. Gutiérrez, Patricia Baquedano-López, and Carlos Tejeda, "Rethinking Diversity: Hybridity and Hybrid Language Practices in the Third Space," *Mind, Culture, and Activity* 6:4 (1999): 286–303.

92. Yrjö Engeström, "Expansive Learning at Work: Toward an Activity Theoretical Reconceptualization," *Journal of Education and Work* 14:1 (2001): 137.

93. Kris D. Gutiérrez, "Developing a Sociocritical Literacy in the Third Space," *Reading Research Quarterly* 43:2 (2008): 148.

94. Gutiérrez, "Developing a Sociocritical Literacy in the Third Space," 149.

95. Gutiérrez, "Developing a Sociocritical Literacy in the Third Space," 152.

96. Kurt Lewin, "Behavior and Development as a Function of the Total Situation," in *Field Theory in Social Science: Selected Theoretical Papers*, edited by Dorwin Cartwright (New York: Harper & Row, 1951), 238–303.

97. Kurt Lewin, "Defining the 'Field at a Given Time'," in *Field Theory in Social Science*, 59.

98. Kurt Lewin, "Psychological Ecology," in Lewin, *Field Theory in Social Science*, 173.

99. Urie Bronfenbrenner, *The Ecology of Human Development: Experiments by Nature and Design* (Cambridge, MA: Harvard University Press, 1979).

100. Peter L. Berger, Brigitte Berger, and Hansfried Kellner, *The Homeless Mind: Modernization and Consciousness* (New York: Random House, 1973), 63–82.

101. Francis Müller, *Design Ethnography: Epistemology and Methodology*, translated by Anna Brailovsky (Cham: Springer, 2021), 13.

102. Müller, *Design Ethnography*, 15.

103. Jürgen Habermas, *The Philosophical Discourse on Modernity: Twelve Lectures*, translated by Frederick G. Lawrence (Cambridge, MA: MIT Press, 1990).

104. Terry Eagleton, "Culture & Barbarism: Metaphysics in a Time of Terrorism," *Commonweal* 136:6 (2009): 12.

105. Ann Douglas, *Terrible Honesty: Mongrel Manhattan in the 1920s* (New York: Farrar, Straus and Giroux, 1995).

106. Susan Hegeman, *Patterns for Americans: Modernism and the Concept of Culture* (Princeton: Princeton University Press, 1999); Marc Manganaro, *Culture, 1922* (Princeton: Princeton University Press, 2002).

107. George W. Stocking, Jr., "Franz Boas and the Culture Concept in Historical Perspective," in *Race, Culture, and Evolution: Essays in the History of Anthropology* (Chicago: University of Chicago Press, 1982), 195–233.

108. Eric R. Wolf, *Anthropology* (Englewood Cliffs, NJ: Prentice-Hall, 1964).

109. Dell Hymes, editor, *Reinventing Anthropology* (New York: Random House, 1969).

110. Eric R. Wolf, "Introduction," in *In Search of the Primitive*, by Stanley Diamond (New Brunswick: Transaction Publishers), xi.

111. Wolf, "Introduction," xii, xiii.

112. Eric R. Wolf, *Europe and the People Without History* (Berkeley: University of California Press, 1982).

113. Immanuel Wallerstein, "A World-System Perspective on the Social Sciences," *The British Journal of Sociology* 27:3 (1976): 343–352.

114. Boaventura de Sousa Santos, "Encountering Other Cultural Universes on the Brink of Chaos," *Journal of World-Systems Research* 28:2 (2022): 440.

115. Boaventura de Sousa Santos, "Beyond Abyssal Thinking: From Global Lines to Ecologies of Knowledges," *Review* (*Fernand Braudel Center*) 30:1 (2007): 45–89.

116. David Barkin, "Shaping a Communitarian Ethos in an Era of Ecological Crisis," *Frontiers in Sustainability* 3 (2022): 1–14.

117. Robert H. Wiebe, *The Search For Order, 1877–1920* (New York: Hill and Wang, 1967), xiii.

118. Hye-Su Kuk and Rebecca Tarlau, "The Confluence of Popular Education and Social Movement Studies into Social Movement Learning: A Systematic Literature Review," *International Journal of Lifelong Education* 39:5–6 (2020): 591–604.

119. Tricia Niesz, Aaron M. Korora, Christy Burke Walkuski, and Rachel E. Foot, "Social Movements and Educational Research: Toward a United Field of Scholarship," *Teachers College Record* 120:3 (2018): 1–41.

120. Ron Eyerman and Andrew Jamison, *Social Movements: A Cognitive Approach* (University Park: Pennsylvania State University Press, 1991), 165.

121. Tricia Niesz, "Social Movement Knowledge and Anthropology of Education," *Anthropology & Education Quarterly* 50:2 (2019): 227.

122. Gary L. Anderson, Dipti Desai, Ana Inés Heras, and Carol Anne Spreen, *Creating Third Spaces of Learning for Post-Capitalism: Lessons from Educators, Artists, and Activists* (New York: Routledge, 2023), 2.

Part II

CHANGING THE SUBJECT

Chapter 4

Project-Based Learning as a Critical Pedagogy

CHANGING THE SUBJECT

Project-based learning has focused the attention of educational philosophers, researchers, and practicing educators for a long time. However, this kind of learning does not come with one definition nor is it guided by one idea. Instead, it has been defined and discussed in many ways and enacted in different ways in schools and classrooms. John Dewey, in *How We Think*, made it clear that what was involved in reflective thinking and inquiry were two sub-processes: "(a) a state of perplexity, hesitation, doubt; and (b) an act of search and investigation directed toward bringing to light further facts which serve to corroborate or nullify the suggested belief."[1] These dispositions and ways of thinking were and are part of project-based learning. For Dewey and others,[2] having students think in these ways is at the heart of students' investigations of real-world problems in this style of learning, in and out of school. The educational purpose for inquiry-based and reflective learning has to do with the preparation for life outside of school and for the advancement of student motivation and thinking. Building on John Dewey's notation of reflective inquiry, James Hiebert and his co-authors state:

> Allowing the subject to be problematic means allowing students to wonder why things are, to inquire, to search for solutions, and to resolve incongruities. It means that both curriculum and instruction should begin with problems, dilemmas, and questions for students. We do not use "problematic" to mean that students should become frustrated and find the subject overly difficult. Rather, we use "problematic" in the sense that students should be allowed and encouraged to problematize what they study, to define problems that elicit their curiosities and sense-making skills.[3]

For Dewey, that preparation for life outside of school embraces both better motivation and thinking and the advancement of active democratic citizenship, inside and outside of a school. However, today's dominant reality in schooling, either elementary or secondary, reflects few of the characteristics identified by Dewey and others who champion inquiry-based and reflective learning. Instead, a persistent set of practices and structures far removed from these features and guiding ideas about meaningful learning exist and persist in many of today's schools, as they have for the past 150 years. Charles Tilly views the institutional arrangements of contemporary schooling as bearing "strong markings of the nineteenth century":

> In the world of education, we still behave as though the effective way to prepare young intellects for the fight ahead were to divide all youngsters of a certain age into groups of twenty or thirty, place each group in a closed room with a somewhat older person, seat the youngsters in rows of small desks, arrange for the older person to talk to them for hours each day, have them write various sorts of exercises for the older person to evaluate, and require them to speak periodically in class about the exercises they have written, about the material they have read, about general issues the older person has raised.[4]

Outmoded bureaucratic organizational structures have thus lingered in the life of schools and school districts.

John I. Goodlad's study of schooling identified these predictable and persistent sets of daily classroom experiences and practices.[5] Typically, instructional practices involve students listening to a teacher talk, students doing seatwork, and students taking tests that resemble their seatwork. Both of the latter largely involve short answers in response to questions of fact. Other studies conducted at the same time as the Goodlad study revealed and confirmed similar persistent classroom activities, despite dramatic changes in the world surrounding schools.[6] Teachers have clung to these outdated notions about students, learning, and educational practices. Arno Bellack and his colleagues collected and studied forty-five hours of typed transcripts derived from tape recordings of teachers' interactions in classrooms with their students. Hoetker and Ahlbrand provide this summary of the results of Bellack's study and analysis of his data:

> What Bellack observed, then, was that his teachers, despite differences in the sizes, ability levels, and backgrounds of their classes, acted very much like one another. They talked between two-thirds and three-quarters of the time. Their major activity was asking and reacting to questions that called for factual answers from students.[7]

The conclusions noted above were drawn from data collected over fifty years ago. However, earlier patterns like those described by Bellack had been captured in the research literature before the 1960s. A study done by Romiett

Stevens in 1912, for example, like Bellack's, focused on four years of classroom observations captured about what she observed in detailed stenographic notes. Like Bellack, her analysis of these data found that the question-and-answer teaching format prevailed, then, at the turn of the twentieth century. And, as today, Stevens was concerned about its dominance:

> The large number of questions suggests that in actual practice there is very little effort put forth to teach our boys and girls to be self-reliant, independent mental workers. The discrepancy between our theory and practice is nowhere more patent. . . . There is no use in claiming to teach boys and girls how to study, and how to command their intellectual forces by the current practice of keeping them at the point of the bayonet in rehearsal of textbook facts at the rate of two or four per minute.[8]

In the twenty-first century, a similar pattern appears to exist in today's schools and classrooms. Robert Pianta and his colleagues have conducted a study of what they call opportunities to learn in classrooms serving middle-class students in the United States. While not a nationally representative sample, their data confirm the persistence of classroom practices that we have been discussing, patterns similar to those found in studies of classroom teaching during the twentieth century. Here is what Pianta and colleagues have to say about their observations:

> . . . experiences in fifth grade, although highly variable, were geared toward performance of basic reading and math skills, not problem-solving or reasoning skills or other content areas. Few opportunities were provided to learn in small groups, to improve analytical skills, or to interact extensively with teachers. This pattern of instruction appears inconsistent with aims to add depth to students' understanding, particularly in mathematics and science.[9]

Economically poorer students and those of color have no better experiences in general in this twenty-first century than the middle-class students in Pianta's study.

In the third decade of the twenty-first century, today's national educational reform rhetoric does not challenge these persistent practices. Instead, as it happened in the early 1980s, after the studies by Boyer, Goodlad, and Sizer came out documenting the strong presence of these predictable and ineffective practices, attention instead went then and now to standards and test scores. For example, the 1983 national report entitled *A Nation at Risk* called for higher standards, not different standards, not different practices, and not different school structures.[10] Nor did the report call attention to the emerging cognitive science that was then and has now continued to challenge the tired and minimally effective behaviorist ideas that have guided these persistent practices. Recent studies of cognitive development highlight the power of the mind of the child and adolescent and its great capacity for memory, thinking, and reflecting, which is often ignored in the dominant set of predictable behaviorist practices.[11]

The 107th Congress passed Public Law 107–110–115, Stat. 1425, which became known as "No Child Left Behind" (NCLB). Section 1001 of the law reads, "ensuring that high-quality academic assessments, accountability systems, teacher preparation and training, curriculum, and instructional materials are aligned with challenging State academic standards so that students, teachers, parents, and administrators can measure progress against common expectations for student academic achievement. . . ."[12] NCLB created "a seismic shift" that altered the public educational landscape through federal and state policies enacted in the early twenty-first century. Consequently, new ideas found their way into public discourse in the educational arena:

> Education policy at all levels, most notably articulate in the federal No Child Left Behind Act of 2001, placed accountability for student achievement at the heart of the education enterprise and called upon stakeholders to employ evidence-based programming and practices in the service of that aim. Equally important was the new federal insistence on exposing disparities in achievement among students from a variety of demographic subgroups.[13]

Like the focus of *A Nation At Risk*, this act directs attention to standards and their alignment with a number of the other persistent dimensions of schooling. It did not draw attention to challenging and questioning any of these persistent dimensions nor did it encourage rethinking and creating new dimensions, and also the guides for them, in classrooms and schools. Instead, and in effect, the argument for these reforms is to keep the persistent dimensions the same and to better connect, improve, and advance them. According to Mintrop, because of NCLB and its intense focus on an effort to improve test scores, classrooms are characterized by

> rigidity, fragmentation, and deterioration. Low-capacity schools are predestined to bank on short-term strategies that require little added capacity. Common strategies are test preparation activities, content alignment, and concentration on tested subjects, benchmark grades, and students near proficiency. In some low-performing schools, this can amount to a parallel test-remediation curriculum that is different from the regular curriculum taught in less-pressured schools, with the result that students are excluded from intellectually challenging content and learning.[14]

The state of affairs provided the impetus for improving learning in today's schools. In a few places around the United States, creative groups of educators have been challenging and rethinking the persistent practices and structures. Two such innovators, Larry Rosenstock and Rob Riordan who co-founded High Tech High in 2000 were asked:

> What should students know in the 21st Century?" Their response, never published, but passed from e-mail to e-mail by many teachers, was a two-page article entitled, "Changing the Subject." Their answer, in brief, as that in the age of ubiquitous and accessible information, it matters less what students know than what they can do. "When we learn—really learn—we transform the content, the self, and the social relations of teaching and learning," they wrote. "This is what it means to change the subject.[15]

Changing the subject will involve considering three current realities about teaching and learning. Each argues for a dramatically different narrative of schooling and the efforts undertaken by children, youth, and their teachers in these places, not for more of the same. First, a persistent set of practices and structures have dominated schooling since at least before the turn of the last century. Second, these persistent practices and structures are in general ineffective for motivating and encouraging the kind of thinking required in the twenty-first century. Third, there is much research and science about human learning and development to support a more powerful set of learning practices and structures in classrooms in schools. Seymour Sarason has called these practices and structures "the regularities of schooling." These innovators have been inventing new guides, practices, and structures for education for the children and youth in their educational settings, which differ in ways from these regularities in the way that Sarason suggested school change would have to occur:

> any attempt to introduce a change into the school involves some existing regularity, behavioral or programmatic. These regularities are in the nature of intended outcomes. It is a characteristic of the modal process of change in the school culture that the intended outcome (the change in the regularity) is rarely stated clearly, and if it is stated clearly, by the end of the change process it has managed to get lost.[16]

The suggested alternatives to existing regularities have to be examined, discussed, and debated in the educational settings where they might be considered to spur new meanings for action in those places. Referring to scientific results represented in books or in research studies, Dewey noted, "Enlightenment, clarity and progress can come about only as we remember that such results are sources to be used through the medium of the minds of educators, to make educational functions more intelligent."[17] Reimagining education in the light of such thinking, twenty-first century educators again embraced project-based learning. The approach emerged as an experiential, hands-on, and student-directed activity as a corrective to conventional schooling soon after the second industrial revolution. The intent was to provide opportunities for reflective inquiry, following Dewey, both inside and outside the classroom. Dewey's notion of the school as a "social laboratory" influenced educational policy a century ago when the United States underwent a great transformation in its educational history toward mass schooling. Dewey's thought took place amid rapid population change as the country shifted to urban modernity and industrial life. The high-school movement where the focus was "schooling for life" accompanied this change.

Project-based learning, which builds on Dewey's work on experiential methods of thought and action, is ultimately delivered within a student–teacher relationship. The structure of this relationship and that of the high school itself were shaped by an industrial culture that developed when the dual revolutions of technology and information processing were transforming the country. During the transition from craft to mass production, high schools were tasked with renegotiating and reframing occupational and world orientations in light of dramatic technological changes. The recent postindustrial transition brought about a similar dual revolution in the spheres of technology and information, which transformed societal institutions and reoriented education in today's world.

TOWARD PUBLIC PEDAGOGIES

The challenges of the recent technological revolution in the Anthropocene, the age of "the great acceleration," again shifted the emphasis of education toward students actively using what they know to explore, negotiate, interpret, and create. Unlike conventional schooling, project-based learning challenges students by acknowledging their roles as participants engaged in producing knowledge. Students also perceive the value of experiencing this form of learning and are rewarded through the responses of others to their projects. Through an amalgam of knowledge, skills, teamwork, and communication, project-based learning challenges routine schooling by helping youth to develop habits of mind associated with personal and occupational success.

As "public pedagogies–spaces, sites, and languages of education and learning that exist outside the walls of the institution of schools,"[18] as Jennifer Sandlin and her co-authors assert, project-based learning can also take place in shops, offices, labs, and in natural settings. Like project-focused classrooms, these venues become "living laboratories" for student knowledge construction, following Dewey. Whether in the school or beyond, youth who engage in project-based learning encounter a public sphere where they may "connect their experiences to specific problems that emanate from the material contexts of everyday life,"[19] according to Giroux. Ira Shor finds such a critical literacy through a praxis that "connects the political and the personal, the public and the private, the global and the local, the economic and the pedagogical, for rethinking our lives."[20] Through experiential approaches that privilege the learner as producer, youth become protagonists engaged in "the struggle for knowledge, a place where consciousness can discover itself, a place where knowledge gives way to creative purposiveness,"[21] as Peter McLaren states.

"Critical pedagogy" forwards democratic instructional practices so that schools may become educational drivers of change in society.[22] Giroux sees public schools and colleges as sites of critical inquiry that, through dialogue and socially responsible teaching and learning, sustain a democratic public sphere. Giroux considers the works of Paulo Freire and Antonio Gramsci on cultural politics for his critique of schooling under neoliberalism. He strives to move education beyond market-based accountability and positivist approaches to classroom instruction. Giroux provides both theoretical and practical elements of a critical pedagogy to reshape the educational landscape in support of core values of freedom and reason. Returning to a notion of the centrality of education to politics, he argues for schooling that motivates students to think and act differently. The promise of democratic schooling thus returns agency to both teachers and students by linking together interests in justice and struggles over identities with knowledge and skills acquisition. Moving beyond merely introducing students to diverse ideas and traditions, Giroux's pedagogy reshapes practices of teaching and learning to expand students' capabilities, enabling them to actively transform knowledge through critical agency and responsibility, following Freire's sense of knowledge creation through reflective inquiry and then engagement with both the world and others:

> Knowledge emerges only through invention and reinvention, through the relentless, impatient, continuing hopeful inquiry human beings pursue in the world, with the world, and with each other.[23]

Giroux's project, then, is toward "linking the knowledge that is taught to the experiences students bring to their classroom encounters,"[24] so that they may lead meaningful lives as active participants in a democratic society.

Through taking a critical look at past and current debates on schooling, Angelina Castagno and Teresa McCarty view education reform as a contested terrain and educational policy as "a situated sociocultural process."[25] For them, educational policy processes establish the knowledge regimes and the power networks that maintain the current system. Policy is thus a social practice that can be described and critically analyzed on the ground through an ethnographic lens. Seeing "policy as processual, dynamic and in motion," critically informed ethnography can locate and probe deeper into the sites where this policy is worked out. Castagno and McCarty thus provide a trajectory for an anthropology of education policy, situating it historically and seeing the field as a potentially "democratizing social science" in light of inequalities both in education and in society. They cite the early twentieth-century efforts of anthropologist Franz Boas, sociologist W. E. B Du Bois, and the newly established NAACP to critically disrupt the racial inequality set in motion by the 1896 Plessy vs. Ferguson case, seeing their efforts as providing a theoretical basis for the later Brown vs the Board of Education case in 1954.

Castagno and McCarty point to the paucity of social scientific analysis on school desegregation immediately after this landmark decision, stating that there were only limited efforts to desegregate schooling in the two decades after the decision. They then link the emergence of an anthropology of education with the establishment of federal governmental initiatives, including the National Institute of Education that funded policy-relevant ethnographic studies of school desegregation, and language and literacy, including bilingual education. Looking forward, Castagno and McCarty hold that a critical anthropological approach using historical, ethnographic, and comparative perspectives can "map the contours of the anthropology of education policy."[26] A critical ethnography of policy studies thus seeks out broader possibilities for democratizing both school and society.

From school boards and unions to textbook decisions and university entrance requirements, each reinforces conditions that assure that conventional schooling remains in place. State governments maintain structural constraints on schools, limiting teachers' autonomy in the classroom. Still, experiential approaches, such as project-based learning both inside and outside of school, provide countless youth the habits of mind to support their forming more engaged self-identities. These approaches combine active and reflective thinking and at the same time incorporate global concerns through inclusive learning. Sustaining such instructional approaches requires ongoing intergenerational support from both teachers and parents. Restricted intergenerational engagement, by contrast, only furthers the current fragmenting of cognitive, affective, and instrumental functions that are essential to healthy socialization. In Stanley Diamond's words:

> The segregation of these functions from each other is, itself, a function of the shrinking circumference of direct personal ties within and between the generations of our society. Correlatively, as the area of socialization diminishes, the area devoted to increased specialized education of an impersonal character increases. *We may, therefore, be nurturing a fundamental paradox, namely, that inadequately socialized persons are and will be drafted by modern society for increasingly narrow and technical training.* This is not merely a split between the so-called "two cultures" but represents a fundamental division at the very

> heart of our society involving all of us—parents and children, teachers and pupils, workers and managers, scientists and artists, engineers and housewives.[27]

Project-based learning, in the classroom and beyond, mitigates this trend by increasing direct interpersonal ties between mentors and learners. Intergenerational ties are also strengthened, as experiential approaches typically involve critical modes of inquiry and presentations of learning that assist youth and their parents to understand and appreciate the outcomes of these practices.

Roger Dale and Susan Robertson suggest that the shift toward neoliberal market arrangements undermined the social contract between the individual and the national society in education and beyond.[28] Consequently, the idea of citizenship moved from the state as the mediator of rights to "the practices of citizens defined as members of communities."[29] Rather than national citizenship, the new social contact emphasizes engagement and participation for active citizenship within a community. Democratic schooling can then motivate students and their families toward becoming active citizens with concern for the good life in their neighborhoods. Robert Crowson and Ellen Goldring hold that this "new localism" in public education signals a turn away from the previous social contract toward the centrality of neighborhood and community, stemming in part from a mistrust of national standards, achievement rankings, and centralized educational reforms.[30] Following Richard Schragger's typology,[31] localism takes distinct forms. The first are "deep community"[32] efforts to protect familial belief systems amid secularization.

Second, "contractarian community,"[33] is based upon loose ties of voluntary association and action on behalf of instrumental ends. A middle ground exists as the "dualist community,"[34] a place-based effort that requires strong ties to actually create community by civic engagement through dialogue and other participatory approaches.

Public pedagogies can support this latter form of localism by providing students opportunities for reflective inquiry about community conditions that challenge a community's quality of life. These range from social and economic disparities in education and inequalities in access to health care, to resource scarcity and mobilization following pandemics, food and housing insecurities, and ecological threats. Such vital concerns continue to influence residents seeking ways to reverse social and political disorders in their communities, frequently through civil society movements. Yet, there remain limitations and constraints to collective action at the community level. While movements for expanded constitutional rights move forward, institutional power that is hegemonic, or dominant, may constrain other efforts, such as climate action and environmental justice.

Civic communities of practice then bridge the gap to resolve such dilemmas by mobilizing resources through dialogical and reflective methods.

Constituents facing challenges in their everyday lives thus bring their grievances and demands into political arenas. However, despite such collective action, contemporary disorders and inversions are often consequences of structural obstacles, including information asymmetries, where any effort to change necessarily requires accountability. Demands for greater accountability often involve contending strategies forwarded by locals and experts on how best to bring about an essential transparency. Reaching common ground will then require increasing social knowledge through strategic information. However, access to such information is often limited by social class, ethnic, and cultural disparities that shape constituents' repertoires of practice.

Contemporary life is dichotomized by a public sphere for advancing the common good and a domestic sphere based upon private interests, often with contrasting and conflicting images of family and neighborhood. In traditional neighborhoods, cultural knowledge frames the ties of kinship, patronage, and clientism that are everyday practices in the private realm. As these private interests frequently structure work and family life, residents may welcome progressive initiatives around their common concerns, yet remain apprehensive when requisite changes disrupt long-standing norms. This ambivalence likewise applies to institutional efforts for cognitive reorientation of learning through experiential instructional approaches. The school and society, the latter represented by the domestic sphere, are often polarized by divergent and at times competing repertoires of practice. Even as progressive measures are introduced through dialogue, constituents may hold competing attitudes, norms, and practices in their domestic spheres. Their hesitation about progressive initiatives may in fact result from hearing opposing and at times confusing positions voiced by influentials.

Through the lens of cultural politics, Antonio Gramsci's view of the two types of actors in the change process is instructive. The first, "organic intellectuals," are conscious members from within a community, and the second are "traditional intellectuals," such as teachers, clergy, and professionals who reproduce the dominant culture.[35] Disputes frequently occur between leaders representing these two contrasting intellectual positions. Certain locals with close ties to informal networks may even push back against cosmopolitans employed by formal institutions who endorse programmatic change for better schools, housing, health, and social services. Local efforts aligned with grassroots leadership are usually more successful than those striving to implement policy directives from the top. Collaborative measures can entail a broader-based strategic review of outmoded institutional missions and procedures. A problem-solving approach thus provides transparency and involves community-based stakeholders as participants in a deliberative process.

In the educational arena, deliberative approaches can ensure that more adolescents successfully enter postsecondary institutions and find work in the new knowledge economy. Although as progressives forward broad-based efforts for twenty-first-century learning, others holding institutional power with hardened attitudes continue to endorse conventional schooling.

Parents embracing traditional views of education, work, and family life may also question progressive approaches, even while seeing how their children's life chances are altered by new economic relations. A major challenge is "familism,"[36] emphasizing the primacy of the extended family over individual interests. While this cultural norm based upon supportive interpersonal relationships safeguards the traditional domestic sphere, unyielding attitudes can deter younger generations from the autonomy required for mobility. It will then be crucial to seek appropriate ways to convince parents that experiential approaches can free their children to explore new occupational techniques and gain requisite habits of mind.

Accordingly, organic intellectuals have served as cultural brokers on behalf of progressive measures, typically in communities where many distrust state power or are suspicious of outside influentials. As arbiters between traditional and modern worldviews, local thought leaders seek an appropriate discourse to inform families about accessing public goods. Despite confusing information, many tradition-bound constituencies have come to understand socially marginalizing legal and administrative barriers to accessing public services and have learned how to negotiate services for themselves and their families. A similar approach will be required to inform how certain instructional methods link to work practices familiar to the parental generation. Conversations with parents could then identify the value of such pedagogies to both their children and the good life of their families.

Anthropological engagement likewise moved toward a participatory mode by assisting in the process of knowledge co-production to yield democratizing and self-liberating innovations.

This involves engaged, experientially based "pedagogies of praxis" to give anthropology relevance not obvious in its conventional disciplinary practices. Participation in and with people and communities relies on an understanding of pathways of power that impact their lives. Engaged anthropology tightens the relationship of academic work, community problem-solving, and democratizing pedagogies. Critical to this kind of praxis is both the capacity for reflection and a toolkit of communicative skills for negotiating complex interpersonal dynamics in the field. Anthropologists are increasingly focused on collaborative and continuous learning on behalf of knowledge-based work, specifically through such engagements with communities, globally. There, ethnographers document how long-held cultural norms elevated in the folk tradition, broadly understood, can facilitate educational and occupational transitions within tradition-bound families and communities. Murray and

Rosalie Wax talk about differences between an atomizing premise underlying modern schooling and the ethos of a folk community:

> What an individual does in school, and later, in his vocation, is an achievement—his individual achievement—deriving from his own initiative and effort, and of benefit only to himself and his immediate family. Contrary to this ideology is the normative system of a folk community that confronts an alien society. For in this system the individual may excel only when his excellence enhances the position of his brethren. If this achievement were to derogate them before others, then it would be incumbent on him to conceal his talents.[37]

The contrasting assumptions of traditional enculturation in the family, community, and modern schooling may be an impetus for schools to incorporate more informal learning of organic elements of a community and its lifeways into their arguably abstract, rational and, to many youth, unexciting and deracinated curricula. According to the Waxes,

> Where in folk society, there is a great stress on the function of language to promote consensus and maintain the integrity of the community . . . in the urban middle class world and its schools, the stress is on language as a vehicle for imparting "rational" knowledge to strangers.[38]

Bureaucratic efficiencies and relevance to technological imperatives moved modern schooling to abandon informal learning characteristic of community-informed "little traditions," instead favoring conventional formal frameworks of a "great tradition" based upon individual achievement motivation. Regardless, formal approaches to organizing, classifying, and enculturating knowledge are frequently mediated by less formal ontologies, as stated in a report by the National Academy of Sciences:

> It is a basic function of science to label and classify the phenomena that are observed and organize them for study in a particular domain. . . . Often, however, it proceeds more automatically and implicitly, as when so-called folk categories from everyday life (informal understandings of phenomena, such as cognition, emotion, perception) play an unrecognized role in behavioral science.[39]

Following from this, a pedagogical dimension that may move tradition-bound students and their parents toward acceptance of modern education is the folk revival. This is a craft-based movement that upholds artisanal values of the little tradition in times of technological change and social transition. Walter Benjamin understood that industrial modernity diminishes the authenticity and aesthetic value of design and artistic production in an "age of mechanical reproduction."[40] Folk and craft revivals came about with an industrial revolution based on production engineering, chemicals, and electronics in the late nineteenth century; as unionized mass production work proliferated in the first decades of the twentieth century, and again at mid-century during the transition to postindustrial computerized service and knowledge work. With folk revivals came alternative models of project- and work-based learning and cooperative education established within contemporary "little communities" like Arthur Morgan's experiments in Yellow Springs, Ohio, and at "mountain schools," such as the Arthur Morgan School in the Blue Ridge Mountains and the ones in other parts of Appalachia where teachers, students, and craftspeople have been experimented with socially conscious work practices for nearly a century. Myles Horton formed the Highlander Folk School in Tennessee that pioneered in adult education with the aim of democratic praxis, initially for labor movement leaders in the Depression and then for postwar civil rights leaders, among them Rosa Parks. Highlander's educational approach begins with practice and moves on to theory, informing a pedagogy and curriculum based upon a relational, pluralistic, democratic theory. Dewey's ideas were central to Horton's model of social activism with group problem-solving that helps move participants toward a critical consciousness realized as autonomous decision-making and responsible action. As Barbara Thayer-Bacon states:

> A relational view of democracy does not begin with an assumption of individualism, as classical liberal democracy does, but starts with Dewey's concept of transactional relationships, that individuals affect others and others affect individuals, for we are all selves-in-relation-with-others. A pluralistic view of democracy emphasizes identity and differences without falling into the trap of thinking there is a unitary subject, and without embracing extreme pluralism that emphasizes heterogeneity and incommensurability.[41]

Horton's approach to democracy "always-in-the-making" often relied on oral transmission, sharing experienced-based stories around common concerns in community meetings, and also social evenings where people got to know each other through eating together, dancing, and singing folk songs. Such group activities can reshape participants' stereotypical views of individual problems by examining them through the lens of larger sociocultural and political contexts.

Along similar lines, Gramsci's dichotomy of traditional and organic intellectuals may be instructive in understanding clashes around individual- and group-level concerns within "little communities," including ethnic enclaves.[42] Conflicts between systemic and lifeworld domains will often move into the broader public realm, including education, and can potentially be resolved through contextualizing them across multiple domains. Disputes may involve whether or not to stay the course with dominant modes of learning or embrace alternative approaches to increase students' life chances and opportunities.

Gramsci understood that such conflicts arise, as Robert Wohl restates, "when the older generation gives up its pedagogical responsibilities and fails to prepare the young for the tasks that the existing state of historical development thrusts upon them."[43] Ethnographers have provided examples, worldwide, where those adhering to traditional social arrangements clashed with others forwarding progressive innovations. In education, these conflicts often surround whether learning should be based upon passive or active knowledge.

Like Dewey and Horton, Gramsci understands culture as informing the acquisition of active knowledge, as Kate Crehan explains:

> Culture here is defined as the work of self-knowledge, but not a narrowly individualized self-knowledge, but rather a critical self-knowledge focused on understanding one's relations to others, including one's "rights and obligations" in relations to them and one's place in history.[44]

In civic arenas, intellectuals from beyond the community introduce ideas that enculturate active knowledge through collaborative dialogue. Challenges from within a community are typically voiced by others who recite tradition-bound narratives, honoring top-down approaches that uphold the status quo. Constituents thus hold and operate from diverse world images and beliefs, especially during cultural modernization processes. In this regard, Gramsci asks:

> How many societies does each individual belong to? And doesn't each one of us make continuous efforts to unify his conception of the world in which there continues to subsist heterogeneous fragments of fossilized cultural worlds? And does there not exist a general historical process that tends to unify humanity continually?[45]

A sociocultural fusion thus inheres in all of us, for individuals are unified by multiple historical processes at any present moment. However, while shared problems call for change based on universal outlooks, people everywhere embrace cultural identities derived by fiercely held worldviews and mindsets. Instructional innovations, as with other progressive initiatives, require dialogue to yield consensus, an outcome mediated by diverse and often conflicting historical arrangements. Even hard-won initiatives, negotiated and accepted by diverse constituencies, will meet with resistance from those holding opposing viewpoints or wishing to maintain power.

All too often the direction of education is depicted as either a public or private good. This dichotomy, together with that of learning as either conventional or experiential, causes leadership to overlook historical lessons in the hurried adoption of current breakthroughs and advances. For many students and their teachers, contemporary learning through electronic means sustains pathways of social isolation. Like Fordist innovations a century ago, digital instruction often mass produces ideas, separating individuals from social and emotional processes necessary for deeper learning. Digital learning more often depends upon designs that serve managerial and administrative ends, further alienating and segmenting student populations. With an increased distancing of learning, connections between the school and society are likewise threatened with division. As with urban sprawl that grew along with Fordism, an overreliance on digital learning is unsustainable, creating ever-widening social and economic divisions. Promoted as an efficient way to learn specialized knowledge for competitive advantage in advanced technological sectors, digital approaches ignore the majority on behalf of future specialists. Max Weber wrote the following over a century ago:

> No one knows who will live in this cage in the future, or whether at the end of this tremendous development entirely new prophets will arise, or there will be a great rebirth of old ideas and ideals, or, if neither, mechanized petrification, embellished with a sort of convulsive self-importance. For the "last man" of this cultural development, it might well be truly said: "Specialist without spirit, sensualist without heart; this nullity imagines that it has attained a level of humanity never before achieved."[46]

Under this spell, an all-encompassing digital future becomes, not simply a priority, but a vision of necessity detached from historical meaning, yet in desperate search of purpose. Conversely, measured experiential learning paths can close social and cognitive gaps and reorient individuals to sources of resolve and consequence within the school and society. In this way, dialogues may ensue to re-envision learning in the context of historical experiences, thereby reconnecting the young to meaningful lifeworlds.

With this in mind, critical ethnographies and democratizing pedagogies, together, can inform an educational praxis that bridges action and reflection through the design of challenging learning situations for youth within and outside the classroom. Moreover, high-quality projects can also create bridging networks that coalesce students, teachers, parents, and the wider community to engage with and through these activities, and perhaps influence local educational policies. The ensuing experiences and relationships among diverse stakeholders can in this manner breach boundaries of class, ethnicity, and gender to challenge conventional norms of mechanistic schooling. The extended case studies demonstrate how reoriented educational spaces inside and outside of school provided developmentally sensitive domains for growing up, exploring occupational practices, and leading a productive life. Woven throughout is the narrative of sustainability, especially the equity dimension, and the primacy of fostering accountability and envisioning processes through experience-based learning.

The cases are experiments in writing a critical anthropology of learning that describes intergenerational engagement and mentoring through project-based activities within classrooms and laboratories, in natural settings and on city streets. They illustrate innovative approaches to assuring youth better life chances in an increasingly precarious economy and society. Each is framed through the lens of community-based transformational research that considers structural factors and power relationships contributing to social and economic disparities within a community. This research approach, based upon principles of personal and group change, holds researchers responsible for collaborating with communities and groups in co-constructing research. Ethnographic and other forms of data are produced in the process. The results enhance both accountability and constituents' ability to carry out subsequent studies rooted in and shaped by the locale. In each case study area, the educational arena changed as schooling regimes adapted to the needs of the new local industries by including project-based learning inside and outside of class where youth participated in projects as fledgling artists, engineers, planners, writers, teachers, and scientists. Each case explores how project-based learning enacted as a critical public pedagogy holds the potential, together with the production of new knowledge, for personal development, creativity, and social transformation. The case studies thus bring an ethnographic dimension to understanding the multisided dynamic and developmental processes that frame the debates over twenty-first-century learning.

NOTES

1. John Dewey, *How We Think: A Restatement of the Relation of Reflective Thinking to the Educative Process* (Boston: Heath and Company, 1933), 9.

2. Dewey, *Democracy and Education*; Phyllis C. Blumenfeld, Elliot Soloway, Ronald W. Marx, Joseph S. Krajcik, Mark Guzdial, and Annemarie Palincsar, "Motivating Project-Based Learning: Sustaining the Doing, Supporting the Learning," *Educational Psychologist* 26:3 (1991): 371.

3. James Hiebert, Thomas P. Carpenter, Elizabeth Fennema, Karen Fuson, Piet Human, Hanlie Murray, Alwyn Olivier, and Diana Wearne, "Problem Solving as a Basis for Reform in Curriculum and Instruction: The Case of Mathematics," *Educational Researcher* 25:4 (1996): 12–21.

4. Tilly, *Big Structures Large Processes Huge Comparisons*, 1–2.

5. John I. Goodlad, *A Place Called School: Prospects for the Future* (New York: McGraw-Hill, 1984).

6. Ernest L. Boyer, *High School: A Report on Secondary Education in America* (New York: Harper & Row, 1983); Larry Cuban, *How Teachers Taught: Constancy*

and Change in American Classrooms, 1890–1980 (New York: Longman, 1984); Theodore R. Sizer, *Horace's Compromise: The Dilemma of the American High School: The First Report from a Study of High Schools, Co-sponsored by the National Association of Secondary School Principals and the Commission on Educational Issues of the National Association of Independent Schools* (Boston: Houghton Mifflin, 1984).

7. James Hoetker and William P. Ahlbrand, "The Persistence of Recitation," *American Educational Research Journal* 6:2 (1969): 145–167.

8. Hoetker and Ahlbrand, "The Persistence of Recitation," 153.

9. Robert C. Pianta, Jay Belsky, Renate Houts, and Fred Morrison, "Opportunities to Learn in America's Elementary Classrooms," *Science* 315 (2007): 1796.

10. National Commission on Excellence in Education, *A Nation At Risk: The Imperative for Educational Reform: A Report to the Nation and the Secretary of Education* (Washington, DC: United States Department of Education, 1983).

11. John H. Flavell, "Cognitive Development: Children's Knowledge About the Mind," *Annual Review of Psychology* 50 (1999): 21–45.

12. Elementary & Secondary Education, Title I—Improving the Academic Achievement of the Disadvantaged, Sec. 101. *Improving the Academic Achievement of the Disadvantaged* (2002). Title I of the Act is amended to read as follows: Title I—Improving the Academic Achievement of the Disadvantaged Elementary and Secondary Education Act of 1965 (20 U.S.C. 6301 et seq.). Washington, DC: Public Law, 107–110—JAN. 8, 2002, 115 STAT. 1425.

13. National Academies of Sciences, Engineering, and Medicine, *The Future of Education Research at IES: Advancing an Equity-Oriented Science* (Washington, DC: The National Academies Press, 2022), 11.

14. Heinrich Mintrop and Gail L. Sunderman, "Predictable Failure of Federal Sanctions-Driven Accountability for School Improvement—And Why We May Retain It Anyway," *Educational Researcher* 38:5 (2009): 358.

15. Jean Kluver and Jeff Robin, *Changing the Subject: Twenty Years of Projects from High Tech High* (San Diego, CA: High Tech High, 2021), 11.

16. Seymour B. Sarason, *Revisiting the Culture of the School and the Problem of Change* (New York: Teachers College Press, 1996), 96.

17. John Dewey, *The Sources of a Science of Education* (New York: Horace Liveright, 1929), 32–33.

18. Jennifer A. Sandlin, Brian D. Schultz, and Jake Burdick, "Understanding and Exploring the Terrain of Public Pedagogy," in *Handbook of Public Pedagogy: Education and Learning Beyond Schooling*, edited by Jennifer A. Sandlin, Brian D. Schultz, and Jake Burdick (New York: Routledge, 2010), 1.

19. Henry A. Giroux, "Neoliberalism as Public Pedagogy," in Sandlin et al., *Handbook of Public Pedagogy*, 495.

20. Ira Shor, "What is Critical Literacy?" *Journal of Pedagogy, Pluralism & Practice* 4:1 (1999): 1 (Cambridge, MA: Lesley College), http://www.lesley.edu/journals/jppp/4/index.html.

21. Peter McLaren, "Afterword: Public Pedagogy and the Challenge of Historical Time," in Sandlin et al., *Handbook of Public Pedagogy*, 650–651.

22. Henry A. Giroux, *On Critical Pedagogy* (London: Bloomsbury Academic, 2020).

23. Paulo Freire, *Pedagogy of the Oppressed*, translated by Myra Bergman Ramos (New York: Continuum, 2000), 72.

24. Henry A. Giroux, "Cultural Studies, Public Pedagogy, and the Responsibility of Intellectuals," *Communication and Critical/Cultural Studies* 1:1 (2004): 66.

25. Angelina E. Castagno and Teresa L. McCarty, eds., *The Anthropology of Education Policy: Ethnographic Inquiries Into Policy as Sociocultural Process* (New York: Taylor & Francis, 2017), ix.

26. Castagno and McCarty, editors, *The Anthropology of Education Policy*, x.

27. Stanley Diamond, "Epilogue," in *Anthropological Perspectives on Education*, edited by Murray L. Wax, Stanley Diamond, and Fred O. Gearing (New York: Basic Books, 1971), 302.

28. Roger Dale and Susan Robertson, "Capitalism, Modernity and the Future of Education in the New Social Contract," *Teachers College Record* 111:14 (2009): 111–129.

29. Dale and Robertson, "Capitalism, Modernity and the Future of Education in the New Social Contract," 124.

30. Robert L. Crowson and Ellen B. Goldring, "The New Localism: Re-examining Issues of Neighborhood and Community in Public Education," *Teachers College Record* 111:13 (2009): 1–24.

31. Richard C. Schragger, "The Limits of Localism," *Michigan Law Review* 100:2 (2001): 371–472.

32. Crowson and Goldring, "The New Localism," 6.

33. Crowson and Goldring, "The New Localism," 8.

34. Crowson and Goldring, "The New Localism," 9.

35. Marcia Landy, "Culture and Politics in the Work of Antonio Gramsci," *boundary 2* 14:3 (1986): 49–70.

36. Maciel M. Hernández and Mayra Y. Bámaca-Colbert, "A Behavioral Process Model of Familism," *Journal of Family Theory & Review* 8:4 (2016): 463–483.

37. Murray L. Wax and Rosalie M. Wax, "Great Tradition, Little Tradition, and Formal Education," in *Anthropological Perspectives on Education*, 7.

38. Wax and Wax, "Great Tradition, Little Tradition, and Formal Education," 15.

39. National Academies of Sciences, Engineering, and Medicine 2022, *Ontologies in the Behavioral Sciences: Accelerating Research and the Spread of Knowledge* (Washington, DC: The National Academies Press, 2022), 35, https://doi.org/10.17226/26464.

40. Walter Benjamin, "The Work of Art in the Age of Mechanical Reproduction (1935)," translated by Harry Zohn, in *Illuminations*, edited by Hannah Arendt (New York: Schocken Books, 1969), 217–251.

41. Barbara J. Thayer-Bacon, "An Exploration of Myles Horton's Democratic Praxis: Highlander Folk School," *Educational Foundations* 18:2 (2004): 7.

42. Henry A. Giroux, "Rethinking Cultural Politics and Radical Pedagogy in the Work of Antonio Gramsci," *Educational Theory* 49:1 (1999): 1–19.

43. Robert Wohl, *The Generation of 1914* (Cambridge, MA: Harvard University Press, 1979), 195–196.

44. Kate Crehan, *Gramsci, Culture and Anthropology* (Berkeley: University of California Press, 2002), 76.

45. Antonio Gramsci, *Letters from Prison, Volume 2*, edited by Frank Rosengarten, translated by Ray Rosenthal (New York: Columbia University Press, 1994), 82.

46. Max Weber, *The Protestant Ethic and the Spirit of Capitalism*, translated by Talcott Parsons (New York: Scribner, 1958), 18.

Chapter 5

Crafting Communities of Learners

SWORDS INTO PLOWSHARES

Structural transformations have taken place in San Diego since the Cold War receded that, although regional in scope, clearly impact the local and global commons. The region is key to the Pacific Rim economy. New forms of enclosure and displacement came with population changes and urbanization pressures in San Diego, affecting housing and services. The New Urbanism design movement in the city's affluent areas adjusted according to these pressures. With the waning of the Cold War, national and regional elites transformed the militarized domestic economy in various postwar defense production areas in a move toward turning swords into plowshares, with regional economies embracing high technology and biotechnology industries.

The federal government enacted a Base Realignment and Closure (BRAC) process that shut down operations at hundreds of military installations, providing funds to finance environmental restoration and mitigation activities while altering the economies of surrounding communities.[1] Following BRAC guidelines, a former San Diego Naval Training Center at Point Loma was transformed into an urban village with a new identity, Liberty Station, through partnerships with the San Diego city government, nonprofit agencies, and corporate sector enterprises. While the BRAC guidelines transformed the navy base into Liberty Station with its hotels, restaurants, and shops, regional elites never fully shifted from the militarized domestic economy. The revenue was simply reorganized into new channels of the military complex, and not necessarily into plowshares. Liquidating portions of these bases, in San Diego and across the nation, was a windfall for real estate developers, architects, and contractors. The cleanup of toxic waste was at issue and the privatizing of revenue was dependent on removal of that waste. Environmental remediation efforts were linked to regional employment concerns, so as to maximize the military dollars spent locally to offset the financial impact of losing long-standing Pentagon largesse.

LIBERTY STATION

The post–Cold War transformation of Liberty Station, the former Naval Training Center in San Diego, into a new neighborhood in Point Loma, at the marine–urban interface within 200 meters of San Diego Bay, is instructive in this regard. The place was converted from a single-use military area to a mixed-use area, with both a cultural and an educational district, including schools developed within a larger regional economy of high technology and biotechnology.[2] Surrounded by green landscaping, Liberty Station is now an urban village with service, retail, and commercial businesses; parks; residential units; hotel sites; marine science laboratories; and an urban waterfront. A hilly peninsula bordered on the west and south by the Pacific Ocean, Point Loma is a seaside community of 38,000 within the city of San Diego. The peninsula is comparable to an island because of the aquatic borders and the urban landscape to the north. The southern one-third of the peninsula is entirely federal land, with residential uses comprising about 90 percent of nonfederal land.

The soundscape includes military and commercial aircraft, Coast Guard and navy activity, vehicular traffic, and the operation of a nearby water-treatment plant. San Diego Bay is a major port of entry, especially for ships traveling north from Mexico. The navy operates several bases with direct access to the bay, including Naval Base Coronado, Naval Base San Diego, and Naval Base Point Loma, with associated submarine bases. The U.S. Coast Guard also actively uses the bay for daily operations and maneuvers. Water quality is affected by the industries around the bay and by commercial, governmental, private, national, and international ships and boats. The city of San Diego operates a nearby wastewater treatment plant that treats one hundred eighty million gallons per day of sewage and deposits the treated effluent four miles offshore at a depth of four hundred feet. Meanwhile, the area contains multiple species of wildlife, both in the federal lands at the southern end of the peninsula (managed in part as an ecological reserve) and in the developed suburban areas. More than 300 species of birds have been observed in Point Loma, which lies on the Pacific Flyway migration route.

From the 1940s, San Diego was a key shipbuilding and aircraft assembly region dominated by the defense industry, with General Dynamics, General Atomics, Rohr, and National Steel and Shipbuilding bound to massive federal government contracts.[3] The San Diego Bay is home to one of California's five major ports and is the base for the U.S. Navy Pacific Fleet. The Naval Electronics Laboratory moved into Point Loma after the war and expanded its radio communications and sonar work to include basic research on electromagnetic energy and sound in the ocean. In 1956, the Pentagon awarded Convair (later a division of General Dynamics) the contract to build the air force's first intercontinental ballistic missiles; with 32,000 employees in the mid-1950s, Convair was San Diego's largest employer. The aerospace and defense industry grew throughout the postwar decades, together with the regional population, with the county growing from 556,000 in 1950 to nearly 2.5 million in 1990.

However, in the mid-1990s, during the wave of post–Cold War consolidation, General Dynamics shut down its Kearny Mesa plant, together with the twenty-seven-acre Missile Park, which provided recreational activities for employees and their families, and 60,000 workers lost their jobs within eighteen months of the closure. The signal event—a plant closing of this magnitude—had a profound effect on the region's business culture. Being so closely tied to defense contracting and the broader military economy, with decades of federal largesse, San Diego had virtually no access to venture capital funding or equity banking to support science-based innovation or entrepreneurship.

Two decades after the end of the Cold War, San Diego reinvented itself as a high-technology hub, with both academic research and industrial production in telecommunications and biotechnology. The region's 1,500 high-tech companies, together with the University of California, San Diego, Scripps Institution of Oceanography, Sanford-Burnham Medical Research Institute, and Salk Institute for Biological Studies on the Torrey Pines Mesa, form a network of regional science and technology-based innovation clusters that fuel a knowledge economy focused on commercial applications of dynamic fields such as genome sequencing, climatology, oceanography, and information technology.[4] Considerable collaborations between San Diego's business and research communities formed over these decades fostered a web of relationships and co-investments that facilitated access to capital.[5] Science and engineering professionals and technoservice workers moved into the North County suburbs, where new industries were transforming a military-, tourist-, and real estate–based economy into a global high-technology and biotechnology production region.

THE SCHOOL AND THE CITY

San Diego's regional elites never fully abandoned the metropolitan–military complex, but simply shifted revenue into a mixed-use, capital-intensive economy. A knowledge economy, both capital driven and resilient, grew together with the region's economic transformation. Liberty Station is the site of public charter schools, which are central to the emergent knowledge economy consequent to the regenerative modality of post–Cold War globalization, together with the concern by stratified economic sectors to assure a secure future for their children. The knowledge base in the form of educational infrastructure reflects this twofold, coordinated interest. As with other school districts, nationwide, San Diego schools followed federally funded systemic change initiatives, such as "No Child Left Behind" and" Race to the Top," which established benchmarks and timelines for accountability and influenced school district decision making. Schooling also began to change as a consequence of the skill demands of knowledge work in firms that put down roots in post–Cold War San Diego. Beginning in 1998, the San Diego Unified School District introduced a seven-year systemic reform effort to improve students' literacy skills. The district's *Blueprint for Student Success*, while focusing initially on implementing new programs in literacy and mathematics, initiated a broader instructional reform effort that brought about dramatic changes in school-based leadership and teacher professional development to improve student outcomes.[6] With the goal of social justice and equity in mind, the district-led reform policies assumed that instructional practice and student achievement growth were linked.

The post–Cold War restructuring of San Diego's economy toward a high-tech knowledge economy influenced the Greater San Diego Chamber of Commerce's Business Roundtable's endorsement of the school district's *Blueprint for Student Success*. Since the mid-1990s, the Business Roundtable actively engaged in educational reform, specifically toward a change in district leadership, "pointing to gaps in student performance and unqualified graduates entering the local workforce."[7] Although some opposed the whole-district content-driven reforms, the Greater San Diego Chamber of Commerce, the Chamber's Business Roundtable, and the San Diego Times-Union, together with local philanthropies and educational researchers, endorsed the *Blueprint*. The William and Flora Hewlett Foundation donated $13.5 million over three years, the Bill and Melinda Gates Foundations donated $15 million over five years, and the Atlantic Philanthropies granted the district $5 million to support the *Blueprint*.[8]

Five years after the reforms were introduced, the *Blueprint*'s success in improving student outcomes represented both time and money well spent by the business and philanthropic communities in pursuit of educational leadership and instructional reform that would support the dynamic technology clusters in the region.[9] These educational priorities, influenced and capitalized by elites for administrative sustainability within the schools, brought about a core redistribution of this form of capital in the region. The Chamber's Business Roundtable also spearheaded the San Diego Charter School Consortium to foster the introduction of public charter schools to the area. Together with federal BRAC policies, the Roundtable provided the impetus for the newly planned Liberty Station to house unique charter schools designed upon ecologically informed principles. The development of Liberty Station and the introduction of charter schools on the site are market oriented and based on smart growth. These initial experimental charter schools—tech-savvy, humanities-friendly, and located amid EPA Superfund environmental remediation efforts—embody and promote ecological principles of sustainable regional management.[10] As educational experiments, the schools are well organized, scientifically and technologically focused, and also sustainably designed. The schools operate within a state- and district-approved educational framework that is adaptive to the changing demographics of a neoliberal urban market economy.

High Tech High School, a public charter school in Liberty Station, was designed with the new knowledge economy in mind.[11] According to Rob Riordan, a cofounder of High Tech High:

> The Business Roundtable discussions led to starting a charter school, but also for a local, pragmatic reason: at the time, there was a cap on work visas for technicians from abroad, and there was a dearth of local talent for the emerging high-tech industry. While there was recognition of the need for change in local business and school administrative circles, the top-down reforms generated a great deal of resistance on the ground. Moreover, the reforms prioritized basic skills and standardization, such as the literacy initiative, over essential skills, such as collaboration and communication, and teacher autonomy. The charter avenue allowed High Tech High to sidestep dysfunctional labor-management dynamics and focus on collaborative design and execution of authentic work.[12]

High Tech High opened in September 2000 in a newly renovated 38,500-square-foot facility, formerly a U.S. Navy technical training center building, with substantial support from Gary Jacobs, at the time a software engineer and senior education specialist of Qualcomm, Inc., a semiconductor, and telecommunications equipment company in San Diego. According to Larry Rosenstock, a cofounder of High Tech High:

> Without question, High Tech High would not exist without Gary Jacobs. Gary was deeply engaged for years in an initiative that began with his insight to make the case that San Diego very much needed a new kind of school. He organized meetings with several key civic leaders who met and planned for years. He and his wife, Jerri-Ann, gave the first critical and substantial financial support to the first High Tech High (which is named after them). But it does not stop there. Gary Jacobs has chaired the board that oversees the sixteen High Tech High schools. He is also on the board of the High Tech High Graduate School of Education. Gary Jacobs has provided steady, and visionary leadership to both organizations. In short, there would be no High Tech High were it not for Gary Jacobs.[13]

Since the original school's founding, six additional schools have been established at the same location, creating a village of three high schools, two middle schools, and two elementary schools. One of forty-eight high schools in the San Diego Unified School District, High Tech High has 501 students, with a 92 percent minority enrollment., majority Hispanic. The school's twenty-three seminar rooms, seven labs, and a large high-ceilinged open area, known as the "Great Room," were designed to support educational program elements, including team teaching, integrated curriculum, project-based learning, community-based internships, frequent student presentations, and exhibitions.[14] The building's flexibility allows students to work individually and in large and small groups. Vast expanses of glass contribute to the sense of transparency; classrooms and common spaces are bathed in natural light to create a sense of "visible learning." Wall and ceiling spaces are dedicated to exhibiting student projects, reinforcing the students' "ownership" of the milieu. The building's design features inform High Tech High's culture of learning based upon a set of educational design principles—namely, equity, personalization, authentic work, and collaborative design.[15]

In a variety of documents, Rosenstock offered the following comment regarding technology at High Tech High: "Technology is not studied as a subject; rather technology tools, both 2-D and 3-D, are ubiquitous and used for producing—that is, making, shaping, forming."[16] Rosenstock's background as a vocational-education (carpentry) teacher, and his reference to technology as a set of "tools" used to "produce," suggests the influence of his background and this metaphor on high tech at High Tech High's pervasive displays of artifacts in the High Tech High building and campus. These exist in conjunction with the school's substantial collection of archived digital media content found on the school's evolving website. In addition, teacher and student portfolios offer solid evidence that technology has played a significant role in various "productions" at High Tech High. What is less apparent, and the subject of a related study on the creation and use of digital portfolios at High Tech High, is the extent to which technology is either perceived or used to support student learning.[17]

At High Tech High, students work together in projects that integrate science, mathematics, the arts and humanities.[18] Integration is frequently accomplished through the use of important or essential questions. Throughout, students are encouraged to develop habits of mind, including evidence, perspective, connection, supposition, relevance, or significance, and then to reflect critically on their learning and personal growth through this lens. As a form of active inquiry, the five habits of mind, or "thinking dispositions," originally put forward by Ted Sizer,[19] can also be framed as questions for both learning and assessment:

How do you know what you know? (Evidence)

From whose point of view is this being presented? (Perspective) How is this event or work connected to others? What causes what? (Connection)

What if things were different? (Supposition) Who cares? Why is this important? (Relevance)[20]

Each student has a faculty advisor who provides ongoing support and helps the student to develop a customized learning plan. Students connect to the world beyond school through field studies, community service, junior- and senior-year internships, and consultation with outside experts; they exhibit that work in professional venues. Assessment is performance based, with all students developing projects, solving problems, and presenting findings to community panels. High Tech High students are required to complete a substantial senior project and a personal digital portfolio to archive and share their learning and to make annual presentations of learning in lieu of exams.[21]

High Tech High teachers employ a variety of approaches to accommodate diverse learners, acting as program and curriculum designers.[22] They work in interdisciplinary teams to design the courses they teach, and teachers have ample planning time to devise integrated projects, common rubrics for assessment, and common rituals, such as a presentation of learning, by which all students demonstrate their learning and progress toward graduation. High Tech High's curriculum is a hybrid of teachers' expert knowledge and what is referred to as *indigenous invention*[23]—namely, a school culture that is created through local knowledge and experience and where the learning environment and styles of inquiry are place-focused and rooted in community action. Based upon new conceptions of learning, cognition, and development, curriculum and standards for school and enrichment programs are co-created by teachers and other local individuals with a stake in children's learning. Project-based learning, especially in high school, will often involve the students as collaborators in the creation of curricular goals and outcomes, derived from their lived experience.[24]

High Tech High has formed school-based learning communities to improve inequitable educational outcomes. Grounded in principles of improvement science that illustrate continuous improvement efforts within schools, High Tech High focused on enabling leadership through collective action for system transformation, notably tackling chronic absenteeism and supporting the college application process. Beyond, student-focused improvement efforts, Riordan discusses the need for continuous adult learning:

> High Tech High founders realized, from the beginning, that a new way of doing things would necessitate new approaches to professional development in the schools—in particular, creating a robust learning environment for the adults. Initially, this was a matter of scheduling ample time for adult learning. The initial schedule at High Tech High, which prevails in all 16 High Tech High schools to this day, provided that teachers would arrive an hour before the students each day, and would meet in various configurations to discuss curriculum,

> students, issues the school faced, and questions educators were pursuing in their own practice. Eventually, this focus on adult learning led to two developments: the creation of a California-approved teacher credentialing program, and the creation of a graduate school of education, offering a master's program in educational leadership. This graduate school served as the incubator for various improvement and dissemination efforts. From the beginning, the High Tech High Graduate School of Education curriculum focused on reflective practice, with action research as the means of linking theory and practice. As such, the graduate school of education was an outlier among post-secondary programs in teacher training, as even action research was somewhat of an outlier in post-secondary academia. The way we put it originally, the curriculum was 20% theory and 80% practice, as opposed to the conventional ratio of 80% theory and 20% practice. As you can imagine, that didn't go over well with our colleagues on accreditation visiting teams. We had to learn to side-step that discussion, and so we adopted a different mantra, describing our approach as "100% theory embedded in practice."[25]

High Tech High Graduate School of Education's Center for Research on Equity and Innovation established improvement hubs and convened a network of High Tech High improvement schools to stimulate dialogues around challenges to college-going, engaging parents and students as co-partners in the project.[26] These efforts enacted a time-honored "diffusion of innovation" approach that built teachers' capacities through improvement networks within and across schools. The ensuing network-based dynamics enabled the more activist teachers to become first adopters and then "evangelists" for continuous improvement on behalf of student success across networked schools. Throughout was a call to "organize for improvement," notably when introducing a new curriculum or technology, building staff capacities, investing in an improvement hub, and developing analytic capacity, all on behalf of cultivation practitioners' agency for change. High Tech High's continuous improvement efforts provide a framework for system transformation within the school community that supports out-of-classroom project-based learning, while providing a mechanism for how such innovations can then inform improvement in classroom instruction. In Riordan's words: "Improvement science emerged as a means of linking theory and practice—and treating educator experience as a 'text' worthy of articulation, exploration, analysis, and sharing with the world—that was acceptable to the academy."[27]

STUDENTS OF CONSEQUENCE

High Tech High students engaged in a long-term project that used San Diego Bay as an outdoor laboratory to understand regional urban ecology, as the school is located in the ongoing redevelopment of Liberty Station. The school's proximity to the bay affords easy contact with several wetland and armored bay habitats. Ninety percent of the bay's historic wetlands have been filled in, drained, or diked. Seventy-four percent of the shoreline has been armored, providing habitat for open-coast hard substrate species in a traditionally soft-substrate estuary. As the city's redevelopment efforts continue to progress, the students' study of the complexity and fragility of urban-bay ecosystems contributes to the city's ongoing discussion and decisions. Teachers and high-school juniors developed strong local partnerships with academic, industrial, and nonprofit-sector stakeholders in a continuing community-based conservation program for the bay. Using this setting as a field laboratory, students began to understand the interconnection between human activities, such as fishing, boating, and military exercises, and local marine life, and then provided decision makers with a set of perspectives, recommendations, and original research findings from their field studies.

The study's faculty designer, Jay Vavra, grew up in San Diego and returned home after a range of experiences in biology abroad and throughout California. He graduated from Stanford University and earned a PhD in marine biology from the University of Southern California. Vavra also worked as a biologist in the fields of biotechnology and zooarchaeology, and as an environmental consultant. He led High Tech High students on two expeditions to Tanzania to carry out projects on global conservation, specifically the African bushmeat trade, using DNA barcoding techniques to help local wildlife-protection officials fight poaching. Vavra also worked in Mozambique on biodiversity assessment with Edward O. Wilson, and contributed to Wilson's digital biology textbook, E. O. Wilson's *Life on Earth.*[28]

In co-designing the study, the first cohort of high-school juniors placed all of the components necessary to understand the bay on the floor, organized them with the class, and then linked them with thread. The activity involved each of the students, who were required to communicate with one another what they knew about the bay from various perspectives. From this original schematic, photographed from above and archived digitally so that subsequent cohorts could carry out a similar design process, students set out to understand the bay as a mix of both human construction and nature's slow reclamation of the marine ecosystem.

In its first eight years, High Tech High San Diego Bay Study[29] introduced 450 High-School juniors to original research in ecological assessment and engagement in writing and reflection regarding their urban ecology. The first study focused on describing the wildlife and reflecting on nature at the nearby Boat Channel, and then it expanded to include detailed field studies of the entire San Diego Bay that would result in a series of published field guides.

Inquiry-based research projects served to integrate a holistic perspective of earth's natural and human systems, specifically the complexity and fragility of the urban-bay ecosystems to better understand the interconnections between human activities and local marine life. Student researchers use photography, interviewing, mapping, drawing, and journaling, together with ecological surveys of species biodiversity and DNA barcoding for identification of species, including invasive species, to develop an understanding of their surroundings and self-awareness of their place within them. Through learning in the field and in the biology classroom, students begin to acquire, according to Vavra, "the ability to make fundamental connections, an awareness of complexity, and the necessary compassion to understand and care for the environment."[30]

In their humanities classes, students produced illuminated journals; reviewed and produced nature photography, poetry, and reflective writing to investigate historical perspectives on the environmental crisis; and debated issues related to stewardship and conservation. Their readings included Thoreau's *Walden*[31] and John Steinbeck's *The Log from the Sea of Cortez*.[32] As they reflected on their interaction with the bay, Vavra and Tom Fehrenbacher, their humanities teacher and co-designer of the project, posed the following questions to the students: "What is our place in broken relationship with our environment? Can we accept ourselves as part of nature?" Considering that these are overarching questions about humanity's place in nature, Fehrenbacher understood that the humanities were essential to both cultivating students' insights concerning these larger issues and helping to answer questions raised in their field studies and reflective journals. In his words:

> In working with Jay, I found out that addressing climate change requires contributions not only from those who study nature, but also from those who study humanity. After all, WE are the species causing it. In humanities, with climate change a settled fact, our Socratic Seminars focused more and more upon the nature of humanity itself. We asked an essential question about climate change: "How can we, as a species, while knowing this is happening, leave it largely unaddressed?" If we had not done so, I could not have looked Jay in the eye. Had humanities students not engaged in relevant social study, neither Jay nor I would have found their work authentic. And, by taking up social analysis, students found that our history, our literature, and the best in our culture can tell us a great deal about life, about living simply, and about sustainability.[33]

From their studies and reflections, students produced a series of books on specific topics about the San Diego Bay, including the historical ecology and conservation practices; each volume integrates cartography, humanities, biology, and art. Four books were produced that focused on the issues that reflect

San Diego's most pressing environmental priorities, including those that students identified as most meaningful to their generation. These books—*Two Sides of the Boat Channel: A Field Guide*;[34] *Perspectives of San Diego Bay: A Field Guide*;[35] *San Diego Bay: A Story of Exploitation and Restoration*;[36] and *San Diego Bay: A Call for Conservation*[37]—were praised by local residents, marine scientists, conservation advocates, and policymakers alike. A fifth volume, *Biomimicry: Respecting Nature through Design*,[38] discusses sustainability and provides case studies of biomimetic design as a way to craft solutions to environmental problems. The students used mathematics to test their biomimetic designs for sustainability and reflect upon the implications of these designs for such topics as "natural capitalism," "tenets of sustainability," and "the biological basis of morality." The books, available through major online booksellers, contain forewords by scientist-advocates Jane Goodall and Edward O. Wilson and endorsements by the National Audubon Society, Sierra Club, Wildlife Conservation Society, and Blue Ocean Institute, and local stakeholders, or "Stewards of the Bay," advocates for San Diego Bay conservation and restoration. Students also used the bay as a site for naturalist and scientific observation, focusing on benthic marine invertebrates. Student teams collected and analyzed different taxonomic groups and tried to identify all species within their assigned taxa. They also collected samples for DNA barcoding work done in their lab. DNA barcode sequences were uploaded to GenBank and Bold, both Encyclopedia of Life[39] content partners. The students also created documentaries, combining video footage of interviewees and their original research with multimedia and graphic arts skills acquired in art classes. This community of practice among students and their adult mentors holds the promise of "knowing" the commons and the prospect of sustainability based upon socially defined competence and habits of mind grounded in a sense of accomplishment and responsibility.

Integrating the arts, humanities, and social sciences with active engagement in documenting and restoring an urban ecosystem can help avert the disenchantment and sense of helplessness that frequently accompanies studies of environmental degradation and the adverse effects of climate change. Fehrenbacher sums up the sense of knowing that comes from adults and students working within a community of open-ended practice with respect to both the discovery process and the integration of the "consequences" of interdisciplinary project-based learning:

> With our students we looked for connections, sought relevance and asked questions no matter where they led. In the field guide's uncharted territory, our students found a place to raise their own questions and express their own ideas. We learned a lot from them. What started as a simple look at the local Boat Channel turned into a study of the Bay's ecological history, which then became

a consideration of the environment itself. In some sense, there was nothing planned about it.[40]

Engaging students in helping to create these outcomes through a mentored process that provides enhanced foundational scientific knowledge, critical skills for success in group settings and inspirational community-based experiential activities, significantly enriched what youth learned through conventional curricula. Etienne Wenger[41] sees communities of practice as enacted through a "dual process of meaning making"—namely, through personal participation in social activities and the production of physical and conceptual artifacts that reify experiences. Rob Riordan understands this dual process as enacted in the San Diego Bay Study:

> This dual process, facilitated by Vavra and Fehrenbacher, is precisely what was going on in the San Diego Bay project, where the social relations of production took the form of multiple drafting and peer critique as students engaged in observation, documentation, analysis, and the collaborative creation of the artifacts (photos, essays, calendars, guides) by which they articulated and shared their findings. In other words, there was an instructional technology that supported the intertwining of the relational and creative processes, as Vavra and Fehrenbacher, invoking larger purposes, moved toward the two overarching student outcomes to which High Tech High aspires: self-directed learning, and relational agency (the ability to navigate systems and relationships). The immediate pedagogical roots of that technology, for Fehrenbacher, reside in the writing process reforms that emerged in the 1960s; the broader older professional roots (for High Tech High) may be seen in the architect's studio as prototype.[42]

The San Diego Bay Study case demonstrates varieties of informal learning outside of school that take place in face-to-face venues, which offer mentoring, apprenticeship, and participation in projects as fledgling artists, engineers, planners, writers, teachers, and scientists. Youth in communities of practice developed over a decade on behalf of San Diego Bay conservation efforts were eager to explore out-of-classroom learning activities that introduced urban environmental issues, such as water quality and wetlands restoration, through student collaboration with peers and mentors on applied projects. Project-based learning especially appeals to adolescents, who are more conscious of their community, by using it as an environment to demonstrate scientific principles. This approach has a long history, dating back to a great transformation in American educational history toward mass schooling and the high-school movement.

Looking again at the case, Riordan asks:

> What were the existing regularities that High Tech High contravened? In other words, what were the structural and cultural enablers of high-quality

> project-based learning? In conventional schools, teachers who do this kind of work do it against the grain. Vavra and Fehrenbacher's work, on the other hand, is the outgrowth of the overall design. It's a natural consequence, not an anomaly. So, we're talking about, not simply a pedagogy, but a re-imagining of school as an institution—all in face of the broader, oppressive content-based, compliance-oriented systems of standard-setting and testing. Among the existing regularities that we confronted: external, occasional pedagogical design; teacher as lecturer/transmitter; elective systems of course-taking; siloed, discipline-based curriculum; heavy student load per teacher; teacher isolation; departmental organization; and tracking.[43]

Project-based learning involved students as collaborators in the creation of curricular goals and outcomes using San Diego Bay as an outdoor laboratory to understand regional urban ecology. This community of practice made up of students and adult mentors supported the goal of knowing the commons and the prospect of sustainability at the marine–urban interface. The learning community was sustained through socially defined competence and habits of mind developed in carrying out urban ecology projects that integrate a holistic perspective of earth's natural and human systems. Over a decade, various projects introduced high-school juniors to original research in ecological assessment, engaged writing, and reflection on the bay and its estuaries. Teachers and students designed their learning experiences to develop an understanding of their surroundings and self-awareness of their place within them. Through learning in the field and the classroom, students began to make connections about the complexity of the local environment and the need to care for it.[44]

NOTES

1. Christopher M. Schnaubelt, "Making BRAC Politically Palatable," *The RAND Blog*, March 16, 2017, https://www.rand.org/blog/2017/03/making-brac-politically palatable.html.
2. Judith Comer-Schultz, "History and Historical Preservation in San Diego since 1945: Civic Identity in America's Finest City" (PhD dissertation, Arizona State University, 2011).
3. Roger W. Lotchin, "The Political Culture of the Metropolitan-Military Complex," *Social Science History* 16:2 (1992): 275–299.
4. Lynne G. Zucker, Michael R. Darby, and Marilynn B. Brewer, "Intellectual Human Capital and the Birth of U.S. Biotechnology Enterprises," *American Economic Review* 88:1 (1998): 290–306.
5. Michael Sable, "The Impact of the Biotechnology Industry on Local Economic Development in the Boston and San Diego Metropolitan Areas," *Technological Forecasting and Social Change* 74:1 (2007): 36–60.

6. Julian R. Betts, "The San Diego Blueprint for Student Success: A Retrospective Overview and Commentary," *Journal of Education for Students Placed at Risk* 14:1 (2009): 120–129.

7. Amy M. Hightower, "San Diego's Big Boom: Systemic Instructional Change in the Central Office and Schools," in *School Districts and Instructional Renewal*, edited by Amy M. Hightower, Michael S. Knapp, Julie A. Marsh, and Milbrey W. McLaughlin (New York: Teachers College Press, 2002), 78.

8. Julian R. Betts, Andrew C. Zau, and Kevin King, *From Blueprint to Reality: San Diego's Educational Reforms* (San Francisco: Public Policy Institute of California, 2005), 3.

9. Public Policy Institute of California, *The Success of San Diego School Reforms Could Serve as a Blueprint for the State*, Research Brief, Issue 104 (San Francisco: Public Policy Institute of California, 2005).

10. United States Environmental Protection Agency, "Superfund Site Information: San Diego Naval Training Center," https://cumulis.epa.gov/supercpad/cursites/csitinfo.cfm?id=0902730.

11. Charles Taylor Kerchner, *The Emperor's Clothes: Traditional and Avant Garde at High Tech High* (Claremont, CA: Claremont Graduate University, 2012).

12. Riordan, personal communication.

13. Larry Rosenstock, personal communication.

14. Bob Pearlman, "Designing, and Making, the New American High School," *Technos-Bloomington* 11:1 (2002): 12–19.

15. Larry Rosenstock, "High Tech High Design Principles," 2017, https://www.hightechhigh.org/about-us/.

16. Larry Rosenstock, "Technology at High Tech High," 2004, http://www.bobpearlman.org/BestPractices/TechnologyatHighTechHigh.pdf.

17. Brian Donnelly, "Digital Portfolios and Learning: The Students' Voices," Doctoral Dissertation. University of California, Davis, 2010.

18. Tara S. Behrend, Michael R. Ford, Kathleen M. Ross, Edmund M. Han, Erin Peters Burton, and Nancy K. Spillane, "Gary and Jerri-Ann Jacobs High Tech High: A Case Study of an Inclusive STEM-Focused High School in San Diego, California" (OSPrI Report 2014-03), George Washington University, Opportunity Structures for Preparation and Inspiration in STEM (2014) https://ospri.research.gwu.edu/sites/ospri.research.gwu.edu/files/downloads/OSPrI_Report_2014-03.pdf.

19. Theodore R. Sizer, *Horace's Hope: What Works for the American High School* (Boston, MA: Houghton Mifflin, 1996).

20. Brent Duckor and Daniel Perlstein, "Assessing Habits of Mind: Teaching to the Test at Central Park East Secondary School," *Teachers College Record* 116:2 (2014): 1–33.

21. David Stephen and Eve Goldberg, *Profile: High Tech High Network—Student-Centered Learning in Action* (Quincy, MA: Nellie Mae Education Foundation, 2013), https://www.nmefoundation.org/getmedia/aa1d06c3-dd1d-460b-b99f-5b283d03e008/ PROFILE-HighTechHighNetwork-NMEF.

22. Stacey L. Caillier and Robert C. Riordan, "Teacher Education for the Schools We Need," *Journal of Teacher Education* 60:5 (2009): 489–496.

23. Paul E. Heckman and Viki L. Montera, "School Reform: The Flatworm in a Flat World: From Entropy to Renewal through Indigenous Invention," *Teachers College Record* 111:5 (2009): 1328–1351.

24. Robert C. Riordan, "Change the Subject: Making the Case for Project-Based Learning," *Edutopia*, January 17, 2013, https://www.edutopia.org/blog/21st-centuryskills-changing-subjects-larry-rosenstock-rob-riordan.

25. Riordan, personal communication.

26. Anthony S. Bryk, "Infusing Improvement into a Charter Network: High Tech High," in *Improvement in Action: Advancing Quality in America's Schools* (Cambridge, MA: Harvard Education Press, 2020), 101–124.

27. Riordan, personal communication.

28. Morgan Ryan, Gaël McGill, and Edward O. Wilson, *E. O. Wilson's Life on Earth* (Durham, NC: E. O. Wilson Biodiversity Foundation, 2014).

29. Jay Vavra, "The San Diego Bay Study: Community-Based Conservation," in *Proceedings of the Pacific Division, American Association for the Advancement of Science* Volume 28, Part 1 (Ashland, OR: American Association for the Advancement of Science, Pacific Division, 2009), 72.

30. Vavra, personal communication.

31. Henry David Thoreau, *Walden*, edited by J. Lyndon Shanley (Princeton, NJ: Princeton University Press, 1971).

32. John Steinbeck, *The Log from the Sea of Cortez* (New York: The Viking Press, 1951).

33. Tom Fehrenbacher, "Logs from San Diego Bay," *UnBoxed* 13 (2015), http://www.hightechhigh.org/unboxed/issue13/logs_from_san_diego_bay/.

34. Students of the Gary and Jerri-Ann Jacobs High Tech High, *The Two Sides of the Boat Channel: A Field Guide* (San Diego, CA: BuenFeVA Press, 2004).

35. Students of the Gary and Jerri-Ann Jacobs High Tech High, *Perspectives of San Diego Bay: A Field Guide* (Providence, RI: Next Generation Press, 2005).

36. Students of the Gary and Jerri-Ann Jacobs High Tech High, *San Diego Bay: A Story of Exploitation and Restoration* (El Cajon, CA: Sunbelt Publications, 2007).

37. Students of the Gary and Jerri-Ann Jacobs High Tech High, *San Diego Bay: A Call for Conservation* (El Cajon, CA: Sunbelt Publications, 2009).

38. Students of the Gary and Jerri-Ann Jacobs High Tech High, *Biomimicry: Respecting Nature through Design* (San Diego, CA: High Tech High, 2014).

39. Richard Blaustein, "The Encyclopedia of Life: Describing Species, Unifying Biolog," *BioScience* 59, no. 7 (2009): 551–556; Edward O. Wilson, "The Encyclopedia of Life," *Trends in Ecology & Evolution* 18, no. 2 (2003): 77–80.

40. Tom Fehrenbacher, "Logs from San Diego Bay," http://www.hightechhigh.org/unboxed/issue13/logs_from_san_diego_bay/.

41. Etienne Wenger, "Communities of Practice and Social Learning Systems: The Career of a Concept," in *Social Learning Systems and Communities of Practice*, edited by Chris Blackmore (London: Springer, 2010), 180.

42. Riordan, personal communication.

43. Riordan, personal communication.

44. This chapter, dedicated to the memory of Jay Vavra, is based on a long-term ethnographic study of High Tech High, part of a larger study of project-based learning in high school carried out by researchers at UCLA and University of California, Davis, and funded by the Sally and Dick Roberts Coyote Foundation. I also wish to thank the following teachers who graciously shared with me their work with their students in their classrooms during and after school hours, on exhibition night and during their students' presentations of learning: Mark Aguirre, Stacey Caillier, Amy Callahan, Tom Fehrenbacher, Jean Kluver, Jeff Robin, and Jesse Wade Robinson. I am also grateful to the many students at the Gary and Jerri-Ann Jacobs High Tech High and at the High Tech High Graduate School of Education who generously shared their projects and reflections on their learning. I am especially grateful to Larry Rosenstock, Rob Riordan and Ben Daley who spent many hours with me sharing their visions of twenty-first-century learning.

Chapter 6

Cultivating Intergenerational Mentoring

A COMMUNITY IN TRANSITION

Pacoima is a low-income, working-class community of 103,000 located in the northeast San Fernando Valley in the City of Los Angeles, with a population density of 9,870 per square mile. Pacoima covers seven square miles at the base of the San Gabriel Mountains. About 87 percent of Pacoima's population is of Hispanic origin. Immigrants make up about 46 percent of the population; 23 percent are noncitizens. The median household income is $64,000 per year; 21 percent of the population lives below the federal poverty level. Forty-five percent of residents hold a high-school degree and 24 percent are linguistically isolated. Of the 17,232 housing units, 62 percent are owner-occupied homes.

Pacoima was established as a rail side stop for Southern Pacific Railroad passengers in 1887.[1] From the beginning of the twentieth century to the Second World War, the area remained a community of small farms, vineyards, and orchards. During and following the Second World War, aircraft and other assembly plants turned mostly inexpensive property into land valuable for both residential and commercial purposes. During the war, restrictions were relaxed to permit African Americans to purchase homes in Pacoima in order to provide a workforce for the industry in the northeast San Fernando Valley. After the war, the area became a blue-collar community of mostly single-family homes with a predominantly African American population. Low-incomc black families also began moving to Pacoima in the early 1950s, with the completion of public housing projects.

Jobs were plentiful for both low-income factory workers and middle-class working families; however, change came during the 1980s and 1990s in Southern California's economy as aerospace-related and automobile manufacturing and consumer product light manufacturing left the northeast San

Fernando Valley. These changes had an enormous impact on local employment in Pacoima and surrounding communities. It is estimated that half of minimum wage and lower-wage manufacturing and warehousing jobs moved from the area since the signing of the 1994 North American Free Trade Agreement (NAFTA), which removed most barriers to trade and investment among the United States, Canada, and Mexico. A large number of good-paying jobs with benefits left the area with the closure of the General Motors assembly plant in Van Nuys, Lockheed's aircraft plant in Burbank, and the Price Pfister plumbing fixture plant in Pacoima. Lack of jobs eventually led to the displacement of African American families from Pacoima. Because of newly available inexpensive housing, Pacoima attracted recent Latino immigrants, many of whom pooled their earnings to rent or buy a house.

Pacoima sits at the edge of the San Fernando Valley where it rises toward the San Gabriel Mountains. Pacoima translates as "rushing waters," a name given to the area by the San Gabrielino Indians who helped farm the extensive mission gardens after the founding of Mission San Fernando Rey in 1771. A stream of water flows into the area from the nearby mountain canyons. True to its name, Pacoima experienced extensive flooding twice in its 125 years, during the Great Flood of 1891 and again in 1938 when 29 inches of rainfall swept away homes and took the lives of residents. Flooding has been controlled somewhat by an extensive flood plain managed by the Army Corps of Engineers at nearby Hansen Dam. However, inadequate storm drainage causes water to back up on many streets during light rains.

Situated at the wild land–urban interface, the area's natural environment, particularly the brush and grass covered foothills to the north and east, block and modify winds. They capture Los Angeles basin smog and other air pollutants during most of the spring, summer, and early autumn. In the late autumn and winter months, the winds reverse, blowing dust, plant particulate, and manmade pollutants off the hills onto the neighborhoods below. A flood basin, nearly three miles in circumference, comprises much of the area. Fall and winter winds blow across the flat terrain of mostly dirt and rocks, bringing dust and sand into residential neighborhoods.

Pacoima Beautiful, a community-based environmental health and justice organization, is dedicated to the creation of a healthy, environmentally safe, prosperous, and sustainable community. It began in 1996 and became a nonprofit under 501(c) 3 in December 1999. A grassroots organization, Pacoima Beautiful was initially formed to help residents clean up but then grew to promote environmental health and justice education, leadership development, and advocacy skills to residents. The organization has an agenda of civic engagement on behalf of environmental awareness and community building and employs an action research approach designed to enhance the quality of life in the community, together with the cultivation of an aesthetic sensibility.[2] The various projects undertaken by Pacoima Beautiful to identify toxic substances, understand the health implications of potential toxic risks, and ameliorate those risks resulted in a long-term community capacity to improve the local environment.

Since its beginnings, Pacoima Beautiful has evolved from a volunteer beautification committee composed of five individuals to an environmental justice and environmental health, community-based organization that operates through the support of a policy board consisting of residents as well as professional advisors. Pacoima Beautiful's mission is "to empower the Pacoima community through programs that provide environmental education, advocacy and local leadership in order to foster a healthy and safe environment." The mission was carried out in the organization's early years through three programs. The Community Inspectors Program taught residents how to identify the sources of toxins and pollution in the community, to understand the potential risks to health, and to find simple solutions to reduce the risks in the community. The Youth Environmentalists Program engaged approximately 150 local high-school youth, annually, through environmental education projects, leadership development, skills training, and college readiness activities. The Safer Homes for a Healthy Community Program helped residents to create healthy homes for their families in order to reduce and prevent environmentally related health problems, including lead poisoning and asthma triggers, such as mold and moisture.

Pacoima Beautiful brought together residents, university environmental health scientists, environmental and other organizations, university service-learning classes, and representatives from governmental agencies to address environmental issues in the community. Being able to partner with experts in various fields served to expand the capacity of Pacoima Beautiful and build a valuable knowledge bank. As environmental health and environmental justice became a more prominent part of the work of Pacoima Beautiful, it was necessary to enlist the support of those who knew how to address the issues. Pacoima has long suffered from environmental neglect that can likely be blamed for the high rates of environmental health risks in the community and the numerous sources of pollution throughout the area.[3]

During the post–Cold War decades, Pacoima underwent a transformation from a heavy aerospace and defense industrial landscape to a service and transportation economy, leaving a set of toxic conditions. Pacoima is home to seven U.S. Environmental Protection Agency (US EPA) Comprehensive Environmental Response Compensation, and Liability Act (CERCLA) toxic release sites, commonly known as Superfund sites, four of which are active and three that have been remediated at the time of this case study. Community concerns focus on the cumulative impacts from contaminants, such as lead in paint and in the soil, emissions from freeways, commuter planes, diesel from trucks and equipment, older gross emitting cars, landfills, and the widespread use of toxic chemicals throughout the community.

ENGAGING YOUTH IN A DIVIDED CITY

In the late 1990s, when Pacoima Beautiful began focusing on the community's youth, Los Angeles County had 1,350 named street gangs, with at least 150,000 gang members, about one-quarter of the nation's active gangs.[4] Gang crime proliferated across the region, including gang-related homicide, which represented 40 percent of all homicides in the county. Throughout the decade, the northeast San Fernando Valley's gang landscape had changed with Central American immigrants moving into traditional Mexican neighborhoods. New gangs increased territorial rivalries with more opportunities for adolescent gang members to kill one another to establish their identities. The area's Latino gangs were responsible for homicidal gang wars that sprayed semiautomatic gunfire in the streets, robberies, drug dealing, assaults, and random shootings. In 1997, a neighborhood park in Pacoima controlled by the violent Humphrey Boys gang that for years had terrorized surrounding neighborhoods was "reclaimed" from gang control by municipal agencies. The Los Angeles Police Department's Foothill Division increased anti-gang patrol efforts, and the City Council appropriated funds to build a pool in the park.

In a divided city, with wide disparities in income and educational attainment, ethnic and class polarization, and long-term shifts in business and industry following the end of the Cold War,[5] Pacoima Beautiful set out to develop environmental work activities so that older youth could become involved in its partnership efforts with local government, including public housing and planning agencies, on behalf of community building. The opportunity that presented itself was linked to a planning initiative that had been mounted to confront the emerging threat to the viability of Pacoima's commercial areas as a result of the redevelopment along Van Nuys Boulevard, a major north–south spine that runs the length of the San Fernando Valley. In 1997, Los Angeles Mayor Richard Riordan initiated the Targeted Neighborhood Initiative (TNI) program to focus public and private sector resources to revitalize communities throughout the city. The program provided $3 million to develop, implement, and sustain public improvements in a neighborhood over a three-year period. The initiative was intended to assure that the local community was fully mobilized and invested in the effort. The decisions about how, when, and on what to spend money were to be made by the people who lived and worked in the community and not the bureaucracy. Van Nuys Boulevard in Pacoima was selected to be a TNI neighborhood. The Community Redevelopment Agency (CRA) was charged by the local city council member to spearhead the efforts. The efforts were unsuccessful because they failed to get any public participation. In 1998, the CRA invited Pacoima Beautiful and two other organizations to help design and implement a successful process.

Out of the process came Pacoima Partners, a group of residents, merchants, and heads of community-based organizations, to serve as a committee to implement the TNI project and to help guide the revitalization of a 1.5-mile stretch of Van Nuys Boulevard. The commercial strip included storefronts, industrial parks, offices, vacant lots, residential homes, apartments, and three miles of sidewalk. The area had run-down buildings, some totally abandoned, and many with boarded-up windows. Other buildings showed visible cracks from the 1994 Northridge earthquake. With few exceptions, the area was barren; telephone poles dominated the skyline. There were seventy-eight trees on the three miles of sidewalk, but they were small and thin with few leaves, and many trees were dead or damaged. There were fewer than a half dozen bus stops with benches in the corridor, and only three that were covered, though most of the community used the bus system. The types and quality of services found on other parts of Van Nuys Boulevard were not found in Pacoima, and residents did not patronize the shops that were there.

Pacoima Partners used the approach developed by the National Main Street Center, a program of the National Trust for Historic Preservation, to encourage public participation in preservation efforts.[6] This team approach involved a network of community members, who worked to change Van Nuys Boulevard in Pacoima into an economically viable commercial district. Pacoima Partners served as a liaison to community groups, businesses, and public officials, as they were responsible for introducing the commercial revitalization concept to the rest of the community and for representing the community's interest to governmental agencies and private investors. The project developed three program components. First, the commercial rehabilitation program would provide loans to business and property owners interested in expanding, improving, or relocating in the area and would promote the image of the Pacoima Town Center to prospective businesses and investors. Second, the commercial façade improvement program provided a target area for the renovation, repair, and improvement of approximately 30 storefronts along the boulevard. Third, the streetscape program would develop an urban design plan to provide for an entry gateway, coordinated street signs and banners, street trees, planters, decorative sidewalks, coordinated signage, bus shelters, improved bus information, and pedestrian lighting. The program also would implement a streetscape design by installing street furniture and managing a cleanup campaign that would place trash cans along the boulevard, steam clean the sidewalks, and repair broken curbs and sidewalks.

Another initiative began in the spring of 2000 when the City of Los Angeles allocated $3 million to improve Van Nuys Boulevard, through the existing TNI. The goal was to make improvements to the streetscape and facades of Van Nuys Boulevard in Pacoima to begin a revitalization process. Pacoima Beautiful helped to link the streetscape and tree-planting initiatives by volunteering the youth hired under the CRA grant to work on the TNI project. The project served as a way to link the tree-planting project to the streetscape improvements and provide the youth with meaningful, relevant work. Fifteen youth were recruited; for most, this was their first job. The youth met after school at least two days a week and as needed on Saturdays. Pacoima Beautiful hired additional staff to make sure the youth were well trained and safe on the job. It was difficult to maintain a cadre of fifteen youth throughout the program for various reasons, including that of parents who did not want their children out on the streets, particularly after dark.

The first assignment was for the youth to set rules for themselves. Through consensus-building activities, the youth decided on a group name, "Pacoima Beautiful Youth Environmentalists," or PB YES. They designed a logo and business cards. They also wrote the following mission statement: "Pacoima Beautiful Youth Environmentalists are educated to provide the community with the tools needed in order to have a cleaner and safer environment here in Pacoima." The youth concluded, on their own, that only those youth who complied with the rules would be able to participate in the project. Some youth were asked to leave because they did not follow the rules. The second assignment was to develop a scope of work and a set of tasks with the TNI architects and landscape architects. With architects and landscape architects, the PB YES inventoried and mapped Van Nuys Boulevard. They learned how to do street surveying and to read and interpret architectural drawings. They photographed all of the boulevard's streetscape and mounted and labeled photos on presentation boards. The youth learned how to use a digital camera, how to prepare photos for a presentation, and how to read maps.

Throughout, they were expected to perform at a professional level, including setting and meeting deadlines. The photo boards presented Van Nuys Boulevard in 2000 at the beginning of the revitalization process. The architects and landscape architects have continued to use the boards in their design work. Upon the completion of the project, the boards were presented to the new Pacoima library for their archives. An additional task was for the youth to conduct interviews with residents of Pacoima. The interviews further helped the youth develop communication and writing skills and provided them with a historical perspective of Pacoima, specifically Van Nuys Boulevard. The architects, landscape architects, and youth incorporated what was learned from the interviews into the designs for Van Nuys Boulevard. Pacoima Beautiful staff invited a number of adults working in Pacoima to team with the youth. The teams met merchants on Van Nuys Boulevard to explain the improvement project and to invite them to the community workshop, which they helped to organize. As the youth met residents and merchants, they developed skills in communication and self-confidence.

This phase culminated in a community workshop in October 2000 designed to inform and engage the community in the improvements to take place on the boulevard. To prepare for the community meeting, the youth were teamed with adults and went out to meet all merchants and residents on Van Nuys Boulevard to inform them about the scope of the project. Through these conversations, the community had a chance to see the involvement of its youth. The youth then helped set the agenda, planned the food to be served, set up the meeting space, helped the architects and landscape architects to arrange the presentation boards, and served as greeters at the meeting. Several of the adults who walked with the youth to meet the merchants became mentors to the youth, offering guidance on issues ranging from the value of a college education and courses of study to ways of coping with interpersonal relationships and family conflicts. The mentors included a staff member from the mayor's district office; a local merchant whose family has owned a business in Pacoima since the 1960s; a former army staff sergeant who was operating a local travel agency; a state senator's chief of staff; a single mother of two who lives and works in Pacoima, attended school at night, and was applying to law school; the founder of local nonprofit health clinic in Pacoima; and the director of the local nonprofit workforce development program. The adult mentors were supportive and consistent and took their role very seriously. For example, one youth who wanted to become a midwife was invited by her mentor to meet his wife, an obstetrics nurse at a local hospital, and to spend some time at the maternity center. As many of the youth had poor reading and writing skills, considerable time was devoted to writing and then reading aloud what was written. Through mentoring, the youth gained self-confidence. At the beginning of the project, it was difficult for the youth to speak to adults, let alone look them in the eye. However, this changed through constant exposure to adults. The youth had no confidence that their efforts could have an impact. One youth, who at the beginning of the project was cynical and unresponsive, stated after the October workshop, "I now understand that I can make a difference." He then became the first to volunteer and to help on any project.

Due to various city timelines and the grant deadlines, the CRA, Pacoima Beautiful, the County of Los Angeles, and Pacoima Partners agreed that the tree plantings would be completed by April 2001. This meant that the plans needed to be prepared by the landscape architects, and had to be accepted by the city. BoSS needed to complete the tree wells in time for each of the tree plantings, and the youth needed to go back to each merchant to explain the program, to seek permission to remove existing dead or dying trees, to hand out materials on how to plant and care for trees, and to get signatures from all merchants to agree to care for the trees. In preparation for the actual tree planting, the youth learned how to become tree-planting supervisors. With the assistance of the TreePeople's forestry staff, the youth received instructions on how to correctly plant trees and how to teach others to do the same. In October, the youth participated in a community tree-planting project, at which time the youth became tree-planting supervisors. As supervisors, the youth could then teach others to plant trees. Several tree-planting dates were set from January through May. The goal was to plant all 240 trees between those dates. The landscape architect selected the trees, which needed to meet all the city standards. However, it was then up to Pacoima Beautiful and TreePeople to locate the trees, and this proved to be a difficult task, as the trees selected proved to be popular and therefore were scarce. BoSS's Street Tree Division required that soil tests be conducted along Van Nuys Boulevard to determine the quality of soil and any needed amendments; the youth then collected the soil samples and sent them off to the lab for testing.

In February 2001, the first tree planting took place. In preparation for the plantings, the youth met with merchants on Van Nuys Boulevard to seek their permission to remove sick or damaged trees and explained how to plant new ones and to care for and maintain the trees. The planting took place on a cold and rainy February morning. With the coordination of Pacoima Beautiful staff and TreePeople, PB YES helped to supervise 250 people, mostly youth, to plant fifty-four trees. The youth were then involved in planting ten to twelve trees at a time until April, when Pacoima Beautiful, PB YES, and TreePeople organized a major Earth Day tree planting. Three hundred fifty participants from all over Los Angeles planted ninety-four trees. In preparation for each of the plantings, the youth contacted all merchants to get them to agree to care for the trees; every merchant agreed, without exception. In addition to tree planting, many of the youth have found other ways to get involved in the community. One youth helped to get his government class to become involved in Pacoima Beautiful activities. Another youth began to serve as a member of Pacoima Beautiful's board of directors. All youth assisted a group of college students in conducting a survey of residents living adjacent to Van Nuys Boulevard to determine their commercial-use patterns. When the youth were asked how they felt about working with the college students on the commercial-use survey, all agreed that it was a positive experience for them. The youth said that the training and experience they gained working with the mentors and merchants gave them the self-confidence they needed to meet with the residents.

Through the efforts of a college student intern turned employee, the youth have been able to explore the world beyond Pacoima. PB YES youth took an overnight camping trip to a state park where they went on nature walks and learned to ride minibikes. Throughout the project, they were mentored in community organization, advocacy, grant writing, and environmental stewardship, and they were exposed to a number of job opportunities, including "greening professions," such as landscape design, landscape contracting, urban planning, waste management, and civil engineering. The youth continued to build skills within the PB YES program, and their mentors continued on to other successful collaborations, notably with Project GRAD Los Angeles on behalf of engaging college-bound youth in environmental restoration, urban recreation, and environmental justice activities.

SCHOOLING FOR SUCCESS

Pacoima Beautiful collaborated with Project GRAD LA to increase the number of youth entering and succeeding in college. Nationally, Project GRAD supports teacher training to improve student achievement in reading and math, guides students in developing self-management skills to succeed in school and at home and connects them with counselors to help with academic and personal needs.[7] Youth who want to earn a Project GRAD college scholarship participate in two "college institutes" during school breaks to increase their skills and experience in reading, critical thinking, communication, and problem solving; other factors toward the scholarship include a minimum GPA and satisfactory completion of a math curriculum. The Ford Foundation acknowledged Project GRAD nationwide as "an unusually successful systemic public-school-reform project with strong results in test scores, improved school atmosphere, high-school graduation rates, reduced teen pregnancy and higher college enrollment."[8]

To implement its programs, Project GRAD LA partnered with 13 schools, with over 18,000 students and 1,000 teachers, in the Los Angeles Unified School District in the northeast San Fernando Valley, including elementary and middle schools in Pacoima, as well as administrators, parents, community volunteers, business sponsors, and mentors. At the outset of the collaboration with Pacoima Beautiful, area high schools had a graduation rate of 18 percent and only 10 percent of those who graduate continued their education. Fifty percent of the students were "limited English proficiency" and qualified for school-based lunches. Before the Project GRAD LA program, students consistently performed below state and district averages in standardized testing. In 1994, Pacoima's children were two to three grade levels behind in reading and mathematics. Since the inception of the comprehensive educational reform program in 1999, the number of students reading at grade level has quadrupled to 40 percent. At the time of the collaborative partnership, the reform effort led to outstanding performance on the California Department of Education's Academic Performance Index (API), an indicator of student academic progress and success, with scores ranging from a low of 200 to a high of 1,000. When comparing the six-year growth of API points, from 1999 to 2005, Project GRAD LA schools doubled that of California's statewide average growth for all levels. The unweighted average API score for the elementary schools grew from 419 to 668, middle schools grew from 441 to 612, and San Fernando High School grew from 476 to 585.

In 2003, the first group of 190 Project GRAD LA scholars to graduate from San Fernando High School completed the requirements for college scholarships. Nearly all of the scholars were Latino and the first in their families to go to college. By the time the fifth class of 180 scholars graduated in 2007, scholars from the previous 4 graduating classes, together, had surpassed the four-year college-going rates for California's Latino population. In California, 71 percent of Latino high-school graduates attend community colleges and only 29 percent choose four-year universities. In the years that Project GRAD LA partnered with Pacoima Beautiful, of the 2003, 2004, 2005, and 2006 classes of scholars, 74 percent selected, gained admission, and were enrolled in four-year colleges and universities; only 26 percent of the scholars were attending community colleges. Parental involvement in the schools improved along with these successes; each year, there is an increased number of parents attending classes and volunteering in the schools.

Project GRAD LA college institutes, month-long learning activities for "off-track" students, are a productive use of vacation time from school, as stimulation and enrichment activities are critical to academic and social development. The institutes engender a spirit of excitement, a passion about learning, and a way of seeing college as a pathway for quality of life. They also promote a sense of the "interactivity of learning" through a mix of cognitive, social, and affective activities that develop communicative and cultural competencies, successful learning and organizational skills, time management, and planning. As the main interface between Project GRAD LA and students, the institutes are socializing experiences aimed at instilling a number of qualities. They cultivate an appreciation of the promise of college, support the development of higher personal academic standards, and help low-income students gain a better understanding of college enrollment issues.

The institute structure also supports leadership and intellectual development by cultivating the experience of a learning community outside the ordinary school environment. Institute activities develop interpersonal competencies, such as supportive peer groups, which upon their return to school would reinforce the "scholar" identity through studying together, sharing rides, supporting Advanced Placement coursework, and engaging in intergenerational discussions with parents and teachers. The learning outcomes for the institute included:

- Leadership Development: The threefold outcomes involved Self-Esteem and Self-Efficacy: Scholars will believe that they can achieve beyond what they may see in their everyday lives. They will be resourceful and motivated and be able to overcome setbacks. Scholars will push for higher standards for themselves and expect others to have high expectations for them as well. Students will have an increased level of self-confidence and drive and see themselves as leaders.

- Goal Setting and Decision-Making: Scholars will develop experience in goal setting and decision-making. They have experience in mobilizing a team around an idea and seeing it through to completion. Scholars will develop short-term and long-term goal setting skills and will understand how to use these skills to facilitate their own success.
- Dreaming About the Future and Making It a Reality: Scholars will be curious and aware of the diverse range of career and educational paths. They will be able to see themselves in roles and situations that may not be a part of their daily life experience. They will be motivated to explore these roles as options and position themselves for success.
- Academic Excellence: Students will gain skills and experience in reading and critical thinking.
- Communication and Problem-Solving: They will set a high academic standard for themselves and seek opportunities beyond what is expected. They will also appreciate the importance of group work as well as develop an academic support network of peers.
- College Awareness: Scholars will become familiar with the basic high-school graduation and college admission requirements, understand that it is possible for them to go to college, and where to get more information and resources about college.
- Capacity Building–Civic Engagement and Cross-Cultural Competence: Scholars will have an increased awareness of, and involvement, in their community. They will develop the capacity to excel in new and challenging environments. They will be able to interact effectively in diverse cultural situations.

Beyond cognitive skills, such as exploration and experiment, the institutes, promote affective development, notably motivation, self-esteem, and self-efficacy, and they support emerging peer bonds through the successful completion of a common task in the humanities, the social sciences, and the sciences. Project GRAD LA began to focus on science as an institute modality in response to numerous national and regional studies that have documented the deplorable state of science education at the high-school level, particularly for students of color. One noteworthy example is the "Nation's Report Card: Science 2000," which found that California's science scores declined for both eighth and twelfth graders between 1996 and 2000.[9]

In 2001, Project GRAD LA approached Pacoima Beautiful about putting together a college institute for "Track B" students at San Fernando High School who received little if any opportunities to participate in enrichment activities. The goal was to engage as many students as possible in collaborative environmental research, stressing math and science. This was a very important opportunity for Pacoima Beautiful to begin to share what it had been learning about environmental health and to bring in experts who had been helpful in the past. In putting together the first of the college institutes, Pacoima Beautiful invited university faculty and students and community residents to participate in developing and implementing a curriculum. Faculty from five area universities, the University of California, Los Angeles, California State University, Northridge, the University of Southern California, the University of California, Irvine, and the California Institute of Technology, designed the research questions and outcomes for each study. Agency staff and residents then created a platform of environmental science projects, including hazardous waste, air quality, soil lead, watershed conservation, atmospheric science, environmental advocacy, and mural design, from which the students could choose. An action research approach informed the specific objectives, data collection methods, and the action plan for each project. Each addressed an environmental issue as framed and prioritized by Pacoima residents and provided a vehicle for students to improve their science, math, and language skills.

As the institutes were being created, various stakeholders became interested in having the students address a key environmental concern in Pacoima, Hansen Dam—a flood-control basin and mixed-use recreation area managed by the Army Corps of Engineers and the City of Los Angeles Department of Recreation and Parks—and they were willing to support them in that effort. The next task was to design projects focused on a specific locale, Hansen Dam, that would catalyze students, teachers, and community residents through a "pedagogy of place," as stated by the Rural Challenge and Evaluation Program at Harvard Graduate School of Education:

> In its most simple form, pedagogy/curriculum of place is an expression of the growing recognition of context and locale and their unique contributions to the educational project. Using what is local and immediate as a source of curriculum tends to deepen knowledge through the larger understandings of the familiar and accessible. It clearly increases student understanding and often gives a stronger impetus to apply problem-solving skills.[10]

Pacoima Beautiful staff and San Fernando High-School science teachers then co-designed the projects, which were then revised and approved by a student advisory board. There were three project platforms, with two project teams per platform, with objectives and sets of questions to guide both the research and critical inquiry:

I. Developing Public Awareness About Hansen Dam

A. Survey Team: Interview residents who use the park and find out the importance of the park. Learn about the different types of surveys and create a survey that can be used to raise the issues for the changes you want to see.

Questions:
What is a survey—what is it used for?
What kinds of surveys are there?
How do you create a successful survey that gets the information you want?
What needs to be known about the area before you can put together a survey?
How does it become cultural- and area-specific and not judgmental?
What types of questions should be asked with regard to Hansen Dam uses?
What questions are to be developed that students and teachers can choose from?
Where are the sources of conflict and potential resolution to the conflict?

B. Press/Media/Community Awareness Team: Make a documentary movie to inform the community. Use the media to promote the importance of Hansen Dam and get the community involved.

Questions:
How can the press get involved?
Why would this happen?
How can you plan to get the community involved?
What can the community do if it gets involved?
How will the video be used to raise awareness?
Where are the sources of conflict and potential resolution to the conflict?

II. Environmental Changes at Hansen Dam

A. History of the Lower Lake and its Potential for Expansion Team: Study Hansen Dam to discover the past and plan for the future. Use water testing equipment and study of trees, birds, and other wildlife. Find out history of Hansen Dam and what it will take to make it a better place for the community.

Questions:
What is the history of Hansen Dam?
Why was it a major recreation area with swimming, boating, fishing? What happened? What will it take to bring it back?
Who are the political forces and what do they want to see happen with the lower lake? What kind of testing needs to be done?
Where are the sources of conflict and potential resolution to the conflict?

B. Link Between the Children's Museum and the Lower Lake Team: Create recycled art for display in the Children's Museum, using the TAP/SCRAP art in order to link the Children's Museum to the lower lake. Work with architects to adapt the Children's Museum to the natural environment.

Questions:
What is the argument to be made that will cause such a linkage?
How can we work with architects from the Children's Museum to incorporate building in a natural setting?
Where are the sources of conflict and potential resolution to the conflict?

III. Preservation of Hansen Dam

A. Cultural Heritage and Historic Preservation Team: Preserve the cultural heritage and history of Hansen Dam. Find out who were the original residents. Study the relationships between Native Americans, Spaniards, Japanese, African Americans, and their historical ties to the area. How should their story be told? Learn about historical preservations and how to document cultural history.

Questions:
How do you learn about cultural heritage?
What is historic preservation?
How do you go about finding out who lived in the area, including people and animals?
How do you find out what happened to them?
What are the options on how and where the story can be told?
Where are the sources of conflict and potential resolution to the conflict?

B. Land Use Team: Learn how to research the history of the land and plan for future uses. What are the environmental consequences of development? Learn about the politics involved in the development of the Hansen Dam area.

Questions:
How do you research the history of the land, and the present, proposed and potential uses? What is the current plan for Hansen Dam?
Who are the players that have to be considered when anything happens in Hansen Dam?
Where are the sources of conflict and potential resolution to the conflict?

To carry out these projects, the institute brought together eighty-four San Fernando High-School students, fourteen college students, and seven community residents for one month, beginning with a weekend leadership training session at UCLA. Students then worked on their projects in the community during the morning and attended classes at the local community college in the afternoon with high-school science teachers. College students served as "near-peer" mentors for the high-school students as they implemented the projects; cross-age mentoring also took place between students and adult volunteers from Pacoima Beautiful. The latter reinforced the notion that these projects were only the beginning of service and that youth would need to step forward and continue in their efforts to address severe environmental issues in their home community. A nascent sense of efficacy also developed around students' ability to actively learn science concepts through their projects and then advocate for change.

Each project team included a Pacoima resident serving as a recorder, who brought a community perspective to project activities and then conveyed the information learned from the project to the community. The data gathered from these projects provided valuable information to help improve the quality of life in Pacoima. Reflective activities, including discussion of alternative paths of action based upon the research, took place among university and high-school students and residents. Youth involved with the various projects presented their findings at a "Town Hall" meeting attended by these two co-managing agencies as well as the Santa Monica Mountains Conservancy, a publicly funded parkland preservation organization; the Children's Museum of Los Angeles, which planned to build a facility at the Hansen Dam site; the Audubon Society; local elected officials; homeowners; equestrian groups; school district personnel; and university faculty. The students also cited the need to teach the community the value and importance of the site; enlist cooperation to improve the park's appearance, significantly reducing the trash and graffiti problem; increase residents' use of the site; systematically remove invasive plants and replace them with natives; provide more facilities for individuals with disabilities; build a boardwalk and walkways for easier access to park destinations; increase the number of playgrounds; add wildlife information signs and viewing telescopes; expand water testing to determine safety for human consumption of fish caught in the area; and advocate to preserve the site as a wildlife habitat for the protection of over 260 species of birds found there. They also suggested that young and adult residents become more active in planning for both Hansen Dam and the proposed museum. The elected officials, representatives from the various constituency groups, and the Army Corps expressed their willingness to work with the students, albeit noting that addressing their concerns about the site would be complicated because of funding issues and overlapping jurisdictions. As a result of the meeting, the local congressman put in a request for additional federal funding for Hansen Dam; over $3 million was secured on its behalf.

It was shortly after this meeting that the Army Corps started dumping construction wastes from the swimming pool at the top of Hansen Dam to the lower lake. This outraged residents, who had heard at the meeting that nothing major would be done without community input; they quickly mobilized to have the dumping stopped. The same group of stakeholders came together a few months later to set priorities for Hansen Dam, namely, developing public awareness through needs assessments and media campaigns; monitoring water quality and the status of trees, birds, and endangered wildlife; exploring sustainable ways to link the natural and built environments; documenting the cultural heritage and historic preservation of the site; and demonstrating how the uses of the site impact the land and community and generate potential conflicts over land use issues. Following the meeting, agency staff and students from the original institute consolidated these priority areas into four platforms for study at a second institute that brought fifty-one students together to further investigate ecological issues at Hansen Dam. The students again shared their recommendations with decision makers, and two of them were selected by the Army Corps for possible implementation: natural trail creation and the construction of informational signage.

The activities developed for the institutes focused on introducing high-school juniors and seniors to the promise of college through two pathways: community-based learning projects and mentoring experiences in small learning settings outside of the school. These activities are viewed as ways to reorient students during the last eighteen months of high school toward the transition to emerging adulthood. Through them, youth begin to cultivate career confidence and competence through work with mentors, including university students and professionals. The institutes' experiential activities supported and enriched the youths' extrafamilial worlds through the development of meaningful intergenerational bonds of mentoring and encounters with college faculty and students. As a result, high-school students were exposed to a number of job opportunities in science and technology, including the health and environmental professions and cultural resource management. Once they acquired a repertoire of social and academic skills, the youth engaged with adult residents in support of public environmental issues identified by informed community residents. Exposure to urban social knowledge, including science, technology, and advocacy, served to increase the youth's environmental and community awareness and academic readiness. Collaboratively involving university faculty and students, high-school students, and residents facilitated transmission of social knowledge in support of revitalizing this community. As these efforts focused on social development within a context of neighborhood resiliency, project-based learning not only improved educational outcomes but also served as a catalyst for community building.

Intergenerational efforts demonstrated the capacity of youth to participate in the community-planning process and to become involved in public affairs through Pacoima Beautiful's partnership with local government and nonprofit agencies. Together, these organizations worked to improve the chances of youth by embracing newer forms of project-based learning. Their advocacy efforts for social equity and access to public goods, namely, environmental and educational quality, compelled young people to understand the broad structural contexts that affect their personal, educational, social, and political horizons. The institute activities also inspired them to fully participate in local-level civic networks and organizations that have their basis in voluntarism, social trust, and reciprocity. In this case, recent immigrants and their children began to freely engage in a form of public culture that has until recently only been available to the more privileged families in American society. More inclusive community social formations that include youth created an intergenerational civic arena, through which recent immigrant families in Pacoima came to understand the host society and its unfamiliar language, customs, and social institutions, and also its constraints and opportunities.

NOTES

1. Leonard Pitt and Dale Pitt, *Los Angeles A to Z: An Encyclopedia of City and County* (Berkeley: University of California Press, 1997), 375.
2. Carl A. Maida, "Expert and Lay Knowledge in Pacoima: Public Anthropology and an Essential Tension in Community-Based Participatory Action Research," *Anthropology in Action* 16:2 (2009): 14–26.
3. Carl A. Maida, *Pathways through Crisis: Urban Risk and Public Culture* (Lanham, MD: Rowman & Littlefield, 2008), 187–231.
4. Cheryl L. Maxson and Malcolm W. Klein, "'Play Groups' No Longer: Urban Street Gangs in the Los Angeles Region," in *From Chicago to LA: Making Sense of Urban Theory*, edited by Michael J. Dear (Thousand Oaks, CA: Sage Publications, 2002), 239–266.
5. United Way of Greater Los Angeles, *Los Angeles County Ten Years Later: A Tale of Two Cities—One Future* (Los Angeles, CA: United Way of Greater Los Angeles, 2010), https://www.yumpu.com/en/document/view/45985294/la-county-10-years-later-a-tale-of-two-cities-one-future.
6. Kennedy Smith, Kate Joncas, and Bill Parrish, *Revitalizing Downtown: The Professional's Guide to the Main Street Approach* (Washington, DC: National Main Street Center, National Trust for Historic Preservation, 1996).
7. Project GRAD, *Project GRAD: Working to Close the Academic Achievement Gap* (Houston, TX: Project GRAD, 2003).
8. Ford Foundation, "President's Message," in *2000 Annual Report* (New York: Ford Foundation, 2001).
9. National Center for Education Statistics, *The Nation's Report Card: Science 2000*, NCES 2003-453, by Christine Y. O'Sullivan, Mary A. Lauko, Wendy S. Grigg, Jiahe Qian, and Jinming Zhang (Washington, DC: U.S. Department of Education, Institute of Educational Sciences, 2002).
10. Annalisa Lewis Raymer, *Pedagogy of Place Facilitation Guide: A Workshop for Cultivating and Promoting Place-Based Education* (Lexington, KY: University of Kentucky Appalachian Center and the Appalachian Rural Education Network (AREN), 2001).

Chapter 7

Reinventing Apprenticeships in Learning

AN URBAN LAND GRANT MISSION

The University of California is a public university with a "Land Grant" mission. For over a century, through its Cooperative Extension, the university has brought researchers to the state's farming areas. In this spirit, the university's Board of Regents established the experimental Community Education Resource Center (CERC) initiative in 1999 to place faculty and students in high-need urban communities in the San Francisco Bay Area and Los Angeles County. The stated mission of the Los Angeles Country CERC initiative was "to serve as a conduit that provides resources to the highest need UCLA-impacted communities." Key objectives were to focus on the social, economic, and cultural factors of youth achievement; articulate schools, community colleges, and universities; link research, teaching, and service through an integrative collaborative approach; advance knowledge creation and dissemination on behalf of community-defined needs; and promote co-learning through the shared expertise of higher education and communities. A community engagement model emphasized quality-of-life concerns that improve the educational chances of youth, including health, the environment, recreation, and the arts. Throughout was the need to instill a college-going culture, providing opportunities for service learning, community-based learning, educational services, and community extension programs to support this ambition. Promised capacity-building efforts included asset mapping, program development, information technology, and public relations. The overarching plan was to add value, while respecting a community's unique identity, by advancing the power of systems integration over fragmentation, seeing the community-based organization as the locus of change. Through local CERC conduits, faculty research could be transferred by collaborative design of community-building programs with implementation by community partners.

The CERC initiative in Los Angeles was originally envisioned as a long-term partnership between the university and partnering agencies within selected communities to address embedded systemic issues that inhibit the academic achievement of underserved youth. The effort would extend the university's work in six of Los Angeles' high-need communities: Boyle Heights in East Los Angeles, Downtown Los Angeles, Watts, Inglewood, Cudahy, and Pacoima. The sites would serve as UC hubs in Los Angeles, bringing campus resources to these communities. The initiative would also provide a vehicle to work comprehensively and coherently within communities to devise and implement plans of action. These plans could then begin to address the educational, socioeconomic, public safety, and health-related issues that impact the youth. The sites were viewed as multifaceted, ones that could provide opportunities for faculty and students to engage in service learning and community service, field placement, and research. The initiative thus sought to create a networked community of learners, enabling communities to build the needed infrastructure on behalf of their children.

Before the CERC initiative, the university supported efforts to design and initiate courses, provide technical assistance to the nonprofit sector, and assign faculty and students for direct service contributions. Building on these efforts, the new initiative could enhance the university's curricular development through service-learning courses that brought college students into the targeted communities. The local areas could then serve as loci for applied research, internships, and opportunities for building synergies among urban planners, social workers, teachers, psychologists, health care personnel, and other professionals. The initiative also drew upon collaborating university academic and service units to provide technical assistance. This involved faculty and student support to develop community-building processes and critique community-developed plans. Areas of need included child and adolescent development, housing and neighborhood revitalization, community-based health care and environmental resources, economic development, and job creation. Campus administrators even tried to form a "higher education trust." The trust, in theory, would link the human resources of California's public higher education system, namely, community colleges, comprehensive universities, and research universities, with schools, community-based organizations, and bridging organizations in the nonprofit and philanthropic sector, such as the United Way. Start-up activities focused mainly on building up social capital in the CERC communities.

COMMUNITY SERVICE LEARNING AND COMMUNITY BUILDING

Within one CERC hub, in Pacoima, the university initiated community-based research and consultation that involved the collaboration of lay community members and experts to produce new knowledge for social and educational change and to improve the quality of life. Faculty and students from the health sciences professional schools, the urban planning department, and from the college began a dialogue with representatives of nonprofit organizations and public agencies on child and adolescent development, community health, and the urban environment. Within eighteen months of its founding, the Pacoima CERC was gradually gaining recognition as a strong grassroots umbrella organization seeking to build and strengthen connections between local schools, community-based organizations, and the university.

Through UCLA's Center for Experiential Education and Service Learning, at the time under the leadership of a distinguished ethnographer within the sociology department, two experimental courses were designed to involve undergraduates in service learning in a school, community agency or healthcare setting. Students were involved in the assessment of issues related to health, law, schooling, and the urban environment. A campus affinity group was formed around the theme of health care and community-based learning in transitional communities in the northeast San Fernando Valley. This group brought together faculty members from public health, dentistry, medicine, nursing, and sociology that had existing involvement with community-based programs and agencies in the San Fernando Valley, with particular emphasis on Pacoima. In conjunction with active community partners from this area, including Pacoima Beautiful, Los Angeles Educational Partnership (LAEP), Project GRAD LA, and San Fernando Neighborhood Legal Services, these faculty and staff members became acknowledged as an ad hoc steering committee for the Pacoima CERC. As such, they led the working interdisciplinary network to coordinate and facilitate community-based learning and applied research. The affinity group held planning meetings with representatives from several of the key agencies involved in the CERC initiative. Group members helped develop and modify the two cooperative community engagement courses, more widely known as service-learning courses.

The two courses brought undergraduates from across major fields together for focused community health problem-solving all related to access to health services, environmental health, health law, health planning, health promotion, and disease prevention. Members also helped guide course faculty, graduate assistants, and field preceptors in developing innovative community-based service-learning activities and classroom-based reflective activities. Faculty and staff on the affinity group saw the potential of CERC to build and strengthen connections between local schools, community-based organizations, and the university on behalf of community-based learning.

Under the auspices of the affinity group, two experimental service-learning courses through the sociology department at UCLA focused on healthcare issues in the northeast San Fernando Valley, examining disparities in health care and how critical issues of health care, notably cost, quality, and access, affected transitional communities. Twelve students enrolled in the winter quarter, and seven in the spring quarter for a total of 1,500 hours of service learning. In the first quarter, students completed group participatory action research projects on an after-school science club at a local middle school, environmental health-related issues in conjunction with Pacoima Beautiful, immigration rights and violence against women projects organized by San Fernando Valley Neighborhood Legal Services, and school-based health education and health career readiness organized by LAEP. In the second quarter course, students completed ethnographic research projects at five health-focused organizations. Project-based learning was a central interest among academic and community partners involved in the Pacoima CERC right from the start. Their mutual intent was to connect the academic and the vocational through rigorous upper-division coursework combined with intensive work-based learning in professional settings. Through project-based learning, students helped assess health-related quality-of-life issues in Pacoima and began to observe the various integrated service responses that had evolved in the community over the previous decade.

CHANGING STRUCTURE OF UNIVERSITY-BASED EXPERTISE IN COMMUNITY BUILDING IN THE TWENTY-FIRST CENTURY

In the decade following the 1992 Civil Unrest in Los Angeles, and the 1994 Northridge Earthquake, local universities became more involved in the affairs of their surrounding communities. These involvements were built upon the efforts by established agencies and institutions outside of academia that had sustained ongoing consultative relationships with professionals and residents within these communities well before the urban crisis. In communities across the city, individual university faculty members and their students initially reached out to local professionals engaged in community-building projects often organized by a regional United Way agency in the years prior to the period of intensive federally funded crisis intervention. Service agencies used a crisis intervention model to disseminate knowledge and helping strategies to meet residents' immediate needs and to reconstruct community infrastructure, including schools, health care, and human services. The model, used by the Federal Emergency Management Agency (FEMA) after an emergency declaration, is top-down and short-term with the goal of returning control to civic and organizational leaders as quickly as possible. In communities, such as Pacoima, where the schools had become the centers of crisis response and rebuilding activities, and locales for the re-establishing of trust between professionals and residents as well, a school-based systems integration approach evolved. The incorporation of FEMA-funded activities within the existing network of the United Way-integrated services on behalf of community building only increased localized trust in outside helping efforts.

As the systems integration approach gradually replaced the crisis-driven model, residents began to work side by side with professionals, including university professors and their students engaged in school-based activities, and paraprofessionals linked through collaborative activities in school-based parent center programs. Within this bridging network, exemplified by the Pacoima CERC initiative, all parties worked together toward common school and youth-focused programmatic goals within the existing educational system and local political structure, with the overarching goal of bringing about change from within the system. However, the school-based service settings, where collaborative partnerships had taken place in the years following the multiple urban crises, were facing severe budget cuts as the economic downturn of 2000–2001 began to affect public funding in California.

In the ensuing years, with the eventual dissolution of the more comprehensive outreach initiatives carried out with direct funding by the universities that were established a decade earlier in the aftermath of civil unrest and natural disaster, the trend shifted to academic community-focused partnering. With publicly funded education, health care, and urban services facing severe budgetary cutbacks, the university would contribute technical assistance, research, and professional personnel, but no direct funds, to high-need communities. Although alliances between academic experts and lay advocates within this CERC initiative were strengthened with each successive and successful joint project, the structure of university-initiated consultation on behalf of comprehensive community building changed considerably over the period covered by the above partnered educational initiatives.

Although, after only a few years, the university abandoned the CERC initiative, withdrawing support for a youth-serving community engagement model to one focusing on school-based institutional building with the Los Angeles Unified School District. Shifting resources thereafter facilitated a different set of synergies that would afford opportunities for training, service, and research to help disadvantaged populations with political will, but very limited resources, to confront their economic, educational, health, and other disparities. In retrospect, a thinly disguised retreat from the crisis model necessitated community-based organizations to find their own ways to resolve these problems. Consequently, faced with budgetary crisis, the universities thus pulled back their support for structural change at the community level.

As a result, the more comprehensive university partnership initiatives rapidly devolved, from collaborative support of building community assets, to funding specific projects linked to individual faculty or staff members. Rather than working together with colleagues in multidisciplinary projects on behalf of a particular community, individual faculty and staff members would now have to compete for funds and recognition as a "community partner" aligned with a specific community-based project. In the end, this shift away from collaboration toward competition, between faculty-partnered projects and between communities, for scarce university funds, created a sense of distrust in the nascent university–community partnerships, such as the CERC hubs, that had just begun to emerge in economically disadvantaged neighborhoods. This breakdown of trust was highest among the many nonacademic professionals who had sustained ties to local schools and agencies throughout a decade of civic crisis.

INSPIRING SCIENCE INTEREST

Amid the changed focus toward project-specific collaborative partnerships, two UCLA faculty members who previously worked within the Pacoima CERC initiative developed a model high-school apprenticeship program for underrepresented and educationally disadvantaged gifted students. Partnering with two community nonprofit programs that held college-going as primary missions, the university thus expanded the original school-to-career foundation of CERC initiatives to target higher learning institutions as an incentive to advance, beyond short-term institutes, to include extended learning opportunities. Formal instruction in urban schools has been relatively unsuccessful at stimulating interest in science and math careers by economically disadvantaged and underrepresented minority students. The mastery and motivations of students to study STEM subjects seem to decrease as they progress in school. Out-of-classroom learning programs serve as an exceptional venue for supplementing classroom instruction in science and mathematics and for introducing the process and value of science. Free from the usual school curriculum strictures, out-of-classroom activities can utilize authentic, interesting, and motivating methods to teach math and science.

The *Trends in International Mathematics and Science Study* (TIMSS), an international comparative study, measures trends in mathematics and science achievement at the fourth and eighth grades every four years. In TIMSS 1995, in grade 4, the United States ranked 12th in mathematics and 3rd in science but dropped to 28th in mathematics and 17th in science in grade 8.[1] Looking at changes over time in mathematics at both grades 4 and 8, U.S. average scores have increased over the long term—with higher average scores in 2019 than in 1995—but show no significant changes between 2015 and 2019. In science, U.S. average scores show no significant changes over the long term (from 1995 to 2019) or over the short term (from 2015 to 2019) at the eighth grade.[2] Student attitudes toward science and math mirror student achievement. Many progressive educators believe that the reason for these declines is that formal instruction alienates many potential science and math students. Conventional classroom instruction often emphasizes facts, memorization, and content knowledge due to proliferating content standards and time constraints. Biology students learn more vocabulary words in one year than they would in many foreign language classes. These trends go against the type of science teaching that research suggests will interest students and prepare them for future study. The National Science Education Standards established by the National Research Council emphasize science inquiry as a way to learn both science content and develop an appreciation of the practice of science. Inquiry is central to understanding the work of scientists and is often linked to developing a curiosity about scientific phenomena.

Until recently, formal schooling has emphasized testing and memorization. Since the pandemic, there are more opportunities for inquiry in the classroom. In light of pivot toward deeper learning based upon reflective inquiry, informal science education may provide the types of experiences that will inspire future scientists. Participation in informal learning environments such as out-of-classroom programs is voluntary and thus the learning is more self-directed than in school settings. Researchers have argued that because youth realize a sense of control and perhaps mastery during informal learning experiences, such as project-based learning, the students' emerging identities as "makers" and "learners" may influence their motivation to succeed in classroom work and their choice of a college major. Out-of-classroom science programs offer opportunities that conventional schooling cannot provide including personal attention from adult mentors, a positive peer group, and project-based activities that build students' interests and self-esteem. For children and youth who do not have positive experiences with science and who have few role models of science-using professionals, the impetus to consider a career in the sciences is problematic at best. Out-of-classroom programs are more effective in building enthusiasm for science than conventional instruction because they can introduce youth to the world of STEM, experientially, and make science more relevant to their lives. Developing effective programs then are important steps in the process of creating a more diverse STEM workforce. Such programs can also address the lack of students from economically disadvantaged communities who are STEM majors at the undergraduate and postgraduate levels.

DESIGNING A HIGH-SCHOOL APPRENTICESHIP

Faculty program directors and laboratory scientists set about to design and test a model that incorporated the best practices for out-of-classroom STEM programs. The intent was to connect students with scientists in the context of science inquiry embedded in project-based learning. The program sought gifted students from area high schools with an interest in science and technology and university-based laboratory scientists who welcomed mentoring opportunities with older adolescents. A common assumption was that the program would provide a link between students and science practitioners through training on both ends of the mentoring relationship. Lab scientists and their graduate students could learn to become mentors and through this relationship the students would be able to see farther down the career pipeline. Many students rarely have access to professional researchers who can supplement both their content knowledge and their own experiences with STEM.

Increasing numbers of students enter college with plans to major in the STEM fields; however, their completion rates are lagging, especially among underrepresented minorities. Generating and sustaining interest in a career in the sciences will require engaging students in STEM learning activities outside of the classroom. To further this goal, UCLA School of Dentistry was awarded a Howard Hughes Medical Institute (HHMI) Pre-College Science Education Program grant in 2007 to engage high-school students in a didactic and experiential two-year program to learn cell and molecular biological research. A goal of apprenticeship program was to prepare gifted, underrepresented minority and disadvantaged high-school students for research programs in college by enhancing their interest in science as a career in their final two years of high school.

The apprenticeship was thus designed to adopt an experiential approach to learning life science disciplines through project-based learning in order to increase the pool of underrepresented and/or educationally disadvantaged youth in the STEM fields; prepare gifted students for science and research programs in college through project-based learning; raise awareness on the value of science as a profession within students' families; and involve faculty, graduate students and postdoctoral fellows in mentoring activities. The program linked university faculty, program representatives from nonprofit community organizations that emphasized "college-going" and, through these organizations, high academic performing high-school students from minority and economically disadvantaged families.

The program brought students to the campus on Saturdays to use the classrooms and laboratories and to learn the laboratory techniques necessary for their summer-intensive learning experience, specifically developed as projects to aid students in their analytical work and guide their observations These activities emphasized the nature and practices of laboratory science as the key to preparing students for college-level science study. Enabling students to see connections between the science they do, and the work of professional researchers adds meaning to their work, helps them to consider science as a career. Helping students understand where scientific ideas come from and how they are tested is a key aspect of science learning. These skills are essential for students to get the most out of the hands-on science activities. Preparing students for science careers depends on keeping alive their natural curiosity and motivation to explore scientific problems and to learn new things.

To ensure that the instruction and mentoring were consistent and that projects were of high quality at all of the laboratory sites, a community of practice was developed during the planning stages. This allowed university faculty and community organization representatives to communicate and exchange ideas and materials and to support the mentors who implemented the subsequent lab-based projects. An essential part of the nascent community of practice was face-to-face meetings to maintain consistency across the planning and implementation teams. Planning the apprenticeship program involved identifying and assessing best practices science learning and mentoring modalities; identifying needed professional development and mentor support to initiate the best practice models; developing plans for implementation of the best practices through mentored project-based learning in university laboratories; and developing methods to assess changes in students' attitudes and preparation for science study. The community of practice enlarged as a result of various co-curricular efforts undertaken to support students' social and emotional growth, and the promise of college-going, forging deeper ties among the various campus-based program and academic units. Speakers for the evening talks, and in some cases, lunchtime forums came from Office of the Chancellor, Academic Advancement Program, Community-Based Learning Program in the College of Letters and Sciences, STEM-PLEDGE Graduate Student Diversity Group in the Graduate Division, Office of Residential Life, Institute of the Environment and Sustainability, and Clinical and Translational Science Institute in the Geffen School of Medicine.

Evaluation metrics and instruments that benchmarked the impact of the apprenticeship program on the students, their families, and their research mentors were collaboratively designed with the California Teach Resource Center, and the center-trained postdoctoral researcher who taught the first-year academy. An outcome evaluation monitored the primary goal of increasing the pool of members of underrepresented and/or educationally disadvantaged groups in the STEM fields, as evidenced by college admission and matriculation with a science major. Process evaluation explored secondary goals: increasing inquiry-based learning through mentored laboratory projects; raising the awareness of science as a career in the students' families, providing an affective support structure as they enter college; and initiating mentoring relationships between clinically applied life science researchers and adolescents in their labs.

Program assessment required the design of instruments that benchmarked the program's impact on the students, their families, and their research mentors. The HHMI-sponsored *Entering Mentoring* materials already used at the university's California Teach Resource Center were incorporated into the expanded Preparing Future Faculty program to include graduate students from campus laboratories who planned to teach and mentor high-school students. *Entering Mentoring* instruments were adapted to assess the effectiveness of the mentoring experience for both the high-school students and their mentors and to assess changes in the graduate students' interest in academic careers and their awareness of the challenges facing their mentees. A culture of mentoring gifted high-school students could then develop within campus laboratories that traditionally mentored only advanced undergraduate, graduate, and postdoctoral students.

A logic model was developed to link the program's purpose, goals, objectives, and activities back into ongoing planning and evaluation. The evaluation, as planned, sought to document the impact of project-based learning and supportive mentoring activities. Assessments used pre- and post-tests to measure changes in scientific attitudes, together with qualitative methods, including key informant interviews and focus groups with parents that were sensitive to issues of language and culture. In addition, students' work was reviewed for evidence of understanding. Participatory evaluation activities included interviews and observations of students' motivation and engagement with science, notably how they view science, the possibility of pursuing a career in science, and themselves as a scientist; students' understanding about science including science content, inquiry skills, and the nature of science; concerns in carrying out projects and mentoring in different laboratory environments; effectiveness of the mentorship development innovations; and the quality of the projects themselves.

COLLABORATIONS

In writing the initial proposal and subsequently designing this research-intensive pre-college science education program with a strong mentoring component, UCLA faculty understood that none of the various activities could have occurred without start-up funds from HHMI. These funds helped to develop the first-year Saturday Academy, co-designed with the department of chemistry and biochemistry, which implemented the "Bridges and the Minority Student Development" programs to assist underrepresented minority students in community colleges and on campus to pursue careers in biomedical research. The staff and evaluators from these programs helped to design the logic model and evaluation metrics for the program. University faculty partnered closely with two community organizations, College Bound and Project GRAD, which served underrepresented minority high-school students, to assure that students are educationally and emotionally prepared to accept the challenge of conducting research in a university environment. College Bound and Project GRAD, Los Angeles program representatives played a major role in developing, monitoring, and shaping the future of the program.

Both community organizations had the same goal of providing academic opportunities and services to prepare students for admission to institutions of higher education. Their programs worked with students from elementary school through high school and addressed the needs of underserved children and youth. The programs differed, in that College Bound was a not-for-profit organization that served 1,000 children and their families throughout the greater Los Angeles area, while Project GRAD, Los Angeles, part of a national program located in various cities, was devoted to improving public education and worked with a cluster of twelve public schools located in the San Fernando Valley of Los Angeles. The apprenticeship program directors collaborated early on with program representatives of the two organizations to develop a partnership that provided ideas of how to design the project; qualified students selected by the organization; support and involvement with the families; insight into how to address the students' needs; and contact with the students after completing the program. The apprenticeship program was designed to include concepts, such as the academies held monthly on Saturdays and intensive summer research experience, which the partners previously used in their own programs. As their organizations were involved with the students since elementary school, community partners had the responsibility for student selection. The partners typically selected students who had already participated in pre-college enrichment programs. They then helped design, draft, and negotiate contracts with the students and their parents. Throughout, program co-directors and community program representatives engaged in two-way conversations to inform one another when students had difficulties. Project GRAD and College Bound are nonprofit comprehensive programs that support "college-going." Each provided academic and emotional support to the students together with diverse enrichment experiences. As a result, the high-school juniors they selected came to campus ready for project-based learning activities. In fact, what made the apprenticeship program so compelling for the students was the prospect of hands-on, project-based learning that began in the fall of their junior year and continued through high-school graduation. The two community partners would not have been able to create these research-intensive programs, themselves, and clearly the laboratory scientists could not have designed project-based activities for high-school students on this scale without external funding.

The program developed around the partners' experiences with students as they moved through their last two years in high school, so that the various program activities connected with students' evolving interests and needs. Throughout, the community partners alerted mentors to the students' competing needs and of their activities within other pre-college enrichment programs. This enabled faculty program directors and mentors to adjust to situations that occurred during the program. It was important that mentors understood that these students were involved in many activities that may compete with the apprenticeship program's demands; therefore, the schedule had to be adapted to that of the students, including SAT training and AP course finals. Certain aspects of the program were then refined to effectively address the challenges faced by these youths. However, there were also hard decisions that required attention. In first year, the enrolled students all participated in the program to a high level, except one student dropped out toward the end of the second year to pursue another activity. In the second year, both cohorts presented unanticipated problems. The first involved two high-school juniors copying homework. The program directors discussed the issue with the teacher who worked with the students during their first year in the program, and also with community partners and decided to drop the students from the program. The second concerned a student who felt uncomfortable living in the dorm during the summer. The third, surfacing throughout the program, involved parents' expectations that participation in the apprenticeship program would lead to their child's acceptance by UCLA. When some students were accepted and others were not, parents of those not accepted were disappointed. As much as mentors thought they identified with the needs of these students, it was the representatives from partnering community organizations who had the insights and ability to work through challenging situations with students and their families.

LEARNING LABORATORY LIFE

Over a seven-year period, from September 2007 to August 2014, sixty-eight high-school students enrolled in the two-year apprenticeship program, completed the program, and went on to college with either a partial or full scholarship. The program framework consisted of three linked components: a first-year academy, a summer research experience, and a second-year academy and capstone.

As students entered the program, they attended a first-year Saturday Academy, a didactic class taught by a faculty member in the department of chemistry and biochemistry that included an in-class laboratory session in which students practice molecular biology techniques; a lecture by a research scientist whose lab one of the high-school students would work in over the summer months; a background reading assignment to introduce relevant terminology; and a take-home, kitchen-style laboratory where students can work with their families to demonstrate a particular scientific concept. The academy's goals were to prepare students for laboratory research; to cover in-depth topics relevant to the ongoing research in oral biology labs; and to enhance scientific writing and communication skills. Throughout the course, emphasis was placed on working together, developing and testing hypotheses, and analyzing results. In-class laboratory sessions gave the students the hands-on skills and confidence to enter a research environment. Take-home laboratories exposed basic scientific concepts to the broader community, including model systems such as fruit flies to understand human diseases, enzymes, DNA and biofilms, meant to engage the family to illustrate the scientific concepts that the students are learning about in the program. The take-home laboratory involving fruit flies consisted of students taking home two male and two female adult fruit flies in order to observe and collect data about the developmental process, from egg to adult.

Parents and guardians were thus involved from the start. In the fall, an orientation at the start of the First-Year Saturday Academy was held where parents, guardians, and siblings learned about the apprenticeship program and the challenges of college life by the associate director of the Academic Advancement Program. Next was a presentation by a biology professor about the challenges of freshman science coursework, followed by one on science as a career from a STEM-PLEDGE graduate student. In the spring, at a ceremony before their parents and other family members, students received certificates indicating their completion of the first-year Saturday Academy sessions and their competency to conduct laboratory research. Then, focus groups with parents were held to discuss their expectations and concerns, and also their observations about their youth's progress over the course of the program. Afterward, parents visited a participating laboratory hosted by the lab director who explained the research conducted, the various personnel engaged in research, laboratory safety, the equipment, and student participation in the laboratory. Through the first-year Saturday Academy, students sensed the collaborative nature of science, began to analyze and question data, interacted with laboratory scientists, and became enthused about biology. Toward the end of the academy in the late spring, students prioritized labs that interested them, and the academy teacher designated the students who would work in seven labs for the six-week residential program. As a result of their first-year academy experiences, students sensed the collaborative nature of science, began to analyze and question data, interacted with laboratory scientists, and became enthused about biology, and were well-prepared and ready for their summer projects.

In the summer research experience, under the supervision of their mentors, the students learned and utilized common laboratory techniques to collect data, and they also learned techniques to present data in a meaningful way, culminating in an end-of-summer research presentation before mentors and family members. A parent and student orientation was held in the residence hall on the first day of the summer research experience. The orientation session was designed to introduce residential life on campus to both parents and students, some of whom were living away from home for the first time, and to explain the ground rules of campus life and expectations of the intensive research program that students were embarking upon. Each student was then presented with a personalized laboratory coat, embossed with an HHMI-UCLA logo. This ceremony, in some way, represented their entry into the prospect of science as a career. Prior to any involvement in laboratory activities, students received a three-hour laboratory safety orientation training course provided by the campus Environment, Health, and Safety unit, which included bio-safety training principles, such as proper use of cabinets, fume hoods, and overall lab safety and precautionary procedures. The students then embarked on their projects, and over the seven years, mentors in fourteen different oral biology laboratories designed projects and mentored the high-school students during the six-week residential summer research experience. Intensive lab-based project-based activities were designed to promote scientific discovery, using problem identification, experimental design, data collection and analysis. The summer activities included but were not limited to cell and tissue cultures, PCR, recombinant DNA, gene expression, Western blots, and microscopy. At the end of each week, the program director and co-director met with the students to discuss the progress of their experiments and their interpersonal relations in the laboratory. At the end-of-summer research presentation and family reception, students presented their summer research before their families, mentors, lab directors, and program staff, followed by a reception and luncheon.

On summer evenings, faculty and staff, and the students themselves, provided social and emotional support on behalf of "college-going" in small groups during evening sessions in the dorms. A clinical psychologist on the university faculty worked with the students on their concerns about speaking in public, working in research environments, and forging a balance between academic, social, and family life. Each Tuesday, the program co-director met with students after dinner and introduced a speaker from either the faculty or the university administration, including the associate Vice Provost for Student Diversity and director of the Academic Advancement Program who each summer led students in a discussion of college and career options, together with personal development issues. Each Wednesday, an undergraduate student from the Community-Based Learning Program or a STEM-PLEDGE graduate student spoke with the students about their own career pathway in science. Thursday evenings were devoted to providing students with social activities on campus or in the surrounding community.

The second-year Saturday Academy, conducted by the program director and co-director, began in September and prepared the students during their senior year for a poster presentation at a regional scientific meeting. In support of students' college-going aspirations, the program director and mentors wrote letters to private universities and scholarship programs. The second-year Saturday Academy sessions for the students who completed their summer research were designed to support their efforts to prepare their abstracts and posters and to practice delivering their posters. Through these monthly sessions, program staff better understood their needs for continued mentoring in the visual presentation of scientific data. They also learned about demands placed upon these students during their senior year in high school, including Advanced Placement and dual enrollment college coursework, SAT/ACT exam preparation, and their college applications. Additional Saturday Academy sessions were then scheduled to support students in completing their posters in time for the summer presentations. As a capstone activity, the students presented their scientific posters in a judged student competition held at the American Association for the Advancement of Science (AAAS) Pacific Division meeting each year. Before the Student Award Committee judges and others attending the meeting, they presented their data in written, visual, and oral formats and fielded questions about their research results and their implications during student poster sessions. Many students won student awards of excellence over the seven years. The students' and their mentors' jointly produced abstracts are published in the Proceedings of the American Association for the Advancement of Science Pacific Division and are part of a print and online archive.

OFF TO COLLEGE

Over the seven-year period, sixty-eight students went on to attend college with a partial or full academic scholarship. Students entered forty-six research universities, eight comprehensive universities, eleven historically black colleges, and three community college-university transfer programs in the sciences. These included UCLA, UC Berkeley, UC Davis, UC San Diego, UC Santa Barbara, UC Irvine, UC Riverside, UC Merced, USC, Stanford, MIT, Columbia, University of Wisconsin–Madison, Rensselaer Polytechnic Institute, Washington University–St. Louis, Rutgers–New Brunswick, Tulane, Cal Poly San Luis Obispo, University of Texas–San Antonio, Loyola Marymount, Dominican, Xavier, Spelman, Howard, Bowie State, Kalamazoo, and many other colleges. Forty-one students declared majors in the STEM disciplines, including the life sciences (24), engineering (10), mathematics (2), physical science (1), neuroscience (2), general science (1), and kinesiology (1). The remainder declared majors in the following disciplines: social and behavioral sciences (18); humanities (3); other majors, including business and education (3); three remained "undecided" as they entered college. Seven students enrolled in college pre-med programs. While in college, two former students returned to campus to attend the residential Summer Medical and Dental Education Program; another attended the PREP Program. Each program was designed to provide premedical and predental students with a means to strengthen their ability and readiness to study medicine and dentistry. Emilio Frias, who was selected by Project GRAD, spoke about how an entirely new dimension of the health field opened to him:

> When I used to think of medicine, I would think of a doctor prescribing drugs. This wasn't like the doctor prescribing drugs. This is behind that. This is the people doing the research to get the medicine the doctor prescribes.[3]

Students attributed successful college admissions and scholarship funding, in part, to the high-school apprenticeship program. Twenty-year-old Michaela Scott was part of the program's first cohort, having been selected by College Bound. Looking back, two years after graduating from high school as a UCLA psychobiology major, she described the apprenticeship experience as follows: "Being part of this program prepared me for college, and I believe it set me apart from other college applicants."[4]

Since the outset of the program, the Scientific Attitude Inventory (SAI) and a satisfaction survey were administered to participants. Data on the SAI were collected during the first session, at the end of their summer laboratory experience, and after students presented their posters. The results of the SAI over the last seven years were fairly flat, with no major changes over the course of the program. In all, the SAI characterizes the students as coming to the program with a good attitude and leaving it with ratings at essentially the same level. The satisfaction survey was administered at the end of the summer research experience and at the end of the program when the students were returning home from the AAAS meeting. Again, the students had high levels of satisfaction after their summer research experience, as well as after the experience of presenting their research at the meeting. In general, both SAI and satisfaction scores indicated that the students came to the program with good attitudes and were satisfied with each aspect of the program. This is not surprising since the students were selected by community organizations that, for many years, had contact with and knowledge about these students and their families.

Many students were the first of their families to either attend college, or pursue science as a vocation, or both. Students reported that the informal learning activities, together with the social and emotional work in small groups, influenced both their own and their parents' attitudes toward science and a scientific career. Noemi Zaldivar from Project GRAD said this about working in a lab with university scientists, "I was excited, but I was also scared. But they make us feel really well about everything because they say, 'You learned more in six weeks than I learned in six months.'"[5] Cameron Stevenson-Monroe from College Bound could not picture himself pursuing a career in the sciences. Being a part of the program changed his attitude considerably, in his words:

> At first, I was apprehensive about participating, since I would be doing something, I wasn't very familiar with, and I didn't want to mess anything up. But I've loved the experience. Not a lot of people can say that they've worked next to scientists in a research lab at UCLA. Before, I didn't have the confidence and was too shy to speak up, but now I'm not as nervous and am really looking forward to asking even more questions in college.[6]

Project-based learning thus enhanced students' motivation to achieve, their sense of well-being and of self-efficacy in performing laboratory procedures during their junior and senior years in high school. For Karolyn Blancas, science seemed fairly easy, but it never especially held her interest because everything she learned came out of a book. Because of her grades in science, she was chosen by Project GRAD to participate in the program. Karolyn admitted, "I was never really into science. It wasn't until I got to the lab that I really liked it."[7] The hands-on experience also changed how Project GRAD student Marisol Nunez looked at science. Marisol said. "In chemistry and biology classes, we didn't do much lab work; it's mostly reading books. To be able to be here and follow procedures is a great opportunity."[8] Over the seven years, it was clear that students who embraced science through the apprenticeship program continued to make an impact on their families, their peers, and their communities. Some graduates even returned to campus to speak with the students currently in the program and their family members. The journeys that these students traveled provided an exemplary pathway for others to follow, and the stories that they told were an inspiration to the younger students who were just beginning to make decisions about college and their choice of majors.

Throughout, a community of practice model assured ongoing trust-building between the community and campus partners and between these partners and the families that entrusted their children to them. Realizing a norm of collaboration promised by community–campus partnerships, such as this program, entails ongoing dialogue between community stakeholders and university partners, and between the partners and families in the community, on mutually determined priorities, strategies, and policies. Moreover, these partnerships are optimally sustained around a common mission, that of service. As it takes considerable time to build up community-scale initiatives, universities may be required to demonstrate a genuine level of commitment for the duration, despite periodic budgetary crises, faculty turnover, and gaps in specific project funding. A culture of academic engagement thus commits the research university to reconnect its educational, scientific, and public service missions through collaborative problem-solving and community-based knowledge development. Universities can then be poised to combine academic expertise and local knowledge on behalf of public goods, such as education, health care, and the environment, that affect the quality of life in neighboring communities and increase the life chances of their youth.

NOTES

1. Albert E. Beaton, Ina V.S. Mullis, Michael O. Martin, Eugenio J. Gonzalez, Dana L. Kelly, and Teresa A. Smith, *The Mathematics Achievement in the Middle School Years: IEA's Third International Mathematics and Science Study (TIMSS)* (Chestnut Hill, MA: Center for the Study of Testing, Evaluation, and Educational Policy, Boston College, 1996).
2. Ina V. S. Mullis, Michael O. Martin, Pierre Foy, Dana L. Kelly, and Bethany Fishbein, *TIMSS 2019 International Results in Mathematics and Science* (Chestnut Hill, MA: TIMSS & PIRLS International Study Center, Lynch School of Education and Human Development, Boston College and International Association for the Evaluation of Educational Achievement, 2020).
3. Adolfo Flores, "Teens Get to Work with Lab Scientists at UCLA; Dental School Opens Lab to High Schoolers with Science Aptitude" (*The Daily News of Los Angeles*, August 1, 2008), A4.
4. Brianna Deane, "UCLA School of Dentistry Creates Future Opportunities for Disadvantaged, Gifted High School Students," *UCLA Newsroom*, July 30, 2012, http://newsroom.ucla.edu/releases/uclaschoolofdentistrycreates236900.
5. Flores, "Teens Get to Work with Lab Scientists at UCLA; Dental School Opens Lab to High Schoolers with Science Aptitude," A4.
6. Deane, "UCLA School of Dentistry Creates Future Opportunities for Disadvantaged, Gifted High School Students," http://newsroom.ucla.edu/releases/uclaschoolofdentistrycreates236900.
7. Flores, "Teens Get to Work with Lab Scientists at UCLA; Dental School Opens Lab to High Schoolers with Science Aptitude," A4.
8. Flores, "Teens Get to Work with Lab Scientists at UCLA; Dental School Opens Lab to High Schoolers with Science Aptitude," A4.

Part III

RATIONALITY AND REDEMPTION

Chapter 8

Times of Promise

ADOLESCENCE AND SOCIAL MOVEMENTS

Social movements and adolescence occur at times marking transitions across diverse social fields and are shaped by historical and cultural moments. Each creates a new social space and the possibility of change through the honing of social skill and cooperative action in group life.[1] Erik Erikson contended that in order to enter history, "each generation of youth must find an identity consonant with its own childhood and consonant with an ideological promise in the perceptible historical process."[2] Emergent adulthood then completes the dialectic of childhood and adolescence forming a synthesis when a person makes conscious life decisions and follows a path with the promise of regenerating the social world through partnership, work, and politics. Erikson's conceptualization of the adolescent path toward adulthood falls within a liberal political framework, according to Leerom Medovoi:

> In one respect, the "search for identity" that comprises the stage of adolescence for Erikson reenacts a classical political metanarrative of the enlightened individual entering into full possession of his/her right to self-determination. Much like the ideals of liberty and independence that it incorporates, therefore, "identity" is a normative term and not just a descriptive one. It named an accomplishment and a positive good.[3]

Youth as a transitional stage with self-determination as the goal involves educational and other community settings where such "psychopolitical autonomy"[4] can mature and prosper. An unfolding life course, or life cycle, underlying individual development requires institutions that support youth's life transitions, offering them directional paths for self-recognition and self-reflection. Erikson understood that life history, or psychogenic development, entails understanding of the historical development of the times. For him, healthy development involves the growth of a conscience, which is indicative of a mature ethical orientation, "a universal sense of values assented to with insight and foresight in anticipation of immediate responsibilities."[5] Erikson held an understanding of humanity based upon a theory of psychogenic development in a social context, which considers certain dilemmas of existence as common to the human condition. These include self-awareness and personal identity, the pursuit of freedom, the search for meaning, and for the ways to manage anxieties over one's own mortality. Erikson thus supported a new approach to education based upon the development of self-knowledge, noting the critical importance of adolescence where identity is consolidated. Consequently, collaborative learning approaches better prepare young people to work together with others and form their identities while exploring various life possibilities at this critical decision-making phase. In earlier societies, education and systems of spiritual direction, together, supported character formation along with vocational preparation, providing pathways to emergent adulthood. Contemporary education may likewise influence youth's inner lives and identify formation through support of their generation's common concerns. This requires seeing the value of social dramas and cultural performances displayed in youth-led movements surrounding what is important to their lives. Earlier social movements are instructive in this regard, especially how emergent experientially grounded social knowledge supported participants as they sought a place in the world. Georg Simmel goes even farther, viewing social movements as holding the promise of liberating "the whole of life," a sentiment shared by each younger generation seeking a place in the world:

> When future times write the history of our century, they might well emphasize two brightly shining points alongside the dark shadows that cover the present times: natural science and social movements. They will depict the first in the security of that which it has achieved and the second in the security of that which it will attain. They will describe the liberation of thought coming from the first and the liberation of the whole of life from the second.[6]

Adolescence and social movements are frames within a temporal "liminality," from the Latin, meaning "a threshold," a state of transition between one life stage and the next in the case of adolescence and a move away from a past toward an imagined "not yet there" for social movements. Both provoke "anti-structure," an opening for persons to consider arbitrary and artificial qualities of seemingly fixed social norms and roles, according to Victor Turner.[7] Liminality attenuates rigid social structural distinctions, drawing actors in a social movement culture toward a spatial sense of *communitas*, or an unstructured, egalitarian community that may evoke feelings of ambiguity and disorientation.[8] Adolescent peer culture provides situations where the young form allegiances based upon crosscutting ties that may blur class, gender, and ethnic differences. As with adolescence, movements provide "spheres of articulation,"[9] following Max Gluckman, where people become caught up in a seemingly inconsistent social environment and attendant interpersonal interactions within it. They are then compelled to select their beliefs to adjust to a particular situation, affording a window on alternative values and lifeways. Movements return participants to a "postliminal" state and a consequent need for them to articulate and uphold their changed beliefs and behaviors. Diverse adolescent experiences thus activate situations that evoke shifts in personal knowledge and being, providing precedents for future thought and action.

Social movements likewise reshape participants' social reality through activities that call into question deeply held cultural beliefs and practices. However, any such reorientation of cultural beliefs requires a means to bring about a shift in perspective. Kenneth Burke recognized the importance of offering participants the opportunities to gain, in his words, "perspective by incongruity," or an act of "metaphorical extension," whereby two opposing viewpoints are linked so as to cause initial confusion.[10] This rhetorical device challenges them to search for the meaning of the conundrum, exploring different ways to gain perspective beyond their accepted frames of reference. Reflecting on such opposites affords opportunities for participants to understand the constructed nature of their social worlds and their habitual modes of thought and action within them. Reflective activities thus force a change in perspective, providing a basis for participants to interpret the inconsistencies in their lives initially in the vocabularies they use to rationalize their behavior within their social worlds. Burke understands that programs for change offer a means for resolving the oppositions that blur reality and limit a person's capacity to explore a range of possibilities. Thus, incorporating various tropes, including metaphor, analogy, and perspective, into a movement's rhetoric can support participants' exploration of their commonplace cognitive and perceptual processes. From such tropes, they can then begin to examine their current situations and create new meanings for their lives going forward. Participants learn how to better interpret their habitual actions, gaining distance through rhetorical and dramatized styles of behaviors. These may include using comic irony as a corrective to challenge the seemingly tragic oppositions within their worlds that prompted a social movement. Participants thus practice gaining a perspective by incongruity, seeing a shift within themselves away from obsolete modes of thought and behavior based upon past "constructed" experiences. Consequently, together with others, they can enact a shift toward a future and attendant alternative cultural explanation of self and world. Through these dynamic processes may come, following Burke, a civic or public pedagogy, as Ann George asserts,[11] from which a movement culture may emerge that potentially reshapes participants' self- and world-images.

MOVEMENT CULTURES

Social movements generate personal change processes by emphasizing a cultural dimension to human action that may serve to reorient, as Alberto Melucci's states, "the biological, emotional and cognitive structures on which we base the construction of our experiences and relationships."[12] For Melucci, a movement's symbolic function, in his words "a prophetic function,"[13] holds the promise of changing participants' lives, even provoking a new interiority, as they mobilize to make societal changes. Emotions are key to the unfolding of movement cultures, as unexpected events or information can bring about a moral shock. An initial emotional upheaval motivates participant recruitment and sustains participation, creating emotion cultures that influence a movement's internal dynamics, and contribute to its growth and influence. A movement thus reorients constituents internally, as they discover new ways to strategically cope with the emotionally charged situations that originally activated their participation. Emotions such as outrage and anger, as well as anxiety, are eventually transmuted into cognitive beliefs embedded in a movement's organization. Tilly explains how a social movement also creates a coherent set of "actors," on each side of the conflict, which may be individual "living participants" or a group:

> Real social movements actually consist of sustained interactions among authorities and challengers. Within real social movements, various challengers attempt to create a coherent actor, or at least its appearance. Furthermore, real social movements always involve a symbolically constrained conversation among multiple actors, in which the ability to deploy symbols and idioms significantly affects the outcome of the interaction.[14]

Within these "spaces of contention,"[15] following Tilly, participants may call on symbols and images from significant political, social, or historical events to legitimate their actions.

Social movements advance repertoires of social action captured in a symbolic discourse that participants express while engaging in such purposeful organized behavior. In this way, movements sustain a particular "vocabulary of motives"[16] or the various meanings that orient participants' actions and their consequences, following Burke. Analogous modes of understanding, cognitive representation, and interpretation are fundamental to the human adaptation for culture. Humans have the capacity to relate with the environment through image making. This form of external adaptation rests on the ability to symbolically represent the physical world as a cognitive map. A cultural worldview reflects the shared cognitive categories of individuals that experience and work within a local set of spatial and temporal arrangements.

Culture has the potential to selectively identify and unify these categories as shared cognitive maps, which serve as symbolic domains of an ecologically framed common world. A worldview, or ethos, fuses these cognitive maps of different individuals within a locale by imposing consistency among meanings, as a paradigm for working economically and efficiently with the energetic world. Jerome Bruner recognizes "that there are specific domains of human knowledge and skill and that they are supported and organized by cultural tool kits," and he maintains that reality construction depends upon "the enablement of the symbolic systems of culture."[17]

By unifying the mental and the energetic in a symbolic system, a local society provides the means to reproduce its own self-organization.[18] A symbolic system involves cognitive framing to produce a set of perspectives that are taken for granted and inform individuals' capacity to interpret requisite information and understand their experiences in a wider social context.[19] There are two types of primary frameworks or schemata that render information meaningful and inform interpretive processes, according to Erving Goffman.[20] The first are natural frameworks that identify events as unguided by human agency. The second are social frameworks that understand events as guided by human motives and intentions. Within social movements, there are framing processes whereby cultural schemas derived from previous movements, or movement cultures, are transformed into symbolizing actions and expressions of solidarity.[21]

Anthony Wallace sees common symbolic processes in social movements and the revitalization of societies after crises, such as disasters, rapid technological change, and forced displacement.[22] For Wallace, society is a social field that shapes members through a network of communication, enculturating them for a place in community networks. An outcome of social learning is the "mazeway,"[23] or the set of cognitive maps containing an individual's images of self and world, and adaptive strategies for satisfying personal goals. A mazeway functions to alleviate and reduce stress, however, it is not beyond challenge, especially when crises confront individuals with circumstances beyond their control. A crisis brings about mazeway disintegration or the perception of disorganization and one's displacement from the familiar environment. The social fabric is thus disrupted and a mazeway can no longer reduce stress, nor can people count upon community networks to protect them from disruptive consequences. Many will then experience a form of cognitive dissonance when their self-image fails to conform to the real system because of external pressure placed upon it by a crisis event.

Revitalization is a radical resynthesis of existing beliefs and values, essentially a new culture that arises phoenix-like from a crisis. A revitalization movement comes about when individuals experience a high level of stress and become disillusioned. According to Wallace, revitalization movements operate through phases on behalf of constructing a more satisfying culture, including mazeway reformulation, communication, organization, adaptation, cultural transformation, and routinization.[24] These movements hold the promise of personal and social change, giving rise to a new set of symbols that guide behavior. Revitalization movements organized in the aftermath of cultural trauma help individuals and groups to reframe postcrisis community worlds. People experiencing crises that threaten their sense of coherence and overall well-being, or destroy their sense of place, may eventually regard their local conditions in a new light. Revitalization movements therefore offer the promise of a "new look" or a changed view of the environment and of social conditions.[25] The change in perspective is supported by communal use of recognizable symbols to carry out rituals of renewal and hope within the changed landscape.

A movement may thus reorient cultural worldview, through participants acting together. By learning new forms of experiential knowledge and employing that knowledge in strategies to gain ground toward movement goals, participants may come to share a new cognitive map. A reoriented worldview, based largely upon values of self-control and self-determination, tends to come about through investment in collaborative practices. In this way, a movement's knowledge base and organization are also continually altered and reconfigured through the open-ended experiences of participants. Melucci holds the fundamental role of knowledge in the process of transformation, especially since dominant institutions shape the identities of their constituents by manipulating symbolic systems:

> Knowledge is a crucial resource for new conflictual actors, both because it is a focus of major conflicts (those over the appropriation and control of knowledge and information, and over the instruments of production and circulation of these), and because only in knowledge can the texture of social relationships be disclosed which lie behind the facade of neutrality that the dominant apparatuses seek to impose on social life.[26]

Thus, control of symbols and language itself are thus necessary conditions for a system's survival:

> . . . in order for highly differentiated systems to be able to guarantee their internal integration, it becomes necessary to extend the system's control over the symbolic levels of action, so as to include in its scope the spheres where the meanings and motives of behaviour are constituted. Control can no longer restrict itself to the external regulation of the production/appropriation of resources; it must also intervene in the internal processes of the formation of attitudes.[27]

Melucci finds that contentious processes underlying social movements are recently taking a cultural turn.[28] Movements create alternative symbolic and linguistic schemas as forms of critical knowledge so that participants can come "to appropriate non-manipulated knowledge."[29] For Melucci, "the *form* of the movement is a message, a symbolic challenge to dominant patterns."[30] Cultural reorientation relies on the symbolic representation of new ways of learning and a "symbolic confrontation" in the face of dominant institutional challenges.[31] Movements thus share common characteristics with culture, itself a form of anticipatory behavior specific to humans, for much of our time is spent reorganizing the world to resist randomness or entropy.

What weaves together diverse interests, habits of mind, and practices into a movement culture is a reliance on self-initiated activity that activates experiential forms of knowledge. Effective movements create social environments for experiential learning where participants solve problems by acting together. Within these settings, the change process is relational, rather than individualistic, emphasizing the connectedness and mutual involvement of participants in meaningful activities. Social movements strive to meld knowledge gained through life experience, rather than that of experts, into an institutional ethos that is shared by all participants. While sometimes appearing similar to expert organizational structures, movement-derived institutions differ with respect to their participants' perspective toward a specific issue or situation.

Mutual efforts to reshape personal life require that participants develop creatively engaged habits of mind to engender new styles of aesthetic and emotional expression. Such creative engagement, notably developing the habit of imagination, is requisite for participants to take informed roles in remaking their lifeworlds on reconstituted moral grounds. Mihaly Csikszentmihalyi views creativity as "when a person . . . has a new idea or sees a new pattern and when this novelty is selected by the appropriate field for inclusion in the relevant domain."[32] However, any aesthetic potential for strategic change will require individuals to anticipate outcomes and use foresight and intentionality as cognitive potentials. Social movements thus call forth a phenomenological individualism. As in childhood, individuals organize and reorganize their social experiences as "standard stories," in Tilly's words, "sequential, explanatory recounting of connected self-propelled people and events that we sometime call *tales*, fables and narratives."[33] Movement participants thus shape and reshape themselves through their storied involvement in contentious events. Their shared stories of patterned transactions within the complex interlocking networks necessary to stage events may assist them toward a reoriented self-identity. As understood, a movement provides participants many opportunities for personal stories that come through interactions with others within any number of potentially metamorphic activities. Sharing stories with one another, after a movement action, can motivate participants to reflect on the meaning of that action for themselves, and then embody those stories as precedents for future action, and these precedents may serve as a catalyst for identity formation or reformation.

Instructive in this regard is Paul Radin's psychologically informed ethnography, which conveyed the primacy of recognizing the historical meaning of a culture through ancestral tales—the stories that preserve an inherited history. For Radin, understanding the "cultural functions of thought"[34] requires immersion into a web of social relationships to discover diverse personality types and forms of life. Dewey appreciated Radin's work, as he understood how the affective dimension provided a "developmental continuity between biological, psychological, and cultural phenomena,"[35] according to Peter Dunlap. Dewey wrote in his Foreword to Radin's *Primitive Man as Philosopher*,

> He makes it clear that objects and nature were conceived dynamically; that change, transition, were primary, and transformation into stability something to be accounted for. His account makes clear that effects, emotional and practical, were the material of the thought of real objects, and that thinkers, in their doctrine of an inner "form," stated in rational terms a notion which was expressed mythically by the mass, a notion which has marked affiliations with a persistent strain in the classic philosophic tradition.[36]

Stanley Diamond saw Radin as "one of the last artisans of modern scholarship at the threshold of the computer age."[37] Radin's analysis of Winnebago tribal rites focuses on psychocultural development, notably how expressive symbols and ritual processes promote specific forms of consciousness.[38] Throughout the Medicine Rite, a new member's initiation performed by a medicine man, Winnebago ancestral tradition is embedded within a ritualized recurrence of biocultural regularities, though always represented through the lens of a culture embedded in history. As Diamond states, "For Radin, history was experience, and history could be more fully reflected, more complete, among primitive peoples in primitive cultures than among class-divided, occupationally compartmentalized, state-organized societies, such as ours, wherein experience is dissociated and fragmented."[39] History and biology form the common ground of the Medicine Rite, where the initiate's life history is enmeshed between these two poles. Cultural history resides in telling stories in the form of ancestral myths and employing archaic ritual speech in the event. Natural history is conveyed through the ritual's emphasis on growth and decay, the recurring natural patterns of the biotic world, and the human condition. Personal history is framed as reminders to the initiate that others passed through the ordeal and thereafter achieved success in their lives. The transfer of social and experiential knowledge is thus reinforced through ritual, which communicates, and frequently encapsulates worldview.

RITUAL PRAXIS AND CHANGE

Anthropology, ethology, and neuroscience view ritualization as adaptive behavior in its ability to encode cultural knowledge.[40] A ritual economy is the "process of provisioning and consuming that materializes and substantiates worldview for managing meaning and shaping interpretation,"[41] as Patricia McAnany and E. Christian Wells state. This economic construct emphasizes how people engage with social, material, and cognitive realms of human experience. In this sense, Roy Rappaport views a shared worldview as being materialized through economic and ecological transactions through a cognized model "of the environment conceived by the people who act in it."[42] The model, according to Rappaport, "elicits behavior that is appropriate to the material situation of the actors."[43] Ritual symbols prompt social action because their referents call up polarities between physiological phenomena and normative values, such as reciprocity, respect, generosity, and kindness.

Suzanne Langer characterizes ritual as a symbolic transformation of experiences that no other medium can adequately express. Because it springs from a primary human need, it is a spontaneous activity—that is to say, it arises without intention, without adaptation to a conscious purpose; its growth is undesigned, its pattern purely natural, however intricate they may be.[44] Langer views various expressive acts, notably speech, gesture, song, and sacrifice, as "transformations of experience in the human mind that have quite different overt endings. They end in acts that are neither practical nor communicative, though they may be both effective and communal."[45] For Langer, despite the impracticality or the limited utilitarian value of certain expressive behaviors, there is clearly a logical framing within ritual processes, as a ritual "expresses feelings in the logical rather than the physiological sense . . . it is primarily the articulation of feelings. The ultimate product of such articulation is not a simple emotion, but a complex permanent attitude."[46] Langer sees the symbolic transformation that ritual provides as giving rise to life-symbols, noting that the fantasies we derive from specific experiences form the basis of recurrent acts that become cognitively and culturally encoded.[47]

In a similar vein, Burke understands dramas of living to be recurrent forms of social experience that can contribute to attitudinal change.[48] Attendant with most life crises, as well as cultural and political crises, is the experience of liminality as a gestation process or transitional state. To make sense of the transition, communities may initiate a subjunctive mood of culture, through symbolic action that dramatizes a sense of fantasy, hypothesis, and conjecture. Raymond Williams refers to these community-scale performances as subjunctive action, notably occasions "to enact alternatives to the hegemonic center, which are meaningful and so common to the community."[49] As a means to resolve community crises, participants will engage in performance behavior that dramatizes the transitional state, such as singing, chanting, playing music, dancing, dressing up, feasting, drinking alcohol, using hallucinogens, to live through this passage from a structural past to a structural future. As rituals, social movements frequently enact such familiar styles of performance that employ multiple sensory domains to dramatize and enculturate new modes of being and living. A movement's high ritual density produces feelings of group solidarity and morality that together with group symbols transmute participants' emotional energy in two ways. Emotions are amplified by a group's shared focus on a common concern. Participants' emotions are then redirected through a similar focus of attention toward group solidarity. These performative and communicative styles can then support participants' embodiment of a desired change, within themselves, in their shared experiences and in their relationships with others.

This was the case as disruptive industrial technologies and cultural modernization brought about diverse crises when the United States was arguably a developing country.[50] In the mid-nineteenth century, agriculture was mechanized and modernized as it moved to become a capital-intensive industry. A structural repatterning of rural life upended long-standing norms of class, gender, age, and personal identity in farming areas. Consequently, regional agrarian protests erupted and then progressed into broader national movements within a civil society framework, notably temperance, abolition, and suffrage. Participants included farmers, merchants, shopkeepers, and mechanics, with large numbers of women and children involved in each movement. The various movements emphasized matters that combined personal and social concerns. Amid rapid capitalist development, certain evangelical groups sought to control the conversation about the sudden changes in personal life, work, and class composition. The movements' cultural schemas, or shared meanings and strategies that motivate social action, were concordant with the schemas of evangelical Protestant sects and their attendant national infrastructure within antebellum America. Participants held to an evangelical faith that stressed an individual's sin and its renunciation in public confession. A feature of evangelism was the emotionally charged religious revival. These "protracted meetings" often lasting up to three weeks used specific spiritual practices, such as twice-daily preaching, the singing of hymns, prayer, and spiritual counseling to provoke conversion among unconverted youth, typically men and women under thirty. The young entered the evangelical revivalist landscape with the promise of a "new birth" through conversion, accommodating a wish for a secure place in a community where each member made a personal commitment toward a Christian adulthood following a relatively clear spiritual direction.

The diverse societies and organizations formed through these movements were rationalized interdenominational benevolent efforts to "redeem a nation" by forwarding evangelical concerns that included mass dissemination of bible tracts, temperance newspapers, and texts invoking "the gospel of emancipation" by famous lecturers on the antislavery ministry circuit.[51] A fast-moving style of "confessional politics" swept throughout the country facilitated by a network of aligned voluntary organizations, including benevolent societies and fellowships that in Michael P. Young's words, "promoted a uniquely national purpose—a campaign against national sin."[52] The various schemas co-constructed by movement leaders and participants formed part of a national political consciousness. Individual and collective schemas converged as innovative communal symbols and action strategies complemented participants' deeply personal spiritual concerns and motivations. Within these movements, the personal and the political coalesced on behalf of mutual concerns within a developing country struggling to form collective imaginings on behalf of a national identity. Their extensive schemas reached beyond local issues to encompass the many people enmeshed in existential or life situations at this time, notably experiences with alcohol, equal rights, or antislavery. By contrast, intensive schemas, or those that are tightly controlled in terms of their meanings and motives, were typically embraced by smaller numbers of participants in more localized movements.

During this disruptive modernizing phase, contentious debates surfaced ethnocultural and subnational tensions with participants advocating freedoms based upon sectarian faith, rather than secular precepts. However, there were profound changes in material circumstances that contributed to this discord. Along with commercial agriculture, a transportation revolution opened water-based trade corridors that facilitated a larger exchange network. New religious congregational communities and utopian settlements were frequently situated along the nation's rivers and canal routes that grew with the market revolution, as in the "Burned-Over District" in upstate New York.[53] Devotional and communal settlements became the sites of movement formations as responses to the commercial revolution that brought about new market relations. Amid such rapid societal changes, evangelical movements for reform thus swept through the countryside, calling forth a "second coming" in the new world, although secularizing streams of thought and action would eventually supplant them. How these schemas fused to become repertoires for collective action beyond this era can be understood in the context of secularization. Peter L. Berger indicates that secularization's "original locale" was "in the economic area, specifically in those sectors of the economy being formed by the capitalist and industrial processes."[54] The secular turn arrived as technologically modernized ways of living and working were occurring in industrializing urban regions connected to one another, and to their hinterlands, by steam-powered rail and national communications networks.[55] Social movements that challenged authority under such rapidly changing urban industrial conditions brought about a broader-based national politics through secularized forms and enactments. Through them flowed a hidden stream of movement culture, notably experientially based organizational "repertoires of contention,"[56] as Tilly and Sidney Tarrow note. Such are the tools and scripted actions that motivate participants in a social movement that, when effective, inspire and inform subsequent shared actions, or their praxis.

Like earlier movements, the new urban social movements provided innovative social knowledge and an awareness of previously successful contentious performances, as in theatre where effective scripts and routines inform and modify actors' subsequent performances. Out of a remembered past, generations of industrialized youth enacted movements through repertoires that fused tradition and innovation, impelled by an emergent peer culture. Machine Age youth challenged market forces and mechanized lifeways that brought even more complex cultural modernization processes. Postwar youth built upon these organizational repertoires in movements for social justice, racial and gender equality, and the fate of the earth. Millennial youth created social movements to contend with neoliberal austerity, climate action, and gun violence. For more than a century, participants during each era reinvented movement strategies, blending traditional and contemporary organizational repertoires and their communicative and performative styles. Many reinvented themselves as well through such movement dynamics. Reinvention, a hallmark of craft consciousness, was thus a driver of change and an outcome for each culture of solidarity in Machine Age America and beyond.[57]

NOTES

1. Neil Fligstein, "Social Skill and the Theory of Fields," *Sociological Theory* 19:2 (2001): 106–124.
2. Erik H. Erikson, *Identity: Youth and Crisis* (New York: Norton, 1968), 257–258.
3. Leerom Medovoi, *Rebels: Youth and the Cold War Origins of Identity* (Durham, NC: Duke University Press, 2005), 6–7.
4. Medovoi, *Rebels*, 7.
5. Erik H. Erikson, "Reflections on the Dissent of Contemporary Youth," *Daedalus* 99:1 (1970): 164.
6. Georg Simmel, *Simmel on Culture: Selected Writings*, edited by David Frisby and Mike Featherstone (London: Sage Publications, 1997), 288–289.
7. Victor W. Turner, *The Ritual Process* (Chicago: Aldine, 1969).

8. Victor W. Turner, "Dewey, Dilthey, and Drama: An Essay in the Anthropology of Experience," in *The Anthropology of Experience*, edited by Victor W. Turner and Edward M. Bruner (Urbana: University of Illinois Press, 1986), 43.

9. Eric Sheppard, Vinay Gidwani, Michael Goldman, Helga Leitner, Ananya Roy, and Anant Maringanti, "Introduction: Urban Revolutions in the Age of Global Urbanism," *Urban Studies* 52:11 (2015): 1955.

10. Kenneth Burke, *Permanence and Change: An Anatomy of Purpose*, third edition (Berkeley: University of California Press, 1984), 69.

11. Ann George, *Kenneth Burke's Permanence and Change: A Critical Companion* (Columbia: University of South Carolina Press, 2018).

12. Alberto Melucci, "The Global Planet and the Internal Planet: New Frontiers for Collective Action and Individual Transformation," in *Cultural Politics and Social Movements*, edited by Marcy Darnovsky, Barbara Epstein, and Richard Flacks (Philadelphia: Temple University Press, 1995), 288.

13. Alberto Melucci, "The Symbolic Challenge of Contemporary Movements," *Social Research* 52:4 (1985): 797.

14. Charles Tilly, *Big Structures Large Processes Huge Comparisons*, 30–31.

15. Charles Tilly, "Spaces of Contention," *Mobilization: An International Journal* 5:2 (2000): 135–159.

16. Kenneth Burke, *Permanence and Change*, 21.

17. Jerome Bruner, "The Narrative Construction of Reality," *Critical Inquiry* 18:1 (1991): 20.

18. Richard Newbold Adams, *The Eighth Day: Social Evolution as the Self-Organization of Society* (Austin: University of Texas Press, 1988), 81–82.

19. Erving Goffman, *Frame Analysis: An Essay on the Organization of Experience* (New York: Harper and Row, 1974).

20. Goffman, *Frame Analysis*, 22.

21. Michael P. Young, "A Revolution of the Soul: Transformative Experiences and Immediate Abolition," in *Passionate Politics: Emotions and Social Movements*, edited by Jeff Goodwin, James M. Jasper, and Francesca Polletta (Chicago: University of Chicago Press, 2001), 104.

22. Anthony F. C. Wallace, "Paradigmatic Processes in Culture Change," *American Anthropologist* 74:3 (1972): 467–478.

23. Anthony F. C. Wallace, "Mazeway Disintegration: The Individual's Perception of Socio-Cultural Disorganization," *Human Organization* 16:2 (1957): 23–27; Anthony F. C. Wallace, "Mazeway Resynthesis: A Biocultural Theory of Religious Inspiration," *Transactions of the New York Academy of Sciences* 18 (1956): 626–638.

24. Anthony F. C. Wallace, "Revitalization Movements," *American Anthropologist* 58:2 (1956): 264–281.

25. Anthony F. C. Wallace, *The Death and Rebirth of the Seneca* (New York: Knopf, 1970); Anthony F. C. Wallace, *Rockdale: The Growth of an American Village in the Early Industrial Revolution* (New York: Knopf, 1978).

26. Alberto Melucci, *Challenging Codes: Collective Action in the Information Age* (Cambridge: Cambridge University Press, 1996), 223.

27. Melucci, *Challenging Codes*, 92.

28. Alberto Melucci, "Liberation or Meaning? Social Movements, Culture and Democracy," *Development and Change* 23:3 (1992): 43–77.

29. Melucci, *Challenging Codes*, 224.

30. Alberto Melucci, "An End to Social Movements? Introductory Paper to the Sessions on 'New Social Movements and Change in Organizational Forms'," *Social Science Information* 23:4–5 (1984): 830.

31. Alberto Melucci, "Social Movements in Complex Societies: A European Perspective," *Arena Journal* 15 (2000): 81–99.

32. Mihaly Csikszentmihalyi, *Creativity: Flow and the Psychology of Discovery and Innovation* (New York: HarperCollins, 1996), 28.

33. Charles Tilly, "The Trouble with Stories," in *The Social Worlds of Higher Education: Handbook for Teaching in a New Century*, edited by Ronald Aminzade and Bernice Pescosolido (Thousand Oaks, CA: Pine Forge Press, 1999), 257.

34. David B. Zilberman and Susan Layton, "Culture-Historical Reconstruction and Mythology in the Anthropology of Paul Radin," *Dialectical Anthropology* 6:4 (1982): 276.

35. Peter T. Dunlap, "The Unifying Function of Affect: Founding a Theory of Psychocultural Development in the Epistemology of John Dewey and Carl Jung," *Educational Philosophy and Theory* 44:1 (2012): 54.

36. John Dewey, "Foreword," in Paul Radin, *Primitive Man as Philosopher*, second revised edition (New York: Dover Publications, 1957), xx.

37. Stanley Diamond, "Paul Radin," in *Totems and Teachers: Key Figures in the History of Anthropology*, edited by Sydel Silverman (Walnut Creek, CA: AltaMira Press, 2004), 52.

38. Paul Radin, *The Road of Life and Death: A Ritual Drama of the American Indians*, Vol. 5 (Princeton: Princeton University Press, 1991).

39. Diamond, "Paul Radin," 61.

40. Charles D. Laughton, John McManus, and Eugene G. D'Aquili, *Brain, Symbol, and Experience: Towards a Neurophenomenology of Human Consciousness* (New York: Columbia University Press, 1993); Roy A. Rappaport, *Religion and Ritual in the Making of Humanity* (Cambridge: Cambridge University Press, 1999).

41. Patricia A. McAnany and E. Christian Wells, "Toward a Theory of Ritual Economy," in *Dimensions of Ritual Economy*, edited by E. Christian Wells and Patricia A. McAnany, *Research in Economic Anthropology*, Volume 27 (Bingley: Emerald Group Publishing, 2008), 1.

42. Roy A. Rappaport, *Pigs for the Ancestors: Ritual in the Ecology of a New Guinea People*, second edition (New Haven: Yale University Press, 1984), 238.

43. Rappaport, *Pigs for the Ancestors*, 239.

44. Suzanne K. Langer, *Philosophy in a New Key* (Cambridge, MA: Harvard University Press, 1951), 52.

45. Langer, *Philosophy in a New Key*, 49.

46. Langer, *Philosophy in a New Key*, 134.

47. Langer, *Philosophy in a New Key*, 28–29.

48. Kenneth Burke, *The Philosophy of Literary Form: Studies in Symbolic Action*, third edition (Berkeley: University of California Press, 1973).

49. Raymond Williams, "Brecht and Beyond," in *Politics and Letters: Interviews with New Left Review* (London: Verso, 2015), 218.

50. Richard D. Brown, *Modernization: The Transformation of American Life, 1600–1865* (New York: Hill and Wang, 1976); Martin J. Sklar, *The United States as a Developing Country: Studies in US History in the Progressive Era and the 1920s* (Cambridge: Cambridge University Press, 1992).

51. Michael P. Young, *Bearing Witness Against Sin: The Evangelical Birth of the American Social Movement* (Chicago: University of Chicago Press, 2006).

52. Michael P. Young, "Confessional Protest: The Religious Birth of U.S. National Social Movements," *American Sociological Review* 67:5 (2002): 672.

53. Michael Barkun, *Disaster and the Millennium* (New Haven: Yale University Press, 1974), 62–63.

54. Peter L. Berger, *The Sacred Canopy: Elements of a Sociological Theory of Religion* (New York: Doubleday Anchor Books, 1969), 129.

55. Craig Calhoun, "'New Social Movements' of the Early Nineteenth Century," *Social Science History* 17:3 (1993): 385–427.1

56. Charles Tilly and Sidney Tarrow, *Contentious Politics*, second edition (New York: Oxford University Press, 2015), 14–15.

57. Rick Fantasia, *Cultures of Solidarity: Consciousness, Action, and Contemporary American Workers* (Berkeley: University of California Press, 1989), 238.

Chapter 9

The Machine Age

DISRUPTIVE MODERNIZATION

Modernizing dynamics, with new modes of transmitting cultural knowledge, occurred between 1850 and 1930 when the United States and Europe underwent socioeconomic transition and cultural innovation. This time marked the ending of a long-standing social reality based upon agriculture and the ushering in of an "urban modernity"[1] with an industrial-based culture formed on science and technology. Miriam Levin writes of elite-led efforts, from urban planning to the creation of new institutions, which both shaped the society and moved city-dwellers toward an appreciation of this cultural shift: "The new culture of change helped tame the social conflict and stress arising from industrialization, while creating a human-built continuum of time and space out of the very technologies and scientific ideas that fueled industrialization itself."[2] Recognizing the hard realities of skill obsolescence and efficiencies, elites within European nations established social legislation, including mandatory health insurance, and polytechnic institutes and universities to prepare the children of farmers and mechanics to enter science, technology, and engineering fields. The United States similarly established public colleges and universities through the Morrill-Land Grant Act of 1862.

Educational improvements corresponded with commercial, agricultural, and manufacturing ones, with each promising a more rationalized way to manage restructured enterprises in the new capitalist market economy. By the turn of the century, much of the natural and human resources in the United States had been plundered by precipitous industrialization. The costs could be observed most strikingly in urban production regions that had become industrial wastelands. Any number of social movements and movements of thought emerged to remedy these conditions. New institutions, together with a fully modern set of core values and distinctly urban knowledges, were created to reshape American life during the Machine Age. The pace of change accelerated in areas of industrial growth, technological advance, labor organization, and agricultural expansion. The corporation, the hospital, the skyscraper, the settlement house and the union hall, the movie theater and the department store, the public school and the playground were central to this reorientation of American society and thought.

Acknowledging this swift change of direction of what social theorist Georg Simmel characterized as the "whole of life"[3] and the varied circumstances that provoked fragmentation, a new philosophical outlook emerged that would reshape American thought. William James and John Dewey forwarded a life-affirming pragmatic philosophy within a Victorian America struggling to make sense of disruptive societal change and itself. Foremost was the widespread death and destruction wrought by the Civil War and subsequent Indian Wars, and also the Spanish-American War. The experientially based propositions of a nascent American pragmatism provided a counterpoint to the emerging urban industrial order. James understood the self as a phenomenon, a relative construct based upon one's point of view, or perspective, not solely on absolute notions. The self as constituted contains empirical and nonempirical elements, both derived from "pure experience"[4] or "the immediate flux of life,"[5] in James' words. Through a movement of thought grounded in experience rather than in a set of a priori ideals, James and Dewey set out to reconstruct thinking as a via media or middle way between natural science and moral thought.[6] Each conversed with thinkers outside of philosophical circles, broadening the pragmatic outlook with inquiries into art, religion, politics, and education. At the same time, progressive social movements developed a cognitive practice to reconstruct the public realm through social knowledge and a sense of moral responsibility rooted in experience. This form of practice clearly resonated with Dewey's belief in the power of participatory democracy, which included political and moral education to promote communicative, collaborative, and reflective activities. The dialogue between pragmatism and progressive thought arrived amid considerable "intellectual disquiet and cultural revolt,"[7] in Felix Gilbert's words, over the industrial transition, consequent demographic changes, and an impending global militarism.

Long-standing professions, notably medicine, law, science, and engineering, brought an advanced knowledge-based skilled service sector to industrial cities. Professional middle-strata practitioners offered cognitive and technical skills as a public service, while charging client fees, establishing rationalized institutions that gave practitioners greater freedoms in competitive urban markets and in their private lives. Many others in the educated middle strata directed their professional lives toward advancing the less fortunate in a changing society. Progressive educators took up the task of reinventing learning in the public school to meet these new exigencies. The entangled history of modern social movements, which opened up discursive arenas to promote critical questioning, debate, and argumentation in American civic and intellectual life, informed their efforts to enact experience-based pedagogies. In the same way as progressive social politics in city planning, modernist architecture, public housing, and social insurance, educational reform at the turn of the century was international in scope. Like other progressive reformers, Dewey's educational thought was influenced by, in Daniel T. Rodgers' words, "Atlantic Crossings"[8] of social politics and intellectual thought, notably the exchange of philosophical, psychological, and pedagogical ideas between Europe and North America. Social movements advocating civil liberties, equal rights for women and for African Americans, environmental conservation, and organized labor defined the era's core issues. European and American social movements, in reflecting the trajectory toward progressivism since the 1870s, thus accelerated reform and enlarged community contexts for adolescents to pursue individual and group inquiry in and beyond the classroom.

Project-based learning in both school and informal settings enabled youth to gain the skills to encounter, understand, debate, and meet the challenges of structural conditions attendant with the second industrial revolution that brought changes in work and personal life. Educators at the time designed and implemented fully modern experienced-based instructional modes as cognitive adaptations and a "pedagogy of resistance,"[9] following Giroux, to existing schooling regimes. This learning style provided youth with the habits of mind and emotional stability to understand social and economic challenges, and an encompassing technological domination of everyday life. Adaptive pedagogical approaches were designed, implemented, defended, and reinvented amid corporate technologies' capture of occupational groups. These educators, together with other progressives, created new approaches to teaching and learning and reformed schools and other youth-serving institutions to reduce children's risks as they struggled to survive in an increasingly complex world. Educators' efforts during this industrial transition remain instructive, for the skills gained through an amalgam of project-based learning, civic education, and craft knowledge remain valuable as youth confront a precarious present.

NEW SPIRITS

The mass distribution of industrial goods, promoted through a new vocabulary of advertising, engendered a middle-class consumer society. Most Americans underwent a change in their perceptions of the physical landscape, their bodies, and their social worlds. They stood in awe of the Wright Flyer, the Model T, industrial architecture, and the motion picture. They observed a changing urban scene peopled with new kinds of immigrants in slums and on street corners and "new women" working in offices and department stores and enjoying city nightlife and other amusements. Americans discovered, as well, that risk, danger, and catastrophe were concomitant with massive urbanization and technical innovation. Images of civil and industrial strife, of global warfare, of natural and technological disasters, of epidemic disease, of internal migration, and of immigrants from abroad flooded their senses. Most found it difficult to comprehend these distressing images, which, according to Martha Banta, contributed to the underlying anxiety at the core of the culture:

> It was an anxiety over how to determine the identity of things in a society where there were no commonly accepted patterns to guarantee confident recognition. For could knowing what was real be accomplished in a country where many were announcing the belief that cultural homogeneity had been lost? Once, they told themselves, Americans possessed clearcut categories for everything under the sun, including sex, race, class, religion and ethnic derivation. Such matters had been a muddle in America from the start, but people who spoke for late nineteenth century values insisted that there had until recently been a clarity to the national scene which was now blurred.[10]

Many turned to the new forms of mass communication to make sense of these unsettling changes on the American scene. The large-circulation magazine, the Sunday supplements of the popular press, department store displays, and poster advertisements in the city's public spaces, and later the motion picture were the domains where they observed the structural changes wrought by a machine-made world. However, these media only produced fleeting images and a mere surface understanding of the troubling phenomena that were occupying their everyday lives. The new graphic and journalistic techniques used to convey the wonder and terror of life were limiting both to the image makers and to the audience. With a commercial style that proved successful in the marketplace, illustrators and news reporters could deliver only conventional images of Victorian cultural behavior and accepted national values. The forces and ideas that were changing the society were conveyed in the popular media through stereotypical representations. However, the machine became the iconic image that defined the era, as Richard Guy Wilson states:

> For many people the period marked a new age, brought into being by the machine. From the clock that awakened one in the morning, to the flicked switch, the faucet handle, the vehicle for transportation, and the radio and motion picture, machines and their products increasingly pervaded all aspects of American life. Machines were everywhere: their impact went beyond the fact of their physical existence to challenge perceptions of both the self and the world. This new consciousness implied a whole new culture that could be

> built as readily as the machine; history seemed irrelevant, traditional styles and pieties outmoded. The machine in all its manifestations—as an object, as a process, and ultimately a symbol—became the fundamental fact of modernism.[11]

Lacking any practical training in the visual arts or the interpretive tools to understand the new iconography, average Americans accepted these conventional images as a form of truth. Such conventionalized forms could not possibly help them comprehend the sources of their own restlessness and of the anxiety that underlay American society at the turn of the century.

During this time, as Rebecca Edwards indicates: "The ideal of 'separate spheres' for men and women began to fade; young women graduated from high school and even college, took jobs in the corporate world, and led great reform movements."[12] This marked a qualitative change in the life of certain more progressive women as these growing reform movements confronted the emerging power of the nation's urban industrial life, recognizing the industrial machine's hold on people's mental life and its socially derived turbulence. The core ideas of experience, culture, community, and civic engagement, forged through aesthetic and reflective activities, were debated in women's colleges, in their clubs and parlors, and in their professional venues in the Gilded Age cities. These women's explorations of distinctly urban styles of living and knowing would influence many others in the American modernist project, including Frank Lloyd Wright and John Dewey. The New Woman of the 1890s was pivotal to the task of socializing a modern form of consciousness that built upon women's earlier advances toward gender equality. These independent women grew up in middle-class families and were college educated. Many were unmarried; however, all supported themselves in the new women's professions, such as teaching, nursing, librarianship, and settlement house work. As professionals, they discovered a public voice through their work and a common voice through associative activities with their peers. While laying claim to the rights and privileges of middle-class men, these new women were also "the daughters of the new bourgeois matron."[13] Carroll Smith-Rosenberg claims that much of the groundwork for their independent lives and careers had been accomplished by an earlier generation of middle-class women on the verge of self-discovery:

> Attending lyceum talks on feminism and on the problems of the city, made possible by the women's clubs that their literal and fictive mothers had founded, they grew up in a world already filled with separate women's urban institutions. They inherited a consciousness of women's new role possibilities almost as their birthright.[14]

New women established a political base in existing women's voluntary organizations and in the new social settlements that spearheaded many Progressive social reforms. New women realized both support and a coherent ideology in Progressive movements that advocated for their equality, voting rights, and increased civic responsibility. The feminization of government was especially critical to Progressives who sought legislative measures to limit male competitiveness and create more nurturing roles in the public sphere. The modern woman-suffrage movement took shape with the founding of the National American Woman Suffrage Association in 1890. Its leaders Elizabeth Cady Stanton and Susan B. Anthony modeled their campaign strategy for women's enfranchisement on reform politics. The new suffragists developed broad-based coalitions to campaign for state and federal legislative initiatives. They also built numerous local political alliances to educate and involve greater numbers of women. The feminist movement, inspired by Charlotte Perkins Gilman's writings on domestic politics, also promoted new roles for women. Gilman advocated the professionalization of housework and the need to socialize young children in nonsexist attitudes. She also held that with economic independence, women would take a greater role in civic culture. Both movements advocated women's leadership in efforts to revitalize the civil society and to regenerate its constituents.

The American settlement house movement took shape in 1889 when Jane Robbins and Jean Fine started New York's College Settlement. That same year, Jane Addams and Ellen Gates Starr established Hull-House, the most famous American settlement house, in an immigrant neighborhood on Chicago's near west side. The social settlements were efforts by private voluntary organizations to create healthy and productive environments in the teeming immigrant enclaves of industrial America. Social settlements originated in England during the 1880s as a response to industrial conditions. Following the precepts of social philosophers Charles Kingsley, Frederick Denison Maurice, and John Ruskin, clergyman Samuel A. Barnett and a group of university students resolved to live as "neighbours of the working poor, sharing their life, thinking out their problems, learning from them the lessons of patience, fellowship, self-sacrifice, and offering in response the help of their own education and friendship."[15] The students put their social Christianity into practice at Toynbee Hall, the first English settlement, located in Whitechapel, a working-class neighborhood on London's East End. From its beginnings, the English settlement movement assumed a role in the revival of native craft traditions. Architect C. S. Ashbee started a handicraft workshop at Toynbee Hall to teach working-class men and boys how to "apply the principle of art to materials themselves: metal, wood, leather, and paper."[16]

The settlement house movement exemplifies a transatlantic social politics rooted in social Protestantism, although in the United States it was more deeply feminized than the English settlements.[17] It was through Jane Addams' widely read observations on the impact of Hull-House on life in the immigrant neighborhoods of fin de siècle Chicago that the settlement house movement found its voice. In 1888, during her first visit to Toynbee Hall, Addams observed how the English reformers used the settlement as the base from which to introduce educational, cultural, and social reforms to their neighbors. Hull-House residents similarly brought culture to their working-class neighbors who attended the settlement's art exhibitions, lectures, concerts, classes, and clubs at the settlement. The premier American settlement also provided a free kindergarten, a day nursery, and a medical clinic and dispensary to its residents and neighbors. Hull-House, like other social settlements, was a female environment, in that most staff members and residents were women. The settlement's programs thus reflected a feminine social and moral vision, including a music school, drama and choral groups, and even a social club for working people. Its adult education programs included English classes for new immigrants, reading groups centered on the works of Plato and Shakespeare, courses in various trades, and training in dietetics, cooking, dressmaking, millinery work, and childcare.

Arriving in Chicago in 1894 in the middle of the Pullman strike, John Dewey was baffled by the disruptive tendencies of labor and capital that surfaced with the polarizing event. Dewey was awed and at the same time inspired by the dynamism of the chaotic city and its neighborhoods. As a result, he underwent a sea change in his thinking that first year at the University of Chicago. Dewey regularly visited Hull-House, where his wide-ranging conversations with Addams about matters ranging from pragmatist inquiry and the aims of democracy to the labor movement and the Pullman strike influenced the direction of his thought. Charlene Haddock Seigfried suggests that the "pragmatist shift from a detached theory of knowing to an engaged theory of understanding"[18] is a direct consequence of Dewey's experiences at Hull-House. Seigfried sees Dewey's rejection of classical liberal individualism as an outgrowth of "the model of the intersubjective constitution of the individual that Addams develops from examining the relation of personal development to social interaction among the women residents of Hull-House."[19]

Dewey was even elected a Hull-House trustee and visited the settlement frequently, teaching a class or sharing an evening meal. Dewey's pragmatic philosophy was inspired by the relational nature of sociality at the settlement, especially the mutuality characteristic of interactions among Hull-House participants. Shannon Jackson sees a direct influence between Hull-House domesticity, Chicago pragmatism, and the subsequent symbolic interactionist approach at mid-century. These framed "settlement epistemology,"[20] which understands human behavior as being shaped by social relations and the physical environment; notably mutual interpretations of the various meanings derived from interpersonal situations. Jackson writes "Hull-House Affiliates' own performance-centered models of the self, of pragmatic interaction, and of democratic pedagogy trouble a purely presentist notion of performance theory and its self-framing as a late-twentieth century theoretical formation."[21] In 1896, Dewey created the University Elementary School, later named the Laboratory School. Like the settlement house, the school was launched in the same spirit of social scientific experimentation, as Louis Menand suggests: "The Dewey School was a philosophy laboratory, in the same way that Hull-House was a sociology laboratory. It was a place, as Dewey later put it, 'to work out in the concrete, instead of merely in the head or on paper, a theory of the unity of knowledge.'"[22] By the unity of knowledge, Dewey meant knowing inseparable from doing. Katherine Camp Mayhew and Anna Camp Edwards, two teachers at the Dewey School, describe the school's cloth-making activities:

> [The student] followed the wool from the sheep to the rug, patiently contriving his own spindle, his own dye, he saw, that while successive inventions of machines have led to the eventual betterment of social life, the immediate results have often been at the bitter cost of the discarded handworker whose plight illustrates an ever-present social problem caused by technological advance for the children of this school, [industrial history] carried many social and moral implications of unsolved problems of human relationships. Thus taught, the history of work becomes the record of how man learned to think to transform the conditions of life so that life itself became a different and less tortured thing and gradually took on, for some at least, comfort and beauty. Here, for all thinking and socially minded persons, logically follows the goading query—Why not comfort and beauty for all?[23]

This lesson on technology brought students up close to an essential material, that of wool, and the ethical concerns around de-skilling in the transition from hand-to-machine manufacture of textiles, including the possibility of change.

Addams and Dewey each considered the modern city of skyscrapers and tenements as a living laboratory, an exemplary setting for gaining experiential knowledge. Chicago became the nation's most dynamic city of the late nineteenth century, one that transformed both the natural and cultural landscapes of the urban region and its hinterlands. The scale and intensity of American industrial life was most evident in the Middle West, which over three generations moved from a wilderness to an agrarian culture and then on to become the nation's industrial center. For Dewey and Addams, then, their experiments in living and learning were of a piece with the spirit of the city, itself. In *The Spirit of Youth and the City Streets*, Addams alerted educators to the needs of the young in an urban world and their alienation from the industrial order, stating flatly: "We may either smother the divine fire in youth or we may feed it."[24] Her remedies included organized recreation, education based upon practical experience, and participatory arts. Dewey, together with Hull-House settlement workers went on to craft social change by discovering new ways to instill an aesthetic dimension to the school and the society amid a turbulent urban scene.

"THE ART THAT IS LIFE"

Through exposure to and practical training in the arts and humanities, Addams and her associates sought to regenerate a community beset by the ills of industrialism. The premier settlement house thereby instilled creativity and aesthetic judgment but also a heightened social awareness that inspired residents to take direct action. With their neighbors, Hull-House residents lobbied for reform of the Chicago public schools and for more city libraries, parks, and playgrounds. Hull-House's leadership thus identified with Victorian critic John Ruskin's belief that social progress depended upon improving aesthetic perception. That both self and society could be changed through moral aesthetics, or a sensibility that is a blend of art and social awareness, perhaps explain why the Chicago Arts and Crafts Society was founded at Hull-House in 1897.

The Progressives romanticized simple elegance of traditional crafts, from hand weaving to ceramics and the "honest use of materials."[25] Hull-House and other American settlements modeled their crafts programs on the famed London settlement's initiatives. They sponsored workshops and exhibitions of immigrant crafts in the belief that "a more genuinely cosmopolitan society might emerge out of the mingling of old and new."[26] By preserving the best of the newcomers' Old-World traditions, settlement leaders believed that such immigrant gifts would lead to a greater pluralism in American life. The feminine textile arts, such as spinning, weaving, and dressmaking, would also transform immigrant women into artisans, give them a sense of pride in their craft heritage, and perhaps contribute to their adjustment to urban living. Settlement workers also hoped that native-born Americans would come a step closer to accepting the newcomers through a greater appreciation of their potential contributions to the nation's cultural life. American followers of the Arts and Crafts movement, like their English counterparts, most likely underwent a conversion experience in trying to work in the industrial city. Hull-House cofounder Ellen Gates Starr was even inspired to leave the settlement to train as a bookbinder in England under William Morris' associate, T. J. Cobden-Sanderson. Upon her return from London, Starr instituted a bindery at the settlement's crafts program. With Jane Addams, she advocated the practice of craft traditions to uplift immigrant laborers from the degradation of factory work.

The Progressive artist's search for a revival of preindustrial techniques was frequently accompanied by a strong disenchantment with industrialism. In their writings, craft reformers espoused a worldview affirming the simple life of the premodern artisan as a corrective to the perils of the machine civilization. Architect Frank Lloyd Wright confronted this seemingly antimodern vision in his talk on "The Art and Craft of the Machine."[27] In the landmark address, delivered at the first meeting of the Chicago Arts and Crafts Society at Hull-House in 1901, Wright asked his listeners to "reflect that the texture of the city, this great Machine, is the warp upon which will be woven the woof and pattern of the Democracy we pray for."[28] Later in the speech, he would challenge artists to understand and to use the machine as a conventionalizing agent or instrument of cultural progress. Wright's progressive and modernist vision was antithetical to the antimodernism of the craft revivalists. Two years later, Wright and philosopher John Dewey would convince Chicago print craft revivalist and propagandist Oscar Lovell Triggs, who had published polemics against industrial modernity, that machine technology could only enhance individual development, after which he began to espouse the liberating qualities of factory work and of industrial bureaucracy. Jackson Lears sees this conversion of craft revivalists, from antimodernist ideologue to apologist for a new industrialism, as typical of the personal odysseys of many middle-class reformers in fin de siècle America:

> The fate of Arts and Crafts ideology resulted in part from the class position and interests of its adherents. It was not just that they accepted laissez-faire economic assumptions; they were personally involved in business ventures. This accommodation with bourgeois culture grew more apparent as their business concerns grew more dominant. . . . Yet from the beginning craftsmen had shared the anxieties common to a business and professional class. They feared their own weakness in the face of restive workers and "swarming immigrants." Their concern about overcivilization combined with their commercial entanglements and (sic) led them to play conflicting roles. They sought business success at the same time they preached social regeneration.[29]

Progressive reformers but also cultural conservatives, like Henry Adams, sought out premodern paths, such as Asian mysticism, medievalism, and even Roman Catholicism, to deal with their ambivalence about technological modernity. They apparently found the real thing or authentic physical and spiritual experiences embodied in these primitive, ancient, and medieval paths to self-realization. In resolving their personal crises, they experienced a sense of self-fulfillment that served to undermine the ideal of self-denial of their Victorian childhoods. Ellen Starr's conversion to the craft revival led her to search for the real thing in the London bindery and then drew her home to institutionalize the impulse at Hull-House. Middle-class reformers brought an alternative vision back into their occupational lives and thereby helped to reconstruct the national culture. The secular society of the 1920s would promote these new forms of experience as ways for individuals to cope with their psychic conflicts. The Progressives' youthful search for authentic experience would be routinized in both therapeutic paths to life adjustment and media-driven consumerist ideology that promised personal security and self-fulfillment through mass-produced goods.

CULTURAL TRAUMA

As the new century unfolded, there remained unresolved cultural traumas across a land transformed by industrialism and fast-paced urbanization. Many immigrants and refugees, though hopeful even amid urban squalor or harsh rural living, grieved for their lost homelands. There were also individuals from families living in the nation for generations who were still haunted by troubled memories from a legacy of slavery and a civil war that threatened to tear apart the nation. A new internationalism legitimated dispatching young Americans into war zones abroad with the nation's emergence as a world power in the 1890s. Progressive internationalism also compelled the country's entry into a destructive global war fought with lethal industrialized munitions and chemicals that left tens of millions killed or wounded and traumatized by horrific battlefield conditions. Industrialized warfare arose in the Civil War when the country initially experienced this form of combat, which produced prevalent cultural traumas decades after the last battle.

The bloodshed, horror, and grief that came with the Civil War formed the social, cultural, and psychological backdrop for a generation of youth coming of age in its aftermath. The Civil War marked the onset of industrialized warfare, introducing new weaponry like the Spencer repeating rifle, a lever-action firearm, and the Gatling gun, a hand-driven machine gun that could fire more than 200 bullets per minute. Most battlefield wounds were the result of artillery ammunition and bullets, typically in the extremities. One-third of Civil War deaths were a direct result of these wounds. The 620,000 deaths documented by historians did not account for the uncounted deaths of civilians who died during the war. According to Drew Gilpin Faust, these "undocumented" deaths resulted:

> From epidemic disease that was spread by movements of population armies and other population movements as a result of the war, deaths from guerilla conflict, deaths from food shortages, deaths from what we today might call collateral damage, families that lived near battlefields, other sorts of instances of military

action that had an impact on civilian lives so that death—its threat, its proximity, its actuality—became the very most widely shared of the war's experiences.[30]

Beyond bereavement of individuals and families enmeshed in battlefield and home front tragedies, the "work of death," in Faust words, "meaning the duties of soldiers to fight, kill, and die, but at the same time invoking battle's consequences, its slaughter, suffering and devastation."[31] Although for Civil War Americans, "work" also required remembering the war dead, and further, finding the best ways to explain the meaning of the war's devastation to themselves and rising generations. Mark Twain wrote, in 1873, that the war "uprooted institutions that were centuries old, changed the politics of a people, and wrought so profoundly upon the entire national character that the influence cannot be measured short of two or three generations."[32] How survivors confronted these challenges had profound impacts on the course of American institutional life.

In the war's aftermath, a period of social disintegration ensued as the nation began the painful period of Reconstruction, with economic growth centered on industry and mechanized agriculture. A restructured relationship between the government, the market, and the country's business elite came about in these years. Entanglements between public and private sectors created the conditions for the subsequent Gilded Age when consolidation of wealth attendant with economic impoverishment of large numbers of Americans brought about economic precarity in the communities. Consequently, many in the country tried to make sense of the profound changes brought about by the war and its aftermath. Cultural trauma, according to William Hirst, "focuses on the process by which society comes to understand and interpret past suffering."[33] Anglo-American communities in both the North and South were suffering collectively and attempted to make sense of the war as an assault on their lives that left memories of pain, grief, and accompanying anxiety. Their eventual interpretation, however, would only provoke more collective suffering in the following decades, as Jackson Lears indicates, for white people would transform, and thus reframe, the meaning of the war from Emancipation to Reunion:

> The key to that transformation was a revived ethic of martial valor, an ethic rooted in Civil War memories and entangled with a developing discourse of Anglo-Saxon supremacy. By the 1890s, Anglo-Saxon militarism would solidify the connection of the white North and the white South to the exclusion of black Americans.[34]

Within African American communities in the post–Civil War period, there was the collective memory of slavery and its traumatic impacts over generations, heightened by ongoing white supremacist terror, including lynching and torture, during Reconstruction.

Reconstructionist amendments went only so far, as state and local laws enforced racial segregation in the South following the withdrawal of federal troops after the Compromise of 1877. The rights of African Americans eroded with this return to long-standing socially and culturally segregated spatial and institutional configurations in the South. Black disenfranchisement and racial segregation under Jim Crow laws, together with the legacy of slavery, were represented in the schemas and repertoires of black American political and social movements and became the principal basis for framing African American identity, itself. Throughout this divisive period, characterized by enduring cultural traumas, life everywhere became faster, with an increased pace of work driven by factory machines powered by the internal combustion engine, which also permitted railroads to speed through city and countryside. Another signal event that produced widespread cultural trauma would bring the machine into the battlefield, together with chemical weapons produced with the new industrial manufacturing techniques.

At the dawn of the twentieth century, many Americans sought ways to continue to live productively amid the carnage of the newer industrialized warfare that came with the outbreak of the First World War. In summer 1914, war broke out in Europe after a century of peace. Industry and empire had transformed modern Europe during the previous century; however, a growing militarism also spread throughout the continent as nations increased their arms purchases. Political crises, combined with the massive stores of weapons, led to war among them. Many were horrified by the destructive scale of modern technological warfare, with its immense mortars, machine guns, and poison gases. Jan Patocka understood that the war had altered the course of human conflict:

> The First World War is the decisive event in the history of the twentieth century. It determined its entire character. It was this war that demonstrated that the transformation of the world into a laboratory for releasing reserves of energy accumulated over billions of years can be achieved only by means of wars.[35]

Battlefield conditions would also be transformed by these powerful technologies. Those facing combat discovered the squalid conditions of trench warfare. The trenches were always filled with pools of standing water, infested with vermin and lice, and shattered under the constant bombardment of artillery shells. The Battle of Verdun on the Western Front illustrates the horrific costs of industrial warfare where some 40 million shells were fired in areas less than eight square miles. The longest battle of the war, fought for 300 days between February and December 2016, cost 300,000 lives with another 450,000 gassed or wounded. Cal Flyn describes the scene:

> Soon the hellfire would become normality: the constant storm of artillery, a thunder reverberating in the chests and the lungs; the earth a pulverized mass of mud and clay and jellified blood and fragments of bone, flesh, and shrapnel pounded into paste; the corpses piling mutilated in the trenches, entrails spilling—buried in the tide of filth only to be blown bare once more. Men struggling with rifles in mud-soaked uniforms, on their knees in filthy water, stepping on the bodies of the dead. Gibbering men, sent mad by fear, silent men, sinking shell-shocked, to the ground, emptied out by the horror of it, grim-faced men, following orders, fighting.[36]

This paradoxical and strikingly modern image of disaster has haunted Americans since the outbreak of the First World War. This form of imagery was first realized in the landscapes of France, which for many Americans represented aristocratic splendor, but especially after the fin de siècle, the decadence of European civilization, considered by some to be responsible for the war. Landscape and memory each reinforced the horror and the consequences of modern warfare, both in the visual images of widespread devastation to France's cities and in remembrance of the hundreds of thousands of American youths who were maimed or killed in the French countryside, itself a wasteland under the constant bombardment of military artillery.

The destructive scale of modern technological warfare with its immense mortars, machine guns, and poison gases horrified many Americans. Many others saw the war as an indicator of the success of the new form of corporate capitalism that had developed in Germany, America, and other Western economies since the late nineteenth century. For this arsenal was amassed through global financial capital but also the inventiveness and industrial acumen of the newer corporate enterprises. Convinced that the very destructiveness of the new artillery would actually shorten the war, they predicted a speedy end to the conflict. James Dawes understands that, for many at the time, skill was an "analogue for morality,"[37] for there is virtue and beauty in craftsmanship; morality likewise shines through when making something virtually. In this vein, Dawes asserts that "John Dewey, with World War I as his backdrop, tracks how control acquired through 'tools and 'technique' functions as a source for our ethical confidence."[38] Unlike Jane Addams who with many other settlement workers formed the American Union Against Militarism, John Dewey welcomed the war, as David Kennedy points out:

> (Dewey) argued that the war constituted a "plastic juncture" in history, a time when the world was made momentarily more malleable to the guided influence of reason. The war presented an opportunity pregnant with "social possibilities," which were not the direct objects of the martial enterprise, but which it might be made to yield. Dewey therefore looked hopefully to the crisis to bring about "the more conscious and extensive use of science for communal purposes" to throw "into relief the public aspect of every social enterprise," to create

> "instrumentalities for enforcing the public interest in all the agencies of production and exchange," to temper "the individualistic tradition" and drive home the lesson of "the supremacy of the public need over private possession."[39]

Dewey hoped to persuade the many pacifist progressives to embrace America's entry into the war based upon a belief that tethering of science and technology to industrial warfare advances the public good. However, many progressive reformers mobilized in a different manner. They established channels of aid for the civilian survivors of the carnage. Their fund-raising efforts resulted in dispatches of food, clothing, and medical supplies to European victims of war. This coalition of social reformers and woman suffragists, including Lillian Wald and Charlotte Perkins Gilman, also built a peace movement and, in August 1914, organized the first modern antiwar march down New York's Fifth Avenue.

The early decades of the twentieth century witnessed the expansion of diverse institutions in the United States that focused public attention on modern warfare. These included the political use of mass communications and propaganda techniques to instill a sense of patriotism on the home front, the mass mobilization of fighting men through the draft, and a global-scale relief effort organized by the Red Cross to help mitigate the suffering of both soldiers and civilians. The print media played a prominent role in shaping American perceptions of the horrors of modern warfare. Pictures of wide-scale destruction photographed from the air brought the European war into their parlors, and also into their consciousness. Images of war landscapes along the western front of France and Belgium and of the sinking of the British passenger liner Lusitania off the coast of Ireland in 1915 were widely circulated in the American press. The Lusitania incident reduced the distance between the war and the home front, forcing Americans to accept that the European tragedy was a disaster for them as well.

Many other Americans sought to understand the war's destructiveness through both secular and religious imagery. At first, the war in Europe inspired the secular imagination. Upon its outbreak, many compared it to a natural disaster, like "lightening out of the clear sky," because of its rapid onset and spread. Theodore Roosevelt compared its scale to that of "the disaster on the Titanic." Writer Henry James referred to the European war as "this plunge into an abyss of blood and darkness." Wars and other disasters have always inspired the religious imagination.

Catastrophes have been viewed either as forms of divine retribution or as signs of the millennium, and the First World War was no exception. John Milton Cooper Jr. points out that the predominantly Protestant Bible-reading public seized upon eschatological images from the New Testament to comprehend the European disaster, especially that of "Armageddon" or "the vision from the book of Revelation of the kingdom-shattering miracle that would precede the Day of Judgment."[404] These supernatural images of divine retribution and ultimate triumph over evil captured the imaginations of Victorian America. It is no wonder that over 24 million Americans had registered for the draft. By 1918, Americans had mobilized about 4.8 million men in a massive war effort, sending about 2 million troops overseas to Europe along with airplanes and motorized vehicles.

The war left over 320,000 casualties, including over 53,400 who died in battle, 204,000 wounded, and 63,100 noncombat related deaths, mainly due to the influenza pandemic at war's end in 1918. The pandemic infected one-fifth of the world and killed 50 million people, among them 675,000 Americans. War deaths amounted to only 11 percent of the country's population of 106,466,000 in 2020. However, as Hugh Rockoff states: "they had a major psychological impact, not only on the families and friends of those killed or wounded, but on the country as a whole, certainly enough to produce strong reservations about any future involvement in a European war."[41] The American Expeditionary Force to the Western Front of the First World War cost over $22 billion, although Rockoff estimates a total cost closer to $33 billion.

Over 7,000 American soldiers returned home suffering from "shell shock," a diagnosis created by the recently established field of psychiatric medicine. The psychiatrist and anthropologist W. H. R. Rivers observed that "the neuroses of war depend upon a conflict between the instinct of self-preservation and certain social standards of thought and conduct, according to which fear, and its expression are regarded as reprehensible."[42] Rivers made early contributions to the nascent field of traumatic stress studies. He was one of the first to recognize the primacy of the emotional stresses of modern warfare in the etiology of these symptoms. Rejecting a strictly organic interpretation of the war neuroses, Rivers instead claimed that they resulted from the mental strains associated with trench and air warfare. A new "forward psychiatry" developed out of the experience of treating combat stress. Its first principle was the reclassification of war neuroses as mental conditions, rather than as physical injuries resulting from blasts and other noises.

The transition to peace in 1919 found many disillusioned with the idea of global war as a way to settle conflicts between nations. A "lost generation" of young men and women returned to the states, sick of the death, violence and carnage of a savage war, and exhausted from caring for soldiers' torn-up bodies and trauma-ravaged minds.[43] Ambulance driver Harry Crosby wrote home the following after experiencing the battle of Verdun:

> I saw the most gruesome sight I've ever seen. Lying on a blood stained brancard was a man—not older than 20 I afterwards ascertained—suffering the agonies of hell. His whole right cheek was completely shot away so you could see all the insides of his face. He had no jaws, teeth, or lips left. His nose was plastered in. Blood was streaming all over.[44]

The malaise and images of death and destruction in the minds of the young returning from overseas, together with prevalent grief over lives lost in the war and the influenza pandemic, appeared to diminish hope during the country's purported "return to normalcy." Following their chilling experiences of the grotesque brutality of a European war, the young in truth returned home to widespread political repression against African Americans, immigrants, labor militants, and political dissidents.

A nascent sense of hope would still be found among American women who achieved the right to vote with the ratification of the 19th amendment in August 1920. In this polarizing moment, Ann Douglas locates an altered mood, originating among younger artists, musicians, and writers living in places like Manhattan, beyond a Victorian ethos of sentimentality and morality toward a "terrible honesty,"[4545] citing Raymond Chandler's words, that for many would come to define an emerging worldview in the 1920s. By the end of that decade, Ernest Hemingway would capture this frame of mind in *A Farewell to Arms*:

> You died, you did not know what it was about. You never had time to learn. They threw you in and told you the rules and the first time they caught you off base they killed you. Or they killed you gratuitously like Aymo. Or gave you the syphilis like Rinaldi. But they killed you in the end. You could count on that. Stay around and they would kill you.[46]

Though for most other Americans, it would take a catastrophic economic downturn and a second global war, to come to terms with this sensibility. Many in the country just wished to move beyond wartime austerity and malaise and return to everyday life. This was especially apparent in the cities where novel urban scenes, a vibrant youth culture, and a mass society influenced by the new media of radio and the movies, together with generational conflict, would define the times.

DREAMTIME AND DISCONTENT

A revitalized urban modernity and the ever-changing influences of the mass society and its popular culture could be seen in the peer-based social worlds of a rising generation of youth born after the turn of the new century. In

1920s America, adolescent peer culture supplanted home-based family life, with teenagers stepping out at night to embrace new musical styles and dance crazes or to attend Hollywood films where images of postwar situations and role models, especially emergent women, flickered across the silver screen. Lary May noted that Hollywood films and their images "screened out the past,"[47] a cinematic reorientation that moved viewers away from outdated Victorian representations of work, leisure, and sexuality toward those of modern life. A youth culture, representing by the flappers, recently enfranchised young women who broke from traditional role expectations in dress and personal style, and embraced "New Negro" jazz and dance music, would supplant the values of older generations. The younger generation, on another life trajectory altogether, came to regard their parents as embarrassingly conventional compared to the modernist images in their heads and the emancipated rhythms of their bodies.

The high school reflected this separate world of adolescence, one that created a small society, as Robert and Helen Lynd characterize the setting in their mid-1920s study, *Middletown*:

> The high school, with its athletics, clubs, sororities and fraternities, dances and parties, and other "extracurricular activities," is a fairly complete social cosmos in itself, and about this city within a city the social life of the intermediate generation centers.[48]

Peers were far more influential than the family in forming adolescent character and values, which fit right into a surfacing culture of individualism based upon personal style and consumption behavior. S. N. Eisenstadt viewed the critical role played by peers in confronting the challenges individuals face as they mature in advanced industrial societies, as Paula Fass states:

> That confrontation, Eisenstadt contends, is often through adolescent peer groups, that provide many of the emotional supports of the family, especially the security of group identification and approbation, while they train the young to respond to extra-personal performance standards similar to those that function in the larger, less personal social environment. Thus, the peer group provides emotional satisfaction in exchange for responsible action.[49]

The high school thus became the place where the young were socialized within peer-based activities, both formal and informal, to eventually take part in the modern industrial society advancing in 1920s America.

Sigfried Giedion's expression "mechanization takes command"[50] came to define the 1920s, when manufacturing productivity increased dramatically by 72 percent. Machine-made goods, like the automobile, began to replace land as the basis of both personal identity and economic security. Automobiles and airplanes were the "heralds of modernity" during this time.[51] With large-scale bureaucracies organizing more of their daily lives, Americans turned to advertising to help them decipher the transition. Advertising men were the "apostles of modernity" helping to create new fantasies about "the good life" made possible by industrialism.[52] Advertisers were not simply marketing commodities to workers and their families. They were selling the modern styles and ways of life that Americans were accustomed to seeing in the motion pictures and reading about in mass circulation magazines. Modern institutions reorganized everyday life in the nation, dissolving the conventions and assumptions of Victorian America. Industrial life eroded the communal bonds and forms of association of that era. Engineers and businessmen replaced craftsmen, farmers, and merchants as the dominant occupations, and the nascent motion picture industry promoted their epic images and dream landscapes in the new mass culture. An organizational revolution engendered new corporations, bureaucracies, and forms of regulation.

As these institutions reoriented the culture, Americans sought new ways of living to accommodate to the change in work and domestic life and an unfamiliar urban scene. American cities became a source of great expectation for the millions who moved in search of the American Dream, but also of great anxiety for native white Protestants concerned with the prospects of increasing disorder and imputed "immorality" among the urban masses. The metropolis became a contested terrain for a number of troubling issues, including immigration, crime, alcohol use, birth control, mass leisure, and the erosion of the work ethic. Many native-born Americans were late Victorians and especially suspicious of the divergent lifestyles of: "New Negro" communities in New York's Harlem and on Chicago's South Side; New Orleans' and Kansas City's Jazz Age nightclubs; Manhattan's Greenwich Village's bohemia, Santa Fe's artist colonies; and Los Angeles' amoral movie colony. Above all, it was the "New Women" who were writing, painting, and living their lives in these places that especially provoked them. For it appeared to many that these free-spirited, creative, and outspoken women were not amenable to the forms of control and moral order that had guided middle-class family and community life since the Civil War. City life and its disruptive modernist culture drove many late Victorians into the suburbs.

The middle-class desire for a safe place to live was shaped in the late nineteenth century when landscape architect Frederick Law Olmsted viewed homeownership in middle-class suburbs as "a civilizing practice." Members of the middle class at this time were seeking a safe place, and suburban home ownership seemed exceedingly practical. Olmsted took it a step further to represent more than the practical and, in fact, shaped the movement to the suburbs as being ideal. Olmsted declared that within suburban enclaves, members of this class, "stand in the vanguard of our civilization . . . they are our stronghold against agrarian and nihilistic tendencies. They are the best security we have."[53] Seeking a sense of security and safety some distance away from big city life, the suburban areas of the nation's 96 largest cities grew twice as fast as the core communities in 1920s America.

A number of reasons explain why suburbs surrounding older industrial cities chose to emphasize their distinctiveness by incorporating rather than consolidating their relationship with the city by agreeing to annexation. First, legislation passed that facilitated the incorporation of suburban communities, which could now offer residents vastly improved municipal services, including roadways, suburban trolley and train lines, electrical grids, water purification, sewage, wastewater, and garbage removal systems. Second, there was a 150 percent increase in automobile registration over the decade. The automobile was chiefly responsible for the suburban boom, but so were increased wages and lower housing prices.[54] The search for a better quality of life also led affluent, native-born Americans to suburbs far enough away from the foreign and disorderly elements in the city. The country was strongly segregated along racial, ethnic, and class lines. Kenneth Jackson suggests that class-based and nativist assumptions were at work in suburbanites' choice not to identify with the metropolis.[55] Hence, many other middle-class families organized themselves in reaction to these disturbing urban social trends.

The new Eastern and Midwestern suburbs, such as Grosse Point near Detroit, Elmwood Park near Chicago, and Nassau County on Long Island, offered a respite from the Machine Age Babylons that most Progressive Era cities had grown into. Though the anonymity of urban life also afforded a break, albeit for a long weekend, from suburban conformity and neighbors' watchful eyes. In *The Great Gatsby*, F. Scott Fitzgerald's Long Island–based narrator Nick Carraway is both repulsed and captivated by the charmed life in Manhattan of his friends Tom and Daisy Buchanan. While gazing out their apartment window as darkness falls, Nick reflects upon this ambivalence amid the urban bustle and allure of the city's diverse urban scenes:

> Yet high over the city our line of yellow windows must have contributed their share of human secrecy to the casual watcher in the darkening streets, and I was him too, looking up and wondering. I was within and without, simultaneously enchanted and repelled by the inexhaustible variety of life.[56]

Nevertheless, the suburb, with its image of tranquility and order, remained the perfect setting for middle-class families to continue their Victorian domestic arrangements.[57] For the wife, the suburban household provided a comfortable environment to care for children and to attend to her domestic duties. As men's public lives in business and the professions became increasingly controlled from above, they attempted to remain in control of the one place that they could call their own, namely the suburban home. The new entanglements between the burgeoning modernist cities and their suburbs to many seemingly offered a late Victorian relief from urban tumult.

In viewing the hasty growth of "dormitory suburbs," mechanized workplaces, technocratic city planning, and urban degradation of the natural environment, Lewis Mumford challenged what were forwarded as solutions to life in an industrial civilization.[58] Mumford spoke up for a younger generation that expressed discontent with capitalism and industrialism, lamenting standardization in machine production and technology's broader impact on the human condition, including a "devitalization" of urban life. He mourned the loss of craftsmanship and the dignity of labor in automated work and observed that the nation had become an employer-dominated world and that its managerial styles pervaded much of the society. His critique especially centered on Dewey's acceptance of a dominant belief in Machine Age America, namely, that science and technology are benevolent progressive forces within society. To his mind, Dewey forwarded an instrumental logic as a mode of inquiry that resembled technological activity. For Dewey, however, inquiry remained a productive skill requiring technical precision and tools of all sorts, both conceptual and physical. Following Aristotle, knowing itself, was for Dewey a "technological artifact," as Larry Hickman suggests.[59]

Throughout the 1920s, Dewey set out to reconstruct pragmatism as a public philosophy, for he came to view the experience of community as the basis of learning that would help direct participants toward both self-discovery and social reconstruction. He was also developing a naturalist and holistic philosophy privileging an aesthetic and introspective culture of lived experience and consciously informed social practice. Dewey's pragmatism provided a link to nineteenth-century Progressivism and a strong intellectual foundation to ground communitarian ideals of citizenship and a participatory democratic culture in the Machine Age. Along the way, he moved pragmatist thinking beyond that of William James, who held to the transiency and malleability of the human condition but remained steadfastly against either absolute explanations for or solutions to the human predicament. Moreover, amid the finitude of the human project, James believed that individuals could diagnose their lives and infer both knowledge and values from their experiences of the world. Dewey accepted James' notion of a self constructed entirely out of the flow of experience, but he required that any personal knowledge derived from such experience be realistic and consistent with institutional life. The pragmatism of James and Dewey with an emphasis on lived experience, in Martin Jay's words, "was a part of the broader crisis of modernist subjectivity" in European and American thought.[60]

Dewey formulated a philosophical position that saw both subjectivity and collective experience as valid paths toward knowing and understanding cultural modernity and its uncertainties. He held out the promise of reconstructing both self and world through involvements within institutional settings characterized by competing demands and conflicting interests. However, the pragmatic activity of self-diagnosis emerging from such involvements threw Dewey into a quandary with respect to its possibilities for organized social action. In one respect, the individual as a knower is rewarded with sufficient insight for having mastered a complex situation, but such truth is at best tenuous and far from universal in its consequences. For Dewey, the individual's claim to truth, gained through experiential knowledge, can retain its significance only in the light of future consequences. He frequently cited a central tenet of the pragmatic sensibility, in John McDermott's words, "that only consequences can effectively validate the truth or the propitiousness of a claim," to demonstrate how well-intentioned efforts at social reconstruction could yield unsatisfactory and even contradictory results.[61]

As a corrective, Dewey set forth a principle of "creative transiency" that points to life's journey, rather than its destination, as the chief source of significance and personal meaning.[62] Following James, he viewed the experience of relations and connections as the starting point for any endeavor undertaken either for personal knowledge or to infer some truth about the world. In the following passage from *Experience and Nature*, Dewey examines the transactional quality of ordinary experience:

> These commonplaces prove that experience is of as well as in nature. It is not experience which is experienced, but nature—stones, plants, animals, diseases, health, temperature, electricity, and so on. Things interacting in certain ways are experience; they are what is experienced. Linked in certain other ways with another natural object—the human organism—they are how things are experienced as well. Experience thus reaches down into nature; it has depth. It also has breadth and to an indefinitely elastic extent. It stretches. The stretch constitutes inference.[63]

The universality of this capacity to have an experience is a central tenet of Dewey's mature philosophy. Such experience is grounded in and cultivated through people's dynamic engagement with their world. For Dewey, the meaning of a particular element, notably an object, another person, or an event, is gained only in the context of the engagement itself. He proposed a theory of culture that would identify how individuals' interactions with nature and their involvements with one another might occasion what he calls a consummatory experience. Through such ongoing involvements within a relational world, an individual may experience an extraordinary event that could lead to personal growth, spiritual vision, or aesthetic insight.

Dewey considered the aesthetic as elemental to all experience, believing that aesthetic experience must not be limited solely to judgments about artistic phenomena. This primacy of the aesthetic is best understood in his distinction between perception and recognition. Recognition is experiencing something and relating it back to what we already know. It alters neither our understanding nor our feelings for the phenomenon. Perception, by contrast, indicates an active experience of something and a discovery of its inherent character. For Dewey, the intrinsic qualities of phenomena are what most affect the experiencer and thereby influence a person's interpretative understanding. Moreover, only through perception can we achieve new insights that allow us to modify our previous habits or conceptual schemes. It is thus primary to aesthetic experience and grounds both learning and psychological growth.[64] Dewey held that aesthetic sensibility, rather than rational calculation, should guide our lives. He promoted an aesthetic ecology, based upon affective feelings and gestures, especially for educational and political institutions that were transitioning into bureaucracies. Dewey wished to replace their stultifying rationality with an appreciation of the aesthetic drama of the ordinary, in John McDermott's words, which celebrated, rather than denigrated, peoples' everyday experiences of the world.[65] In the following passage from *Art as Experience*, Dewey points out how personal identity and the sense of meaning are shaped through mundane events:

> In their physical occurrence, things and events experienced pass and are gone. But something of their meaning and value is retained as an integral part of the self. Through habits formed in intercourse with the world, we also in-habit the world. It becomes a home and the home is part of our every experience.[66]

Here, again, Dewey considers the interplay of the imagination and the environment in creating a sense of personal belonging. Dewey points out that insights drawn from ordinary experience cultivate habits, which, in turn, contribute to our sense of the world as a home. Through shaping events to our intentions, we learn more about ourselves and of our capabilities of dwelling in the world. With this self-knowledge, we may begin to relate across to the farther reaches of our world, perhaps to embrace other forms of life and culture shaped by different habitual and mental sets.[67] Dewey's critics, most notably Mumford, rejected his impulse to reform society, viewing pragmatism and progressivism as merely modifications of an essentially Victorian worldview.[68] Mumford called for a culture of personality and self-fulfillment based upon creative ideas and more authentic experiences in place of the Victorian culture of character founded upon self-denial and rigidly defined personal roles. He held that creativity and authenticity within a regional culture would both reconstruct a nation being degraded by industrialism and give meaning to the lives of men and women, and further held out hope that a creative release could bring about broader renunciation of a vapid and mindless consumer culture. Though during this same time industrialized work, everyday consumption, and the modernist city had transformed the tenor of life in the Machine Age.

MONETIZING THE MOMENTUM

The second industrial revolution marked the emergence of human systems capable of large-scale intervention into natural systems. The period, characterized by acceleration of the mutual feedback between two forms of knowledge, science, and technology, yielded greater control over the earth's resources. This includes subterranean exploitation of energy and metals; vertical expansion into the atmosphere and oceans for nitrogen, minerals, and food; and control over areas of the electromagnetic spectrum. This last transition brought with it the power to transform the planet into a "whole system."[69] The industrial transition accelerated development of manufacturing technologies, notably of vehicles powered by the internal combustion engine and of chemicals, from industrial and agricultural compounds to pharmaceuticals. Agricultural productivity based upon new husbandry, new tools and chemical fertilizers led to human population increase and better living standards for the middle and working classes. Precipitous urbanization and industrialization during this time impacted the soil, water, air, and other common resources. These advancements reoriented life in the countryside and created complex technological and organizational networks in the cities and their surrounding hinterlands.

The United States sought industrial technologies to move from manual to machine production. Though labor short, industry intensified under the factory system, as did the country's expansion into territories rich in resources but incapable of managing their distribution. Engineers thus created global land and sea infrastructure to move raw materials and distribute finished commodities. The second industrial revolution was based upon navigating greater complexity, administered through systems of capital transactions at differential levels. Markets were modifiers, facilitated by transportation technology, financial instruments, and remote exploitation to meet demand. Technology was controlled by capital consolidation and accumulation and concentrations of economic and political power.

Monetized growth of an industry was a key indicator of success. Company securities publicly traded on financial exchanges, and futures and options trading for agricultural goods on commodity exchanges varied on the basis of monetary value. Economic metrics became the manner used to calculate the value of each good. Commodities earlier negotiated through bargaining transactions became rationalized, through pricing, and objective measurement. Private transactions came to follow a rational pattern of market-based bid and ask prices. Accompanying the transition were standardization practices in the manufacturing of machine tools, steel rails, telegraphy, and the telephone, together with a standardized time system. Money-based economic indicators were used to calculate the well-being, or prosperity of Americans and as measures of forward momentum, or progress, of the society as a whole. The country's monetary policies were originally meant to reform institutions and thereby reframe American life in commercial terms to encourage manufacturing and industry. Over time these policies gave shape to a "monetary logic" that assessed the worth of occupations, institutions, and ways of living. As James Livingston suggests at this time, "the debate over the definition and uses of money was in effect a debate over the conditions of selfhood and self-production."[70] With the assigning of monetary value came new personal, occupational and group identities. As a consequence, significant domains within the country were monetized. Land, labor, capital, and knowledge took on abstract value through a price revolution accompanying the market revolution.

Capitalist development also brought about countless crises in the late nineteenth century, when boom and bust characterized the business cycle of precipitous economic expansion and contraction, and with limited state regulatory controls. Huge sums could be made or lost. The agrarian crisis of the 1880s brought falling prices and narrowing markets, clearly impacting the standard of living and quality of life for rural families. These precarious conditions, especially the sense of economic insecurity, threw the rural family into a conflict surrounding their traditional approaches to farm life, although a subsequent world economic crisis would eventually render many rural and town-based patriarchal and patronage arrangements obsolete. The capitalist global crisis in the late nineteenth century culminated in the Great Depression of 1893. The depression was a signal event brought about by a changing economy characterized by massive productive capacity and great concentrations of financial and productive power. Accelerated by failed firms and falling stock prices, the fin de siècle depression swept away long-standing traditionally based social worlds and cultural practices worldwide. Production, consumption, and distribution systems across the globe were penetrated by or integrated with market capitalism. Remaining were remnants of lifeways that would soon disappear as the new capitalist arrangements turned traditional households, farmsteads, artisanal shops, and entire communities upside down.

American industrial capitalism's resilience depended, in part, upon the ability to weather crises through rationalized institutional practices and adaptive economic behaviors.[71] Businesses centralized, creating formal administrative systems with sophisticated accounting and purchasing departments. The reshaped enterprises swept away thousands of nonperforming firms and inefficient labor practices, consolidating countless industries. Managerial capture of a rationalized workplace and monetary policies regulated the new economy. Together, these dynamics influenced the course of institutional life, from firms to farmsteads, and eventually households. Scientific management moved beyond the industrial workplace, as corporate elites drew upon Taylor's principles to move their white-collar managerial and professional firms toward greater efficiency and productivity. North American corporate consolidation through rational management practices necessitated governmental reform, toward an "associative state"[72] that grew in scale and scope during the first decades of the new century. An associative state denotes a partnership between corporate firms, industrial associations, trade unions and the government, each appropriating scientific management to assure greater efficiencies. As cooperative institutions, each would optimally seek self-governance and commit to nonstate solutions to public problems. The "American system,"[73] as envisioned by then commerce secretary Herbert Hoover's words, would create a new "associational order"[74] that contributed to a rise in living standards and limit industrial conflicts, assuring a "harmonious community of interests."[75] Once in place, the associative state adopted managerial practices of the giant corporations.

Professional–state relations under these arrangements also changed the culture of professionalism, notably within medicine, law, engineering, and the university professorate.[76] All reorganized their disciplinary associations to further professional autonomy, control over entry and practice, and efforts to raise standards of education. Rationalized protocols included university-based training, credential and licensing requirements, reformed both medicine and law. Although their guild power was more limited under centralizing conditions of market capitalism, federal and state courts and regulatory agencies protected their professional powers and market control. Sharing a common organizational structure and rationalized operational procedures reorganized corporate firms, the professions, and public bureaucracies could better articulate their distinct interests under increasing regulation. Structural changes in both the state and society thus accompanied the immense economic growth attendant with the industrial transition. These included an accelerated pace of widespread urban industrial capitalism and an expanded market. The monetization of laborers' services and financial capitalist wealth as income produced a growing tax base. In view of these changes, a new system of national public finance was enacted with a progressive tax on both income and wealth transfers.[77]

Consequently, associative state agencies began to employ the same calculative rationality guiding the large-scale industrial corporations in their governmental revenue-generating and regulatory practices.[78] The various structural rearrangements brought about by the second industrial revolution disrupted family, schooling, apprenticeships, and work, which over generations were rooted in distinctive patterns of urban and rural life. Many experienced, firsthand, the ills of industrialism especially the misuse of natural resources in extraction and production methods, seeing their distressing consequences on their physical environments of both the cities and the countryside. As it progressed, the industrial transition engendered a new way of economic life, namely Fordism, which denoted a set of sociotechnical processes designed to increase production efficiency that would come to define an era.

Fordism describes the mass production techniques used to manufacture consumer goods.[79] The key to Fordist production was standardization. Ford originally produced only the black Model T. Building one model meant lowered costs and increased control over every aspect of production. Ford engineers designed single-purpose machines that required relatively few skills to operate. Workers were trained to perform only a few simple operations, while the machinery determined the pace of work. Although the assembly line was an efficient production method, it was the chief source of worker discontent and alienation. Ford plants experienced high rates of worker turnover, and this proved costly to the company because new employees had to be trained before they could take their position on the line. To counter both turnover and the threat of unionization, Ford offered workers an unprecedented five dollars per day and reduced their workday from nine to eight hours. As workers became progressively "de-skilled," turning over control of the productive process to the manufacturer, they were rewarded with greater purchasing power and more time to enjoy the mass-produced commodities of the Machine Age. In 1910, there were fewer than 500,000 automobiles and trucks on American roads. Ten years later, nearly 10 million were registered, and there were over 26 million motor vehicles by 1930.

During these years, electrification of American homes created a market for mass-produced household goods, like the washing machine, the mechanical refrigerator, and the vacuum cleaner. The telephone and the radio also became indispensable to life in Machine Age America. By 1925, there were 571 radio stations and 2.75 million receivers linking listeners to a network of mass entertainment and advertising. However, telephones and radios also contributed to greater individualism and separation between people in a period of considerable uprooting and resettlement. The phone and the radio were also relatively passive media, requiring only a simple movement to operate, like turning a switch or lifting a receiver. Like the assembly line, they seemed to render individuals more passive in their social relations, expectant of control from above, and dependent upon the guidance and direction of experts.

The creation of a class of urban manufacturing and clerical workers with surplus income drove a consumer revolution. The family everywhere was commodified as a household, becoming a market-targeted consumer unit. Local, state, and federal governments taxed households based upon income. Through retail marketing and advertising, the various commodity merchants sold families "necessary goods," or "consumer durable goods"[80]—washing machines, sewing machines, radios, and for many automobiles—that were taxed as well. Influenced by media and family-based social networks, certain commodities became viewed as necessities throughout the lifecycle from social events, such as proms, parties, showers and weddings; to stages of life, including pregnancy and childbirth; infancy, childhood and adolescence; and adulthood, including caring, healing, dying, and death itself. Many working families then quickly adopted media-influenced notions of belongingness as depicted in advertisements. In both city and suburb, families embodied these commercialized sentiments through shopping for, preparing, storing and consuming branded manufactured and agricultural commodities. Progressive efforts in support of family life included the United States Children's Bureau, established by Congress in 1912, which immediately produced booklets for parents entitled *Prenatal Care* (1913) *Infant Care* (1914), and *The Child from* 2-6 (1915) and on through to adolescence.[81] Progressives also enacted compulsory public schooling to assure appropriate socialization of the young for productive roles within the new industrial society. These initiatives provided the many millions of recent immigrant families and rural families from domestic farmsteads to advance their own well-being and their children's life chances in industrial cities.

TRADITION AND MODERNITY RECONSIDERED

Successive waves of mass immigration from abroad and internal migration from agrarian regions expanded employment in the industrial sector from 2.5 million in 1880 to 10 million workers in 1920. Along with manufacturing, agriculture was industrialized, transforming land, labor, and capital in the countryside, bringing disorienting conditions, including displacement and uprooting of populations followed by relocation in modern cities. The changed rural areas they left behind were the prototypes of suburban life just outside the geographically expanding, dense metropolises. Even before the Great Depression of 1893, the Southern United States underwent shifts from traditional agriculture and cottage industry to more capital-intensive modes of production. The agricultural village and small town were thus altered, and industrial cities in the New South became central to the region's economic growth. Throughout the 1890s, African American and white southerners moved into industrial cities of the New South or migrated to the North.

African Americans then migrated in larger numbers from the rural South to cities in the North and Midwest at the time of the First World War. Between 1915 and mid-1917, nearly a quarter million black southerners moved North to take jobs left unfilled by men who had been drafted or by immigrants who had returned to their families in Europe. The first wave of the "Great Migration" continued until 1930, with more than 1 million people leaving the South, moving as far west as Los Angeles.[82] Around 1910, the struggle for African American rights was carried out by the National Association for the Advancement of Colored People (NAACP). W. E. B. DuBois and a group of middle-class educated black professionals had earlier formed the Niagara Movement to counter Washington's conservative political and economic agenda. Members of the Niagara Movement and of the allied Constitution League founded the NAACP as a legal action organization to support the interests of a college-educated African American elite.[83] The organization drew widespread support from white Progressives, including settlement movement leaders Lillian Wald and Jane Addams. By 1914, the NAACP had 6,000 members in 50 branches, and its publication, *The Crisis*, edited by Du Bois, boasted a circulation of more than 31,500. By 1919, the organization's membership had increased to 91,000; however, it had failed to win the support of poorer African Americans in the North and the South. New York City's Harlem in the mid-1920s would have a population of 150,000, making it "a city within a city" that created an artistic renaissance, albeit with an African American population divided between upholding their African heritage and embracing the American mainstream.[84] Du Bois understood how this dichotomy resulted from a double consciousness, or, in his own words, "the sense of always looking at one's self through the eyes of others, of measuring one's soul by the tape of a world that looks in amused contempt and pity."[85] Clearly there was the necessity of forging a character commensurate with a precarious life situation, especially in the face of racial stereotyping encountered in interpersonal relations and in negotiating with the wider society. For Du Bois, "The problem of the twentieth century is the problem of the color-line."[86] However, it would take another three decades before the American civil rights movement mobilized African Americans and many others to confront racial inequality and forward social justice.

During these same decades, larger and more powerful fast-moving ships carried mostly unskilled laborers displaced from the European countryside to American shores. Between the 1880s and the 1920s, shifting demographic processes catapulted Americans into the global community. In the mid-1880s, immigrants from southern and eastern Europe began to arrive in this country. From 1880 to 1930, 4.5 million people emigrated from Italy; 4 million came from Austria, Hungary, the Balkans, and Greece; with another 4 million arriving from Russia and Poland. By 1910, 38 percent of America's foreign-born population originated from these countries. Transnational migration within North America also increased after the turn of the century. Between 1900 and 1919, 185,000 Mexicans crossed the border to work primarily as casual laborers in the cotton, rice, and beet-sugar fields of the Southwest, particularly Texas, California, Arizona, and Colorado. Over the next decade, this number increased to about 750,000. Mexicans also worked for the railroads and in the mines, with some moving on to take industrial jobs in the Midwest. By 1930, 1.5 million English-speaking and French Canadians had crossed the border. Most Canadians came to work in the forests of the Great Lakes and the Far Northwest, but also in the textile mills of Northern New England.

The demographic movements across national borders provoked feelings of nationalism and nativism in the old stock population and polarized the country around ethnocultural differences and their manifestations in city life, and in the countryside, itself facing troubling economic consequences. The postwar decades were years of extremes, characterized by a boom in industrial cities where the consumer goods were produced and marketed, but hard times in rural areas where conditions in farming, coal mining, and the textile industries remained depressed throughout the 1920s. Many rural families, both poor whites and African Americans, who could no longer get by in the countryside, moved to the cities that accounted for fully one-half of the nation's population. Amid the turbulence, newcomers to industrial cities called upon their earlier civic and communal traditions, as artisans and agricultural workers from a fast-disappearing rural society to devise new forms of urban culture, including ethnic-, work- and faith-based mutual aid initiatives as strategies for survival.

As the nation's urban reform agenda moved to a civic culture forged in the image of business, the Progressive impulse was institutionalized in the national government during the New Era of the 1920s. At the same time, many of the Progressives who had tirelessly pushed for legislative action before the First World War became depleted of their political energies after the war. Progressive leaders, who espoused an older liberalism, also became disenchanted with the way their reforms had been routinized by new liberal politicians. The urbanized nation with its mass culture was sweeping away the Victorian past and its evangelical traditions that helped shape the Progressive worldview. The old reformism espoused by native-born Americans from the small town was giving way to a more urban liberalism that drew its strength from diverse groups within the Machine Age metropolis. Throughout the decade, the Progressive legacy endured in the governmental agencies and nongovernmental organizations established in previous decades. For professionals and agency staff who still retained a progressive impulse, a formidable task was that of assimilating millions into a modern urban industrialized society.

Coming into industrial cities were those displaced from the American countryside and agricultural workers from faraway homelands that had recently arrived on American shores. All sought opportunities for factory work and an escape from rural economic precarity. Native-born Americans and immigrants, alike, became modern by their adaptation to urban life and its rationalized institutions, notably an industrial work schedule and, for their children, the public school. Working people gathered together in a traditional religious and fraternal sphere within ethnic neighborhoods surrounded by modern commercial districts that contained factories, offices, schools, union, and meeting halls that formed a public sphere. They were directed by Progressive reformers on appropriate ways to use these spaces to renegotiate their social worlds and reform their identities after a disorienting set of circumstances attendant with uprooting and relocation. Modern economic institutions that centralized and coordinated production and distribution activities transformed the world, while social movements mediated the crises wrought by these profound changes. In their efforts to give voice to people either thrown together or displaced by industrialization, Progressive movement organizers created institutions attendant on a modern vision of individual freedom and voluntary association, such as trade unions, cooperatives, and political parties. Within these organizations, individuals experienced and came to understand the project of modernity as they learned concrete social knowledge to replace old beliefs or obsolete skills. Late nineteenth-century social movements became, in Alain Touraine's words, "The bearers of global meaning, of an image of the society, and they are not confined to the limited world of protest, claims, and reforms."[87]

Samuel Hays and Robert Weibe, respectively, characterized the Progressive reform movement of the time as "a gospel of efficiency"[88] and "the search for order,"[89] from conservation in the countryside to regulating the city and its disorders through organized school systems, social settlements, juvenile courts, and supervised parks and recreation venues. Similar to Taylor's scientific management practices, Progressive reforms were individualistic change strategies that directed those recently encountering modern industrial life toward accepting their new situations in the city. Progressive leaders thus became ministers of reform with a moral vision partly derived from youthful influences of evangelical Protestantism and the social Christianity of the abolitionist movement. Progressive programs of reform adopted an evangelistic style characteristic of religious modes of status passage as a remedy for individuals caught up in disruptive social and cultural circumstances. Reform-oriented organizations' missions were those of regenerating the demoralized so that they could gain control over their own destinies, much like earlier Protestant sects, where the change process was endowed with both spiritual and communal meanings. Arthur Vidich and Stanford Lyman see that "regeneration not only was a religious matter but also bestowed upon the chosen a right to vote and to hold office. It meant acceptance into the community, equality of status and respectability, and civic standing."[90] The spirit of the Progressive project was to socialize fully modern roles and careers that would reorient and direct individual lives amid the chaos of industrial life.

Progressive movements thus enlarged the public sphere, for out of their secular revivals came a reconstructed moral economy. The new routines and identities that would organize life in modern society were frequently constituted and then endlessly negotiated within the public spaces of reform organizations. These were the meeting rooms of settlement houses and union halls, public school classrooms and auditoriums, hospital and clinic waiting rooms, and municipal courtrooms. Immigrant laborers, displaced artisans and farmers, and their children gathered in these places to learn and negotiate the styles of social inquiry and political discourse requisite for a transition to technological modernity. They emerged from these organizations regenerated as trade unionists and activists, citizens and clients, students and professionals. Many men and women would embrace a secular calling, or career, but also a new personal identity through which to view and understand the modern world and their place within it.

Along this course toward modernity, there was a growing divide between urban industrial values and defenders of traditional cultural values that would eventually give way to the material capture of the American household. Steadfast traditionalists upheld an outdated ideal of the household, admonishing opposing positions, at times even castigating critics as disrespectful of the home as the family's "sacred" domain. These attitudes were especially prevalent in the countryside, where the population was declining together with the drop in demand for farm products after war, leading to hundreds of thousands of farm bankruptcies. In small towns, traditionalist' resentments and envy of city life took on anti-urban sentiments and gave rise to a religious fundamentalism and an antiscientific worldview, notably against Darwinian evolution. Amid the rancor, the wealthiest triumphed, while the traditionalists argued incessantly on behalf of a household that no longer existed, seemingly unmoved by the reality of a family life captured by governmental, financial, marketing, and advertising sectors for at least a generation. In actuality, the traditional family was caught up in the worldwide economic upheaval that altered the farmstead, the village, and agrarian life itself. By 1920, only one-quarter of the country's workforce was agricultural, with the urban factory and office as scientifically managed workplaces unseating the farm as the center of the economy. Many rural youth were leaving their farms for city work. For those remaining in the countryside, Progressives supported efficient management of rural family life in the transition. The U.S. Department of Agriculture (USDA) Cooperative Extension was created by the Smith-Lever Act of 1914 as an outreach effort to educate farmers on advances in agricultural practices and technology.[91] The USDA Bureau of Home Economics, established in 1923, supported the work performed by homemakers and developed home economics handbooks for kitchen design, fabric selection, garment and pattern selection, food preservation, nutrition and healthy food choices, and to youth through 4-H clubs. All were originally designed to move rural families toward greater efficiencies. With county offices in every state and in partnership with the public land grant universities, Cooperative Extension educators provided outreach to rural households and farmsteads.

Industrialization and attendant modernizing social and cultural processes clearly affected the life trajectories of newcomers after migration to American cities and the countryside. Though for those arriving in the early decades of the twentieth century, following Reinhard Bendix, "the growth of citizenship and the nation-state" were significant dimensions of modernization.[92] New immigrants and their children, mainly urban Catholics, Lutherans, and Jews, would come to grow national political identities through motion pictures and the broadcast media, beginning with the 1932 presidential election. Franklin D. Roosevelt campaigned and won the presidency because of widespread public discontent with Herbert Hoover and New Era capitalism. Roosevelt captured the public's attention in movie theater newsreels and on the radio where his speeches were listened to by millions. Seeking needed support from traditionalist communities, Roosevelt appropriated certain rhetoric from agrarian populists to justify progressive economic reforms, promising "new deal" for "the forgotten man at the bottom of the economic pyramid."[93]

At the time of the Great Depression, most unskilled and semiskilled workers in mass production industries were not organized in trade unions. In the 1930s, two major events signaled the breakthrough to mass production unionism. The Wagner Act of 1933 led to the establishment of the National Labor Relations Board.[94] This legislative sanctioning of collective bargaining rights and the formation of the Congress of Industrial Organizations (CIO) paved the way for unionization of mass production industries. With the outbreak of the Second World War, defense production boosted corporate profits, but also the number of union contracts. As a result, industrial workers saw vast improvements in their hourly wages and benefits, in seniority provisions, and in grievance procedures. Membership in the CIO more than doubled between 1940 and 1941. By 1945, most of the mass production industries had been unionized.

Roosevelt and his advisors invested heavily in public works projects. New Deal projects were "experimental and incremental—not ideological," as Lizbeth Cohen suggests.[95] Federal infrastructure initiatives followed a rational course; should a specific project approach succeed, then it would move ahead, as in the case of hydraulic efforts. The Works Progress Administration built waterworks and dams in the Tennessee Valley, providing electricity and regional economic development in an impoverished region. In the American West, federal efforts constructed dams along the Columbia and Colorado rivers to control flooding, generate hydroelectric power, and channel water to both farming and urban industrialized areas.[96] Enclosure of the hydraulic energy commons on the Colorado Plateau, Tennessee Valley, and the Columbia River region, a technological transformation facilitated by the federal government, mandated dispossession of communities and their inhabitants and the federal financing of dam building on a massive scale. New Deal hydraulic initiatives paved the way for subsequent wartime strategies on the homefront to support the war against fascism. Hydraulic projects facilitated the growth of wartime production regions to build the tanks, arms, ships, and aircraft that would win the war. The hydraulic power of these regions allowed the United States to become competitive with German industry, plane for plane, bomb for bomb, tank for tank, and ship for ship. Investments in infrastructure were never damaged or destroyed during the war, forming the material underpinnings for a productive postwar recovery characterized by phenomenal economic growth.

Before entering the global war, New Deal crisis managers' decisions within certain strategic arenas, notably employment, unionization, and infrastructure, helped assure an Allied victory in Europe and the Asia-Pacific region. With the peace, the federal government grew increasingly more powerful because of the contractual relationships sustained during the war with the health, energy, defense, building and commercial sectors, all part of the great acceleration that formed the Anthropocene. Private industries profited by producing goods for both households and public entities, the latter being assured by governmental largesse in contracting arrangements. Through such public–private partnerships, American cities, notably New York, Chicago, Seattle, and Los Angeles experienced a high modernity and a sense of technological wonder that was only a promise during the Depression years. With the planned renewal of postwar cities, the nation's industries moved beyond wartime production toward overproduction of cars and other commodities.

Still, traditionalists continued to rail against modernizing trends and to defend a remembered world of the village and small-town life. These social types raging against the current were represented as both comic and tragic rustic characters in early films. Coming of age in the interwar years, their children became avid consumers of the era's commercial goods and the films from Hollywood's motion picture industry. Through popular movies, rising generations viewed modern representations of characters and situations that helped to move them beyond their parents' tradition-bound lifeways and outdated expectations. Peer-based group life, initially during high school, and then in industrial plants and offices, reinforced fully modern ways of living. There, they learned more rational approaches to confronting and resolving personal and social challenges, including traumatic losses and economic shocks. These succeeding generations also soon grew local and national political identities as they suffered together with the rest of the country through the hardships wrought by a worldwide economic depression. Many then took part in a total war against tyranny, serving in uniform on warships and distant battlefields, and in civilian efforts on the home front. Honored and rewarded with social benefits for their sacrifice and service, they were apprised, repeatedly, of the need for steep spending on peacekeeping and for the economic recovery of Western European nations and Japan through Marshall Plan and World Bank reconstruction efforts. They would be swept up in the postwar boom of economic expansion and accelerated growth that brought prosperity to the country.

NOTES

1. Miriam R. Levin, Sophie Forgan, Martina Hessler, Robert H. Kargon, and Morris Low, *Urban Modernity: Cultural Innovation in the Second Industrial Revolution* (MIT Press, 2010).
2. Miriam R. Levin, "Dynamic Triad: City, Exposition, and Museum in Industrial Society," in Levin et al., *Urban Modernity*, 2.
3. Simmel, *Simmel on Culture*, 288.
4. William James, "Does Consciousness Exist?" in James, *Essays in Radical Empiricism*, 5.
5. William James, "The Thing and its Relations," in James, *Essays in Radical Empiricism*, 50.
6. James T. Kloppenberg, *Uncertain Victory: Social Democracy and Progressivism in European and American Thought, 1870–1920* (New York: Oxford University Press, 1986), 27.
7. Felix Gilbert and David Clay Large, *The End of the European Era, 1890 to the Present*, fifth edition (New York: Norton, 2002), 7.
8. Daniel T. Rodgers, *Atlantic Crossings: Social Politics in a Progressive Age* (Cambridge, MA: Harvard University Press, 1998).
9. Henry A. Giroux, *Pedagogy of Resistance: Against Manufactured Ignorance* (London: Bloomsbury Academic, 2022).
10. Martha Banta, *Imaging American Women: Idea and Ideals in Cultural History* (New York: Columbia University Press, 1987), 6–7.
11. Richard Guy Wilson, "America and the Machine Age," in *The Machine Age in America, 1918–1941*, edited by Richard Guy Wilson, Dianne H. Pilgrim, and Dickran Tashjian (New York: Abrams, 1996), 23.

12. Edwards, *New Spirits*, 4.

13. Carroll Smith-Rosenberg, *Disorderly Conduct: Visions of Gender in Victorian America* (New York: Knopf, 1985), 176.

14. Smith-Rosenberg, *Disorderly Conduct*, 176.

15. Herbert Stroup, *Social Welfare Pioneers* (Chicago: Nelson-Hall Publishers, 1986), 78.

16. Eileen Boris, *Art and Labor: Ruskin, Morris, and the Craftsman Ideal in America* (Philadelphia: Temple University Press, 1986), 16.

17. Rodgers, *Atlantic Crossings*, 64.

18. Charlene Haddock Seigfried, "Socializing Democracy: Jane Addams and John Dewey," *Philosophy of the Social Sciences* 29:2 (1999): 207.

19. Seigfried, "Socializing Democracy," 207.

20. Shannon Jackson, *Lines of Activity: Performance, Historiography, Hull-House Domesticity* (Ann Arbor: University of Michigan Press, 2001), 14.

21. Jackson, *Lines of Activity*, 14.

22. John Dewey, "The Theory of the Chicago Experiment," (1936), *The Later Works, Vol 11: 1935–1937*, edited by Jo Ann Boydston (Carbondale: Southern Illinois University Press, 1981), 98, quoted in Louis Menand, *The Metaphysical Club: A Story of Ideas in America* (New York: Farrar, Straus and Giroux, 2001), 320.

23. Katherine Camp Mayhew and Anna Camp Edwards, *The Dewey School: The Laboratory School of the University of Chicago, 1896–1903* (Chicago: D. Appleton Century, 1936), 314, quoted in Kurt Stemhagen and David Waddington, "Beyond the 'Pragmatic Acquiescence' Controversy: Reconciling the Educational Thought of Lewis Mumford and John Dewey," *Educational Studies* 47:5 (2011): 477.

24. Jane Addams, *The Spirit of Youth and the City Streets* (New York: Macmillan, 1909), 161.

25. H. Allen Brooks, "Chicago Architecture: Its Debt to the Arts and Crafts," *Journal of the Society of Architectural Historians* 30:4 (1971): 313.

26. John Higham, *Strangers in the Land: Patterns of American Nativism, 1860–1925* (New Brunswick, NJ: Rutgers University Press), 121.

27. Frank Lloyd Wright, "The Art and Craft of the Machine," in *Roots of Contemporary American Architecture*, edited by Lewis Mumford (New York: Dover Books, 1972), 169–185.

28. Wright, "The Art and Craft of the Machine," 176.

29. Lears, *No Place of Grace*, 93.

30. Drew Gilpin Faust, "This Republic of Suffering: Death and the American Civil War," Address (Washington, DC: The National Archives, January 8, 2008), 4, https://www.libraryofsocialscience.com/assets/pdf/Faust-This_Republic_of_Suffering.pdf.

31. Drew Gilpin Faust, *This Republic of Suffering: Death and the American Civil War* (New York: Knopf, 2008), xiv.

32. Mark Twain and Charles Dudley Warner, *The Gilded Age* (New York: New American Library, 1969), 137–138.

33. William Hirst, "Cultural Trauma: Guest Editor's Introduction," *Social Research* 87:3 (2020): xxxvi.

34. Jackson Lears, *Rebirth of a Nation: The Making of Modern America, 1877–1920* (New York: Harper Collins, 2009), 13.

35. Jan Patocka, *Heretical Essays in the History of Philosophy*, translated by Erazim Kohak, edited by James Dodd (Chicago: Open Court, 1996), 124.

36. Cal Flyn, *Islands of Abandonment: Nature Rebounding in the Post-Human Landscape* (New York: Viking, 2022), 179.

37. James Dawes, *The Language of War: Literature and Culture in the U.S. from the Civil War Through World War II* (Cambridge, MA: Harvard University Press, 2002), 113.

38. Dawes, *The Language of War*, 113.

39. David M. Kennedy, *Over Here: The First World War and American Society* (New York: Oxford University Press, 1980), 50.

40. John Milton Cooper Jr., *Pivotal Decades: The United States, 1900–1920* (New York: Norton, 1990), 221.

41. Hugh Rockoff, *Until It's Over, Over There: The U.S. Economy in World War I*, Working Paper 10580 (Cambridge, MA: National Bureau of Economic Research, 2004), 19.

42. W. H. R. Rivers, "War Neurosis and Military Training," in *Instinct and the Unconscious*, 208. Originally a report to the Medical Research Committee, London, and published in *Mental Hygiene* 2 (1918): 531–533. Quoted in Samuel Hynes, *A War Imagined: The First World War and English Culture* (New York: Atheneum, 1990), 177.

43. Malcolm Cowley, *Exile's Return: A Literary Odyssey of the 1920s* (New York: Viking, 1951); Ernest Hemingway, *A Moveable Feast* (New York: Scribners, 1964).

44. Geoffrey Wolff, *Black Sun: The Brief Transit and Violent Eclipse of Harry Crosby* (New York: Random House, 1976), 48.

45. Douglas, *Terrible Honesty*, 33.

46. Ernest Hemingway, *A Farewell to Arms* (New York: Scribner, 1929), 327.

47. Larry May, *Screening Out the Past: The Birth of Mass Culture and the Motion Picture Industry* (New York: Oxford University Press, 1980).

48. Robert S. Lynd and Helen Merrell Lynd, *Middletown: A Study in Modern American Culture* (New York: Harcourt, Brace and World, 1929), 211.

49. Paula S. Fass, *The Damned and the Beautiful: American Youth in the 1920s* (New York: Oxford University Press, 1977), 121–122.

50. Siegfried Gideon, *Mechanization Takes Command: A Contribution to Anonymous History* (Minneapolis: University of Minnesota Press, 2014).

51. Richard Overy, "Heralds of Modernity: Cars and Planes from Invention to Necessity," in *Fin de Siècle and its Legacy*, 54–79.

52. Roland Marchand, *Advertising and the American Dream: Making Way for Modernity, 1920–1940* (Berkeley: University of California Press, 1985); Jackson Lears, *Fables of Abundance: A Cultural History of Advertising in America* (New York: Basic Books, 1994).

53. Charles E. Beveridge and Paul Rocheleau, *Frederick Law Olmsted: Designing the American Landscape* (New York: Universe, 1998), 99.

54. Robert A. Caro, *The Power Broker: Robert Moses and the Fall of New York* (New York: Knopf, 1974); Clay McShane, *Down the Asphalt Path: The Automobile and the American City* (New York: Columbia University Press, 1994); David Ward and Olivier Zunz, eds., *The Landscape of Modernity: New York City, 1900–1940* (Baltimore: Johns Hopkins University Press, 1992).

55. Kenneth T. Jackson, *Crabgrass Frontier: The Suburbanization of the United States* (New York: Oxford University Press, 1985), 150.

56. F. Scott Fitzgerald, *The Great Gatsby* (New York: Charles Scribner's Sons, 1925), 36.

57. Elaine Tyler May, *Great Expectations: Marriage and Divorce in Post-Victorian America* (Chicago: University of Chicago Press, 1980).

58. Lewis Mumford, "Botched Cities," *American Mercury* 28 (Oct. 1929): 150.

59. Larry A. Hickman, *John Dewey's Pragmatic Technology* (Bloomington: Indiana University Press, 1990), 18–19.

60. Martin Jay, *Songs of Experience: Modern American and European Versions of a Universal Theme* (Berkeley: University of California Press, 2005), 170.

61. John J. McDermott, "A Relational World: The Significance of the Thought of William James and John Dewey for Global Culture," in *Streams of Experience: Reflections on the History and Philosophy of American Culture* (Amherst: University of Massachusetts Press, 1986), 119.

62. McDermott, "A Relational World," 119.

63. John Dewey, *Experience and Nature*, in *John Dewey: The Later Works, 1925–1953*, Volume 1: 1925, edited by Jo Ann Boydston (Carbondale: Southern Illinois University Press, 1981), 12.

64. Mihaly Csikszentmihalyi and Eugene Rochberg-Halton, *The Meaning of Things: Domestic Symbols of the Self* (Cambridge: Cambridge University Press, 1981), 176–177.

65. McDermott, "A Relational World," 123; John J. McDermott, "The Aesthetic Drama of the Ordinary," in *The Drama of Possibility: Experience as Philosophy of Culture*, edited by Douglas R. Anderson (New York: Fordham University Press, 2007).

66. John Dewey, *Art as Experience* (1934; reprinted, New York: Putnam, 1958), 104.

67. Csikszentmihalyi and Rochberg-Halton, *The Meaning of Things*, 39.

68. Robert B. Westbrook, *John Dewey and American Democracy* (Ithaca, NY: Cornell University Press, 1991), 380–387.

69. John W. Bennett, *The Ecological Transition: Cultural Anthropology and Human Adaptation* (New York: Pergamon, 1976), 10.

70. James Livingston, *Pragmatism and the Political Economy of Cultural Revolution, 1850–1940* (Chapel Hill: University of North Carolina Press, 1994), 148.

71. Alan Trachtenberg, *The Incorporation of America: Culture and Society in the Gilded Age* (New York: Hill and Wang, 1982).

72. Ellis W. Hawley, "Herbert Hoover, The Commerce Secretariat, and the Vision of an 'Associative State', 1921–1928," *The Journal of American History* 61:1 (1974): 118.

73. Hawley, "Herbert Hoover, The Commerce Secretariat, and the Vision of an 'Associative State', 1921–1928," 117.

74. Hawley, "Herbert Hoover, The Commerce Secretariat, and the Vision of an 'Associative State', 1921–1928," 118.

75. Hawley, "Herbert Hoover, The Commerce Secretariat, and the Vision of an 'Associative State', 1921–1928," 117.

76. Burton Bledstein, *The Culture of Professionalism. The Middle Class and the Development of Higher Education in America* (New York: Norton, 1976).

77. Ajay K. Mehrotra, "American Economic Development, Managerial Corporate Capitalism, and the Institutional Foundations of the Modern Income Tax," *Law and Contemporary Problems* 73:1 (2010): 25–61.

78. Ellis W. Hawley, *The Great War and the Search for a Modern Order: A History of the American People and Their Institutions, 1917–1933* (New York: St. Martin's Press, 1979).

79. Bob Jessop, "Fordism and Post-Fordism: A Critical Reformulation," in *Pathways to Industrialization and Regional Development*, edited by Allen J. Scott and Michael J. Storper (London: Routledge, 1992), 46–69.

80. Martha L. Olney, *Buy Now, Pay Later: Advertising, Credit, and Consumer Durables in the 1920s* (Chapel Hill: University of North Carolina Press, 1991).

81. Kriste Lindenmeyer, *A Right to Childhood: The US Children's Bureau and Child Welfare, 1912–46* (Champaign: University of Illinois Press, 1997).

82. William Trotter Jr., editor, *The Great Migration in Historical Perspective: New Dimensions of Race, Class, and Gender* (Bloomington: Indiana University Press, 1991); Nicholas Lemann, *The Promised Land: The Great Black Migration and How It Changed America* (New York: Knopf, 1991); Milton C. Sernett, *Bound for the Promised Land: African American Religion and the Great Migration* (Raleigh: Duke University Press, 1997).

83. August Meier, *Negro Thought in America, 1880–1915: Racial Ideologies in the Age of Booker T. Washington* (Ann Arbor: University of Michigan Press, 1963).

84. Nina Baym and Laurence B. Holland, "Introduction," in *The Norton Anthology of American Literature, Volume D: American Literature Between the Wars, 1914–1945*, edited by Nina Baym (New York: Norton, 2003), 1075.

85. W. E. B. Du Bois, "The Souls of Black Folk," in *Writings: The Suppression of the African Slave-Trade, The Souls of Black Folk, Dusk of Dawn, Essays and Articles* (New York: Library of America, 1986), 364.

86. Du Bois, "The Souls of Black Folk," 359.

87. Alain Touraine, *Return of the Actor* (Minneapolis: University of Minnesota Press, 1988), 125.

88. Samuel P. Hays, *Conservation and the Gospel of Efficiency: The Progressive Conservation Movement, 1890–1920* (Cambridge, MA: Harvard University Press, 1959).

89. Wiebe, *The Search for Order*, vii.

90. Arthur J. Vidich and Stanford M. Lyman, *American Sociology: Worldly Rejections of Religion and Their Directions* (New Haven: Yale University Press, 1985), 24.

91. Wayne D. Ramussen, *Taking the University to the People: Seventy-Five Years of Cooperative Extension* (West Lafayette, IN: Purdue University Press, 2002).

92. Reinhard Bendix, *Nation-Building and Citizenship: Studies of Our Changing Social Order* (Berkeley: University of California Press, 1977), 433.

93. Franklin D. Roosevelt, Radio Address From Albany, New York: "The 'Forgotten Man' Speech." Online by Gerhard Peters and John T. Woolley, *The American Presidency Project*, https://www.presidency.ucsb.edu/node/288092.

94. Fantasia, *Cultures of Solidarity*, 45.

95. Lizabeth Cohen, "The Lessons of the Great Depression," *The Atlantic*, May 17, 2020, https://www.theatlantic.com/ideas/archive/2020/05/how-rebuild-nation/611704/.

96. Gerald D. Nash, *The American West Transformed: The Impact of the Second World War* (Bloomington: Indiana University Press, 1985); Roger W. Lotchin, "The City and the Sword through the Ages and the Era of the Cold War," in *Essays on Sunbelt Cities and Recent Urban America*, edited by Robert B. Fairbanks and Kathleen Underwood (College Station: Texas A&M Press, 1990), 87–124; Roger W. Lotchin, "World War II and Urban California: City Planning and the Transformation Hypothesis," *Pacific Historical Review* 62:2 (1993): 143–171; James T. Lemon, "Los Angeles, 1950: The Working Class Thriving on Military Largesse," in *Liberal Dreams and Nature's Limits: Great Cities of North American Since 1600* (Toronto: Oxford University Press, 1996), 191–241.

Chapter 10

Postwar

THE CRISIS OF AFFLUENCE

The era of postwar prosperity followed American industrial mobilization during wartime, including long-term capital investment, expansion of the industrial infrastructure, and consumer production. The postwar economic boom was facilitated by the mass production techniques of Fordism that had been the mainspring of wartime production itself. The Fordist system of production generated a mass market of industrialized goods for societal restoration and economic growth in a world recovering from the massive destructiveness of total war. Together, mass production unionism and military spending propelled the American economy for nearly half a century. By mid-century, military spending increased dramatically, accounting for one half of the federal budget. Sizable government contracts were awarded to aircraft manufacturers and electronic assembly firms in production regions across the country providing high-paying blue-collar jobs with generous wages and benefits. These contracts were one reason why so many industries focused their research and development efforts on military products, such as missiles, guidance systems, bombers, and spacecraft. Because of this federal largesse, massive tracks of land were developed to accommodate the families of those employed in the burgeoning defense sector.

From 1939 to 1945, the annual gross national product increased from $91 billion to $166 billion; 14 million workers entered the work force, with significant employment gains for women, African Americans, recent European immigrants, and Mexican farmworkers (*braceros*). Congress passed the GI Bill in June 1944, which provided educational and other veterans benefits; two-thirds of the returning veterans used educational benefits to develop new occupational skills and professional identities. Significant geographic mobility occurred after the war, with population density increasing in the urban counties of the Northeast and Midwest and with widespread declines in predominately rural counties. Agricultural fields were quickly developed into suburbs, creating new local governments, school districts, healthcare systems, and other amenities, entirely connected by automobile-centered transportation patterns. Fordist market developments took hold regionally as specialized technologies targeted consumer demands and altered previously undifferentiated labor forces to accommodate segmented domestic markets in goods and services based upon diverse consumer interests. Micro-marketing strategies of flexible specialization localized business development and initiated a new interest in a specialized and professionalized knowledge sector that lent itself to a more differentiated mass population of educated consumer-based individualists. The accumulation of material goods and status consciousness based upon these possessions became a fixation in middle-class America.

Postwar Fordism thus emphasized changes in production to address an urban market economy that had become less homogenized. What had been a one-size-fits-all standard of production refocused upon differential and even competitive demands from various consumer markets. Politics and prominent ideologies soon followed these same patterns, reflecting and matching those changes. Some critics sensed that such progress had perhaps taken an unplanned course, for amid the spiraling affluence of a new middle class many were left impoverished in the postwar society. As if to silence these dissenting voices, there were chants extolling a triumphant America victorious against all foreign enemies. A darker side emerged with such credulous boosterism, that of a suspicion of "enemies within," a term marked by fear that designated as "unAmerican" any movement that appeared to contest the escalating patterns of mass production, consumption, and Cold War militarism.

After the war, federal tax measures enacted a peacetime tax system that arguably stimulated capitalist economic growth in mid-century. Governmental officials in pro-business city halls and statehouses were heralds of an era of extraordinary economic growth, championing advances in infrastructure development and mass consumption. Metropolitan life and its amenities, from expressways, airports, and public transit to school and hospital systems, marked a welcome change from the lean years of the Depression and wartime austerity. The economic expansion promised a better life for many working and upper-middle-class families in the nation's sprawling cities and new suburbs with their regional shopping centers. Cold war families embraced the domestic sphere as a haven amid the uncertainties of the times, as Elaine Tyler May points out:

> the home seemed to offer a secure, private nest removed from the dangers of the outside world. The message was ambivalent, however, for the family also

> seemed particularly vulnerable. It needed heavy protection against the intrusions of forces outside itself. The self-contained home held out the promise of security in an insecure world. It also offered a vision of abundance and fulfillment. As the cold war began, young postwar Americans were homeward bound.[1]

Though, as Lizabeth Cohen argues, postwar policies, such as the GI Bill and credit practices were discriminatory, only reinforcing conservative gender and racial expectations, which led to the fragmented residential pattern of postwar American metropolitan life.[2]

A privileged workforce of predominantly white, male union members, their affluence permitted them to move their families into suburban homes filled with every modern convenience. They drove their full-sized cars to jobs at modernized plants that were the envy of the world. The automobile, more than any other commodity, symbolized the material abundance and well-being that characterized "the American way of life" in the postwar era. Since Ford's Model T, the auto had influenced how Americans worked, lived, and fantasized about their futures. In the 1950s and 1960s, it was crucial to the suburbanization of American cities. Interstate highways and the drive-in movies and restaurants that catered to the burgeoning car culture further altered the landscapes and the aspirations of Americans. Many families uprooted to take advantage of the prospect of starting over in the growing Far Western cities. Writing in 1961, Lewis Mumford would specify the automobile, notably its speed, as a way to frame the character of life during the early years of the great acceleration:

> Our present civilization is a gigantic motor car moving along a one-way road at an ever-accelerating speed. Unfortunately as now constructed the car lacks both steering wheel and brakes, and the only form of control the driver exercises consists in making the car go faster, though in his fascination with the machine itself and his commitment to achieving the highest speed possible, he has quite forgotten the purpose of the journey. This state of helpless submission to the economic and technological mechanisms modern man has created is curiously disguised as progress, freedom, and the mastery of man over nature.[3]

The era's prosperity and its culture of abundance changed the course of metropolitan life beyond New York, in cities like Chicago and Detroit in the Midwest, Phoenix, Seattle, and Los Angeles in the West, and in New South cities of Dallas, Houston, Atlanta, Birmingham, Raleigh, Charlotte, and Tampa.

The country's massive public Land Grant universities, especially in the South, and in the Midwest, Mountain, Southwest, and Far West regions, provided an educated workforce for a thriving business civilization, from engineers and managers to accountants and lawyers. NCAA team rivalries in the Big Ten, Pac Ten, Southwest, Mountain States, and Southern conferences were more than in-conference matchups; they were annual secular rituals with game-day traditions. The events ritualized local and statewide commercial boosterism and civic pride. From revived downtowns to campus youth rituals at burgeoning state universities, the nation's business communities saw only growing affluence and prosperity in the country after the postwar recessions over the 1950s. Entrepreneurial and corporate elites in these thriving communities held optimistic views of both their own well-being and that of the country.

The postwar years were times of American economic dominance, particularly in industries such as aviation, electronics, and chemical engineering that had boomed in wartime. By 1947, the United States produced half of the world's manufactured goods. It also provided the world with 57 percent of its steel, 43 percent of its electricity, and 62 percent of its oil. The core industrial regions in the Midwest and on the Pacific Coast were among the first to be reshaped by federally funded postwar reconstruction efforts. These regions were significantly altered by urban renewal, suburban development, and expansion of transportation and communications systems. As regions central to the new global economy, they were a destination for massive flows of raw materials from the rest of the world. These regions' mills and factories mass-produced goods would dominate the world market. The global economy linked New York, Chicago, San Francisco, and Los Angeles as financial centers. These cities were pivotal to the control of data generated by a world economy. They were the hubs of banking and information services, but also of travel and tourism services necessary to a growing international culture, which American industrial workers shared during the triumph of postwar Fordism and its mass production, marketing, and consumption, forming the basis of a thriving business civilization.

YOUTH AND MID-CENTURY MOVEMENTS

Alongside the prosperity brought with Fordism, the American high school had grown to a place of prominence within these regions where secondary schooling systems had endured for decades. Sherry Ortner recognizes that the postwar high school became more visible to the public, representing both middle-class values and "evaluative social categories" within its regulating system.[4] Ortner also views the high school as a place that embodied national aspirations to lift "a large chunk of the former working class (tenuously) into the middle class."[5] In concert with this ambition, the high-school experience converged with the emerging youth culture, situating the high school as a "relative autonomous social space."[6] Within this space, youth created a "social world unto itself,"[7] with distinct cultural norms based upon young people's own self-perceptions:

> The hyperarticulation of the high school social system of the fifties was one part of a broader coming together of a so-called youth culture with its own self-awareness; its own (heavily marketed) styles of clothing; and its own "public culture" of music, movies and television.[8]

In postwar America, television brought the mass society, its entertainment, and breaking news events into people's living rooms. TV commercials with their catchy jingles and slogans framed a popular consensus about commodified ways of living and their prospects for consumer happiness. Those managing the new medium, together with corporate advertising and marketing, understood the potential of a postwar youth culture. Thomas Frank points out that corporate media and advertisers recognized the commercial appeal of "youth" as a concept that transcended any specific age-related demographic, stating:

> What is less frequently recognized is the basic marketing fact that "youth" had a meaning and an appeal that extended far beyond the youth market proper. This point is driven home again and again in the trade press of the era. The imagery and language of youth can be applied effectively to all sorts of products marketed to all varieties of people, because youth is an attractive, consuming attitude, not an age—an attitude that was preeminently defined by the values of the counterculture.[9]

The counterculture itself came about as a revolt against the mass society and would eventually embody the discontent residing beneath the surface of early television's dramatized sense of security and evening newscasters' reportorial authority. Despite the medium's normalizing functions, many television programs only increased viewers' anxieties over distressful images of atomic explosions, drug addiction, violent gangs, and on-the-scene reporting of political assassinations, labor unrest, combat situations, and riots in racially tense cities. Together with tightened work discipline, television reinforced hardened attitudes in a political culture based upon any number of differences, notably race, class, gender, and religion. Staunch partisanship typically involved well-accepted binaries: Democrat–Republican, urban–rural, white collar–blue collar, Christian–Jew, immigrant–native born, men–women, white–nonwhite, and rich–poor. Elite expectations that postwar life would continue along this ideological trajectory, based as it was on such categorical thinking, would begin to unravel in the 1960s.

Michael Harrington saw "a separate culture, another nation, with its own way of life,"[10] that of the poor and disenfranchised who dwelled beneath the surface of the country. Harrington wrote *The Other America* in 1962, originating from his Catholic Worker roots, to characterize the conditions of those living in poverty in an affluent postwar society:

> But the new poverty is constructed so as to destroy aspiration; it is a system designed to be impervious to hope. The other America does not contain the adventurous seeking a new life and land. It is populated by the failures, by those driven from the land and bewildered by the city, by old people suddenly confronted with the torments of loneliness and poverty, and by minorities facing a wall of prejudice.[11]

Harrington showed the contradictions of the 1950s mainstream ideology of upward mobility and dreams of the good life, or at least a better one, documenting the many who were not under "the big tent" of middle-class life and its aspirations during the postwar boom times. Harrington observed these forgotten people on big city streets and subways and along rural routes across the country. While the postwar boom offered many people widespread opportunities for a good life, there remained 25 percent of the country who lived in poverty with minimal levels of education, skill, health, food, and housing that led to their deteriorating quality of life and mental health.

Members of a revived middle class that had experienced the two sides of fortune and misfortune were now engaged in conversations about their own prospects in union halls and town hall meetings and on college campuses with the rising generation. Having witnessed the loss of freedom abroad, constituents of both the working class and the new middle classes no longer wished to tether themselves to an earlier industrial work regimen and to an obedience to authority based upon elite interests. Both the parental generation and their descendents took center stage over the next decades when the polarities of conflicting interests defined all that was to come later in the century and into the new millennium.

To understand the pivotal mid-point of the twentieth century in the 1950s and the appearance of countercultural movements in the 1960s requires historically situating diverse formations and positions. With a changing worldview that situated people in their physical and social environments—their place in the world where quality-of-life concerns were key to their survival—came an awareness of interested stakeholders and the public that inspired a succession of movements. After a crisis-based policy agenda inaugurated during the Depression and the Second World War restructured society, the United States entered a period of prosperity with restored industrial rigor and a vigorous economic growth that renewed the promises of capitalism.

Amid this prospect, a new form of conservatism ascended to challenge the rise of global communism as the alternative to chronic worldwide poverty that resulted from the crises of monopoly capitalism. Conservatives sought to subvert New Deal policies, including Keynesian economics, with a monetarism that foreshadowed subsequent globalization and neoliberal financial arrangements.

By the 1950s, an earlier progressive tradition culminating in the New Deal gave way to corporate conservatism, which sought a new fundamentalism characterized by religious and political conformity. However, a return to an American Victorianism was rejected by the postwar "Baby Boom" generation in near-simultaneous movements of civil rights, free speech, feminism, environmentalism, and a critical approach to society. An ecological logic emerged together with the new social movements that sought nothing less than to reorient the culture itself away from status seeking and wealth management interests toward an increased concern for sustaining the earth's resources. Corporate conservatives viewed the nascent countercultural worldview as an insurgency, conflating it with radicalism, socialism, and communism concomitantly. A counter-insurgency, spontaneously constructed with financing in support of dominant class interests, then called for a "law and order" approach to confront and silence new social movement participants, even those who were nonviolent in exercising first amendment rights. These elite political dynamics incipiently germinated amid the discourse of prosperity and its promises in the 1950s, though not without its discontents.

Fifteen years after the Second World War ended, Congress passed the Labor Management Reporting and Disclosure Act of 1959, amending the New Deal era National Labor Relations Act.[12] The new law established a code of conduct that promoted democratic procedures within labor unions, including members' rights to speak out on union matters. After the bill's passage, discontented workers went on strike in various industries, including film and television entertainment writers, newspaper employees, steelworkers, and farm laborers. There were also struggles for civil rights, antiwar demonstrations, campus upheavals over free speech, and a multi-wave women's movement for equal rights and personal freedoms. At the time, there were transfers of strategic knowledge across these seemingly disparate movements that strengthened their participants' resolve to push forward with their claims. As a social movement's repertoire demonstrated effectiveness in voicing grievances or engendering change, these attendant cultural schemas then informed other movements. Succeeding social movements then created their own situations to question and at times disrupt the status quo and would adopt new vocabularies to frame their actions. Consequently, movement participants discovered experiential knowledge based upon taking action together despite the isolating tenor of postwar life. The movements also brought about an innovative set of behaviors, including styles of dress and demeanor that were worlds apart from any performed in commercial interactions within the business communities.

THE PROMISED LAND

Social movement knowledge frequently draws from established sources to inform the present situation, as with Martin Luther King, Jr. After graduating from Morehouse College in Atlanta, King went north to study at Crozer Theological Seminary in Chester, Pennsylvania, where he became committed to social gospel, nonviolence, and social reform through Mohandas K. Gandhi's teachings. King then pursued doctoral study in systematic theology at Boston University School of Theology, where his "theological convictions, as shaped by his Methodist mentors, informed his vision of the beloved community," according to Richard A. Hughes.[13] There he came to master a canon that white New England divinity students had learned for decades. In Boston, King was introduced to Swiss reformed theologian Karl Barth's neo-orthodoxy, a response to the horrors of the First World War and its aftermath, which gives primacy to human fallibility, sin, and a transcendent deity.

King's position balanced immanence and transcendence, drawing from the ideas of liberal theologians, like Paul Tillich and Reinhold Niebuhr. Niebuhr's earlier views of politics were based upon the "fall of man," original sin, and human fallibility. Niebuhr's reasoning along these lines may have derived from Barth's influence on his earlier thought. Niebuhr eventually turned away from Barth's "crisis theology" and his neo-orthodoxy in the 1930s. At the onset of the Second World War, in "The Nature and Destiny of Man," Niebuhr held that "the evil in man is a consequence of his inevitable though not necessary unwillingness to acknowledge his dependence, to accept his finiteness and to admit his insecurity."[14] Affirming that anxiety is a condition of human freedom and finitude, Niebuhr recognized this existential state as a source of both temptation and creative living. A similar theme is found in "The Serenity Prayer," attributed to Niebuhr who recollected first using it in a wartime sermon at a village church in Heath, Massachusetts, in 1943.[15] The prayer, regarded at the time as a response to Nazi evil, has since been widely adopted by church groups and twelve-step programs. Years later, Niebuhr admitted to perhaps having adapted the prayer, albeit unconsciously, from times when he preached in colleges and youth groups during the 1930s.

Tillich's systematic theory drew upon existential analysis to answer foundational questions about being human. He derived a "method of correlation"[16] bringing together contemporary existential and psychological analyses to understand ontological questions common to theology and philosophy. Tillich's work, *The Courage to Be*, considers three types of anxiety inherent in the human condition: the anxiety of fate and death, or biological finitude; the anxiety of guilt and condemnation, or moral finitude; and the anxiety of meaninglessness and emptiness, or existential finitude.[17] Tillich engaged in socialist politics as a young man in Germany after the First World War and in the early 1930s with the rise of fascism. Following his arrival in the United States in 1933, as part of the intellectual migration of refugee scholars, Tillich espoused a liberalism radicalized by faith to realize universal equality and freedom. King inscribed a copy of *Stride Toward Freedom* to Niebuhr, acknowledging his "great prophetic vision . . . unswerving devotion to the ideals of freedom and justice."[18] When Tillich died in 1965, King stated: "He helped us to speak of God's action in history in terms which adequately expressed both the faith and the intellect of modern man."[19] Upon receiving his doctorate, King returned to the South and took what he learned to lead an organization that would join with other civil rights efforts to confront an entrenched segregated system that had grown over a century since Reconstruction.

The modern American civil rights movement began in the 1950s, a decade that celebrated affluence and was fired by a consumption ethos. In an act of nonviolent resistance, seamstress Rosa Parks, the secretary of a local chapter of the NAACP, rejected the bus driver's order to give up her seat in a segregated bus in Montgomery, Alabama, in December 1955, disobeying a state law that required black passengers to give up their seat to white passengers on full buses. By the 1960s, the decade marking the centennial of the American Civil War and the Emancipation Proclamation, movement leaders portrayed an image of a segregated society and of the poor and disenfranchised. Poorer African Americans in the rural and urban South had been denied the right not only to vote but also to equitably use resources held commonly by all residents of a community, namely, parks, swimming pools, and other recreational facilities; transportation; schools; housing; eating places; and even water fountains. Organizers chose to direct public attention to conditions in the South, appropriating the region's public resources as the arenas of contention.

After four black students from North Carolina A & T College took seats at a "whites only" lunch counter at a Woolworths in Greensboro, North Carolina, on February 1, 1960, 70,000 others participated in such nonviolent sit-ins that year. In 1961, the Northern-based Congress on Racial Equality (CORE) organized Freedom Rides to the South, based on nonviolent Gandhian principles, to test the recent court rulings that mandated the integration of Southern bus terminals. In 1963, King's Southern Christian Leadership Movement (SCLC) focused efforts on the desegregation of Birmingham, Alabama, organizing meetings in the city's largest Baptist church and mass demonstrations to disrupt business in the downtown commercial district. Both CORE and Student Nonviolent Coordinating Committee (SNCC) organized the 1964 Freedom Summer voting rights campaign in rural Mississippi. Inspired by the Algerian-born French philosopher Albert Camus, Robert "Bob" Moses, SNCC's first field representative in rural Mississippi, often stated that the Negro should be "neither victim nor executioner."[20] Like the Algerian masses under French colonial rule that he read about in Camus' works, Moses "experienced the American South as an occupied territory with a terrified, yet submissive population,"[21] in Ron Eyerman and Andrew Jamison's words. He was committed to building an activist community, based upon participatory democracy, privileging the movement over any single individual. Moses was caught between Camus' understanding of rebellion framed by "thought that recognizes limits"[22] and the need to take an activist stance. According to George Cotkin:

> Moses argued against allowing the problematic of action to result in quietism, whereby you "subjugate yourself to the conditions that are and don't try to change them." One must act, but always with the sense of walking a tightrope between asserting one's own rights and not denying the rights of others. In so doing, one battles for "humanitarian values" And, "if it's possible, you try to eke out some corner of love or some glimpse of happiness within. And that's what I think more than anything else conquers the bitterness, let's say."[23]

Moses faced a test of his position when he brought 1,000 primarily white volunteers from northern universities to spend their summer in rural Mississippi to help with voter registration during Freedom Summer. He and the young volunteers were met with racist hostility, black church burnings, beatings, the jailing of a thousand activists, and the murder of three civil rights workers, Michael Schwerner, James Chaney, and Andrew Goodman, that summer. Moses held himself responsible for creating situations where activists could meet brutal vengeance and death. Though Moses claimed he would not place any fellow activists in risky situations that he would not undertake himself. Still the event weighed heavy on him; after a while, Moses would move beyond the Deep South to take part in the national antiwar movement. After fleeing to Canada as a war resister, Moses then went on to teach math and English in Tanzania in East Africa from 1969 through 1976. Under a presidential proclamation granting pardon for draft evasion, he eventually returned to Cambridge, Massachusetts, in 1977 to teach math. Moses was later awarded a MacArthur Fellowship in 1982, which he used to create the Algebra Project to improve children's math literacy, viewing algebra as a "gatekeeper" subject for students' advancement in the STEM fields.

Young black civil rights protesters were early exemplars for a far-reaching political movement. The Southern sit-ins and Freedom Summer were signal events that preceded mid-1960s campus unrest and the broader antiwar mobilization. A well-elaborated civil rights movement culture prevailed in the South in 1960 when SNCC was founded, one that provided a facilitating framework and intergenerational support for the fledgling student movement that initially worked closely with SCLC. SNCC's later embrace of militancy and support for urban rioting spurred internal conflicts that fragmented the organization, rendering it ineffective in mobilizing broad-based support across the civil rights movement. Consequently, by 1968, long-standing community organizers became disenchanted with the militant ideology and no longer viewed SNCC as a force for change, and left the organization, which soon thereafter collapsed.

Despite later developments, SNCC's early culture of solidarity eventually caught on beyond the civil rights movement to inform white college students' protests. Sit-ins and their aftermath ignited a prairie fire, first among discontented and articulate college student leaders like Mario Savio of the Free Speech Movement and Tom Hayden of the Students for a Democratic Society, and their followers, and then among a larger swath of college students engaged in antiwar activities in the North, Midwest, and Far West with social justice at the core. All three student-initiated movements prospered, as their narratives became intertwined. Seeing how CORE, SNCC, and SCLC mass protests in the South, including the March on Washington in 1963, mobilized on behalf of social justice and influenced the passage of federal civil rights legislation, student antiwar leaders at the nation's colleges and universities organized campus-based protest to forward their movement.

"DO NOT FOLD, SPINDLE, OR MUTILATE"

Postwar universities typically assigned their undergraduates critical readings on modern society and the human condition drawn from the expanding social science canon. Public college students discussed ideas from the same texts as their wealthier peers at the private colleges, notably about cities and suburbs, elites and inequality, bureaucracy and rationality, technology and alienation, and the prospects for personal identity and dehumanization in a mass society. Postwar-era social scientists often confronted the rhetoric of an "American Century" appropriated by mass media writers to promote a triumphant, affluent postwar "victory culture,"[242] in Tom Engelhardt's words. Their critiques served as a rationally informed negativity that inspired college students to forge a new synthesis in the dialectics of popular sovereignty within an overly bureaucratized and rationalized mass society. The students would move from the social theories discussed in class to an informed praxis for institutional change on the campus and off in an increasingly militarized industrial society.

C. Wright Mills' writings were arguably most influential to the thinking of protesting college students.[25] In his first years at Columbia University, Mills worked as an applied sociologist with Paul Lazarsfeld's group studying labor leaders and Puerto Rican migration. During that time, he carried on serious long-term intellectually engaged relationships with German exiles, originally Hans Gerth at Wisconsin, and then the Frankfurt School group at Columbia before they returned to Germany, together with the British Marxian thinkers and the various creative people in the New York scene. These influences, together with earlier reading in American pragmatic and European social thought, shaped his thinking as he wrote two widely read critiques of contemporary American society and its culture, *White Collar*[26] and *The Power Elite*.[27] Mill's writings served as a reminder to the young that a white-collar class composed of rational and pragmatic professionals were at the helm, managing the organizations that framed the institutional life in postwar mass society. Despite the rise of this new middle class, the higher circles of elite power, based upon property ownership, remained influential to the control of both state and society through expert-driven rational solutions to social problems and pragmatic policies that characterized the age of information and bureaucratic organizations.

Much of contemporary social science is an abstracted sense of the real, as Mills pointed out in *The Sociological Imagination*, referring to the most prevalent trend in social science as that of abstracted empiricism.[28] The approach assembles data from *talk*, notably opinions, attitudes, knowledge about some thing or experience, which are then quantified and provided to governmental and corporate purchasers and to the lay public. As with art and musical pieces, and literary and art criticism, social science data is bought and sold, rationalized on behalf of exchange value and profit. Similar to contemporary scientific writing, social science often mystifies the ordinary. While the former is abstract talk about hypotheticals, much of the latter is talk about talk. Complex rhetoric has dominated science writing, and, in Mills estimation, the social sciences adopted the same style. By appropriating the conventions of scientific writing, Mills held mid-century social scientists responsible for abandoning the humanistic spirit that characterized social thinkers in the classic tradition, notably Thorstein Veblen, Emile Durkheim, Max Weber, and Georg Simmel, among others.

Mills' studies in the sociology of knowledge, following Karl Mannheim's classic writings, and his critiques of mid-century middle classes and the national elites, led him to redefine social and political change. He moved away from an accepted Marxian hope of a working-class movement that would upend the status quo to identify another agent of change in advanced capitalist societies. Rather than workers, Mills designated "the New Left" as one of young intellectuals and college students who had become increasingly dissatisfied with the postwar order and felt disenchanted by its institutional life. In his words, "To be 'left' means to connect up cultural with political criticism, and both with demands and programs. And it means all this inside every country of the world."[29] Mills viewed the young activists, armed with humanistic and secular knowledge acquired in national universities, as the historical agents that would lead future political and social upheavals. From a mix of critical thinking about American postwar society, chiefly influenced by Mills' work, and the lessons learned from civil rights movement cultures, students derived a movement culture. Through a critical lens, they gained a grasp of their eventual fates after college. While not seeing themselves toiling on factory assembly lines, they envisaged working in a deadening white-collar job in a bureaucratic office cubicle and making payments on a mortgaged "little box" house in the suburbs. To them, the IBM punch card, prevalent throughout their college years, represented the postwar machine-dominated informational and organizational society. The request on the card, "Do not bend, fold, spindle, or mutilate," became an apt metaphor for their presumed position within a computerizing world.

The Berkeley Free Speech Movement (FSM) in the early 1960s viewed the university as a bureaucratic machine, with student activists characterizing their education as a form of mass production, within an impersonal and alienating knowledge factory. The students viewed the IBM punch cards, adopted by university administrators in mass higher education's control revolution as symbolic of the new information technology (IT) that would regulate higher education in ensuing decades. Evidence of disenchantment with the emerging socioeconomic landscape would appear, initially, on research campuses, such as the University of California, Berkeley, according to Eric Foner, the "quintessential Cold War megaversity"[30] at the time of the Free Speech Movement there in 1964–1965. The movement's agenda included civil rights, antiwar, and academic freedom concerns and its leader, Mario Savio, clearly ahead of his times, held a vision of participatory democracy that was forged through a number of radicalizing experiences.[31] He volunteered with a Catholic relief organization on a social action project to build sanitary facilities in the slums of Taxco, Mexico. Savio was arrested during labor protests against the San Francisco Hotel Association for excluding African Americans for non-menial jobs. He also traveled to the South during the Freedom Summer where he helped African Americans in Mississippi register to vote.

Like Savio, many student leaders grew up in families that endorsed working people's rights and progressive union organizing activities and labor actions. Savio was able to integrate tropes of Catholic social justice and civil rights with union-based political rhetoric in his speeches. This was the era when more working-class youth were admitted to public research universities. Many were familiar with the progressive causes espoused by union families that were politicized during the Great Depression and in postwar union movement cultures. As a result, working-class youth may have viewed higher education differently than their more affluent classmates, who also took part in mass demonstrations. The political rhetoric that administrators and faculty members heard, often for the first time on public Ivy campuses such as Michigan, Wisconsin, and Berkeley, was the direct result of entering students who were often the first in their families to gain access to higher education. Working-class parents understood that public higher education was their sons' way out of the factory and their daughters' path beyond constraining domestic roles. These parents also held the hope that their children would bypass the "worlds of pain" attendant with postwar blue-collar family life among the "Silent Majority," notably "their troubled marriages, uncommunicative sex lives, unfulfilling work, and all too costly leisure,"[32] in Lillian Rubin's words. Their sons' college military deferments, for a time, also kept them far away from the rice paddies and jungles of Southeast Asia.

During the 1950s, college freshmen and sophomores on postwar Land Grant university campuses took compulsory Reserve Officer Training Program (ROTC) classes where they were exposed to the drills and commands attendant with military life. These classes allowed collegiate men a glimpse of that life and its subculture, beginning at eighteen, the age when they were required to register for the draft. However, protests against the requirement took place at public universities around the country leading the ROTC program to become voluntary in 1960. Male students nonetheless understood that their local draft boards were awaiting them to either drop out or graduate from college. Many had seen the fate of hometown youth who finished high school and did not attend college and receive a student deferment. Should they not pursue a graduate and professional degree, or marry and have a child upon graduation, their draft status would change from Selective Service Class 2-S, a student deferment, to Class 1-A, indicating availability for military service. The change in draft status generated a letter from the local draft board with an order to report for a preinduction physical examination at an Armed Forces Induction Center that, once passed, would follow with basic training and a likely tour of duty in Vietnam or at a Cold War era military installation abroad. Though it was Vietnam that framed an arena for young white and black Americans to grapple with a common threat, that of traumatic violence and impending death on Southeast Asian battlefields, as Mark Greif points out:

> Once Lyndon Johnson began committing American ground troops to warfare against the North Vietnamese and the guerillas of the Viet Cong in 1965, young men of all backgrounds lived under threat of conscription and passage to Indochina to fight and possibly die in a profoundly unpopular war. Though racial

> integration of the armed forces had been mandated in 1948, Vietnam was the first major US conflict since the Civil War fought without any segregated units. Insofar as the war against anticolonial peasant forces on the other side of the world became something that both white and black young men (and soldiers) could oppose, the new evaluation of the generic "White Man" who was behind it emerged as something blacks and whites could share in rejecting.[33]

For college-educated youth who took part in mass antiwar demonstrations, and for the people of color organizing struggles for freedom and equality in the urban North and the South, military service in what they came to see as an illegitimate war was politically indefensible and morally unconscionable. With their increasing participation in higher education and campus political activities, young women sought personal liberation and gender equality in both the work and domestic spheres inspired by the advances made by the civil rights movement.

EXISTENTIAL FEMINISTS

The Second World War years were pivotal to directional change in gender expectations as domestic labor demands sent men abroad to fight the war and brought women out of farmhouses and city apartments to work in factories. When men returned home from the war eager to get back to their jobs, companies discharged women from assembly lines to again return to the domestic sphere. By the mid-1960s, debates surrounding gender were widespread, with magazine articles, books, and the media advancing women's voices in public matters. Influential were Simone de Beauvoir's *The Second Sex*, published in an English translation in 1953[34] and Betty Friedan's *The Feminine Mystique*, published in 1962,[35] which focused on the lived experience of being a woman. Beauvoir and Friedan shared an existential perspective on how a woman's freedom is constrained by oppression and objectification, conditions that "can be overcome by acts of conscious recognition and choice."[36] At the same time, educational opportunities opened up for young women, and their career options expanded in the growing urban commercial sectors.

The unfolding postwar arena around gender issues overlapped with a mosaic of advancements, restraints, and transitions that led to change in the domestic and public spheres since the early twentieth century. Progressive movements and suffrage battles had been waged against a sterile patriarchal political arena that marginalized and minimized issues involving women's empowerment or their involvement directly in positions of power. It would be fair to say that class mattered, and for the better part of the century, the daughters of wealth held stature and could gain social power and elite educational opportunities. The acceleration of mass higher education and techno-social communication provided a societal impetus to the overturning of patriarchal traditions that grew stagnant in comparison to market individualism and independence. The idea of freedom as self-sufficiency gained ever greater value, together with the notion that competing with men was not only an avenue for potential independence and freedom but also a question of equality-generated collective resolve. Regarding innovations produced by the new American women's movement, Tilly and Sidney Tarrow write:

> Parts of the movement sought a new self-representation through changes in women's dress, language, manners, and collective activities. Although only representative of one branch of the movement, consciousness-raising was a creative tool for development of a new representation of women. Such grassroots activities fostered "sisterhood," women who had previously been demure and retiring learning to speak up for themselves, and new frames of meaning occurred.[37]

Women were thus deciding for themselves what their rights were under equality and precisely what freedom meant for them.

With second-wave feminism, social-class divisions were not considered among the more radically minded, many of whom had come from generations of privilege. Voices arose on behalf of greater opportunities for all women regardless of race or class. Grant money flowed to support higher education for middle-class youth that brought large cohorts of women onto public college campuses. The National Organization for Women formed in 1964 to realize the promise of national employment opportunity for women. The National Women's Political Caucus, founded in 1971 after Congress failed to pass the Equal Rights Amendment the year before, coalesced on behalf of increasing women's participation in political and public life, broadly understood. This pivotal moment, from 1964 through 1971, would soon recede with partisan divisions again gaining their hold, fragmenting the once unified movement. The women's movement was polarized and politicized with divisions. For many women, the pursuit of happiness was a quest for more satisfying career options or at least independent living. For some, the countercultural and antiestablishment fervor was not necessarily against the country's military involvement in Viet Nam; it was whether they had a right to fight in the war. Sexuality also took on a powerful impetus for change at a time of psychedelic experimentation. Certain women joined hands with like-minded men in a short-lived "free love" movement for sexual liberation and sexual freedom, a concept dating back to a turn-of-the-century American Bohemianism.[38]

These divisions were represented by stereotypical and sensationalized images in the press, with much of the movement's earlier moral direction becoming lost within easily forgotten media stories. Diverse radical groups were galvanized around women at the time. A feminized power center in extremist political movements included undertones of militant revolutionary rhetoric and actual violent actions. Women's participation in these groups, from Angela Davis' involvement with the Black Panthers and Patty Hearst's entanglements with the Symbionese Liberation Army, to Bernadine Dohrn's leadership of the Weather Underground, were widely publicized in media stories and created heated disdain across the spectrum. This response was class-driven and politically manipulated to offset consensus among women across the nation aspiring for more liberatory ways of life. The personification of women holding extreme radical views also engendered a reaction among those finding their security in conservative forms of life. Many women would claim traditional values against the breakdown of social norms and then forwarded outmoded ideas of security under essentially the rule of men.

Second, feminism's impact on society was wide and deep. None more so than among women and their fight for identity, freedom, and equality, which continued into the third and fourth waves of the movement. Viewing these waves of feminism as the most important contemporary cultural movement, Touraine characterizes our present "society of communication" as "the time of women."[39] Such a society requires more diversified roles that integrate rationality with emotional work. Touraine states:

> Women's experiences and demands are a central component of societies of communication for two main reasons which are intertwined to the point of inseparability. The entrance of women into the world of culture out of their historic relegation to the order of nature, represents an historic shift in the way in which societies conceive of the limit between society and its natural environment. This shift—in the western historical context—corresponds to the end of the ideal of rational production (the "one best way" of Taylor) as the dominant cultural interpretation of human activity, and the integration of rationality and feeling as necessary and unavoidable dimensions of communication.[40]

Having this as a backdrop, the individuals who set out to understand themselves were often a mix of diverse positions, with few holding absolute membership in any one faction. Both men and women were faced with localized survival and that generally meant seeking their adult identities among peers in a neighborhood, college, or workplace although many also came by themselves to live more freely in vibrant postwar cities. The city was a vortex that drew so many individualists of both genders seeking a place for themselves where economic opportunities thrived and pivotal events were historically and dynamically occurring. Yet all who arrived had come from different family worlds and conditions of life, creating a mosaic of men and women reaching for a kaleidoscopic vision of gender equity, one that was often overtaken by the ebb and flow of city life. Younger women who embraced metropolitan life experienced a newly found freedom as previously closed doors were opening to them. Increased educational opportunities for women and new reproductive technologies altered their life trajectories, and also their sexual attitudes and practices.

Despite this, many outmoded ideas persisted as women entered and competed successfully in male-dominated workplaces. Male identity typically emphasized autonomous individualism attainable through economic self-sufficiency. While women were free to eschew traditional roles and take risks on behalf of this goal, the consequences were not always in their favor, for inequality still worked disproportionately against them as a whole. Their promise of independence was not typically promoted by conservative employment practices of the time, notably the gender wage gap. Discrimination based on maternity and family only deepened gender inequities, with the lack of job-protected maternity leave and routine firing of pregnant women. Eventually, under neoliberal entrepreneurial corporate preconditions for social mobility, freedom became a monetized market pursuit, and sustaining any earlier political idealism became increasingly difficult.

SUSTAINING LIFE

Parallel with the emergent civil rights and social justice movements, constituents concerned about the health and degradation of their natural and built environments organized into an environmental movement. The push toward environmental consciousness was a response to pervasive air, soil, and water pollution that affected health and well-being and the degradation of these commons into a wasteland. There was a growing awareness that the earth's supportive systems were becoming disrupted as a result of industrialization, mass production, surface mining, and a fossil fuel economy. The movement flourished in the late 1960s, when people began to view professional, industrial, and governmental experts more critically and to demystify their premises and procedures. There was also an upsurge of consumer activism focused especially on the various dominant industries and on the regulatory agencies in the federal government, such as the Food and Drug Administration, the Department of Agriculture, the National Highway Traffic Safety Administration and the Occupational Safety and Health Administration.

Attorney Ralph Nader jumpstarted the consumer movement with his book *Unsafe at Any Speed*, published in 1965, that took on the auto industry for resisting safety features, such as seatbelts, endangering the lives of drivers and passengers, leading to the inception of the U.S. Department of Transportation.[41] Movements surrounding such "life political" concerns developed because of an increasing mistrust of the various corporate firms, public agencies, and their professional staffs sanctioned to look after the public interest in the areas of science, technology, and health. This sense of disenchantment was brought on with the consequences of late social modernity, notably a diminishment of personal control as more of people's everyday lives became subject to professional management.

Marine biologist Rachel Carson's *Silent Spring*, published in 1962, alerted Americans to the threat of synthetic pesticides, such as DDT, to the living world.[42] The book grew out of series of *New Yorker* articles that took on the chemical industry for withholding data on the health and ecological risks related to chemical spraying programs. Carson also criticized the federal government agencies for accepting uncritical chemical industry representatives' accounts on the limited risks and hazards of pesticide exposure. *Silent Spring* was instrumental to an activist American environmentalism that went far beyond the mainstream conservation activities of the Sierra Club and National Audubon Society. Five years later, in 1967, the Environmental Defense Fund was created and organized an initial campaign to ban DDT.

In 1969, two signal events would catalyze the movement. That January and February in Santa Barbara, a major spill from a Union Oil platform in the Dos Cuadras Offshore Oil field sent 80,000–100,000 barrels of crude oil into the Channel and onto the coast, killing 3,500 sea birds and countless marine animals, such as dolphins, elephant seals, and sea lions. In June, the Cuyahoga River near Cleveland, laden with industrial pollutants, would again catch fire, this time by a spark from a passing rail car that ignited an oil slick. That year, the National Environmental Policy Act (NEPA) was signed into law, creating the Council on Environmental Quality and mandated a detailed statement of environmental impacts for "significantly affecting the quality of the human environment."[43]

As a counterpoint to these ecological disasters, on Sunday, June 29, 1969, Apollo 11 astronauts Neil Armstrong and Buzz Aldrin landed the *Eagle*, a lunar module, in the Sea of Tranquility, and walked on the moon. The $24 billion Apollo Space Program successfully landed the first humans on the moon and brought them safely back to earth. Apollo 11 carried on board a comprehensive set of camera equipment so that astronaut still photography could occur during this historical mission. As with the previous two human missions, Apollo 8 and 10, photographic training was part of the intensive training the astronauts underwent in preparation for their moon explorations. The iconic "Earthrise" photo taken by Apollo 8 astronaut William Anders on

December 24, 1968, portraying earth peeking out from beyond the lunar surface, changed the perception of the planet and its environment.[44] The Apollo 11 lunar landing created another a new perspective on the planet and again raised public awareness of the earth's environment. Less than a year later, a national event celebrating the earth would deepen that perspective by focusing on the deleterious impacts of industrial development on the environment and human health.

The first Earth Day, held on April 22, 1970, was inspired by the vision of founder Gaylord Nelson, a former Wisconsin senator, who believed that the nation required a day of environmental awareness, education, and peaceful rallies on behalf of the planet.[45] Nelson, a lifelong conservationist, envisioned the event after seeing the effects of the Santa Barbara oil spill and in the wake of news coverage the Cuyahoga River fire and hired the young activist Denis Hayes to organize the event modeled after the Vietnam War teach-ins held by students throughout the previous decade. Nelson and Hayes garnered widespread support, ranging from United Nations Secretary General U Thant to labor leader Walter Reuther and his United Auto Workers (UAW) union, which contributed the major financial and operational support.

The first Earth Day celebration in Union Square in Manhattan brought New York's Mayor John Lindsey and actors Paul Newman and Ali MacGraw to "defend the earth." Across the United States, 20 million people rallied for environmental reform, serving to broaden the base of an environmental movement that was coalescing in the previous decade. Hayes would go on to organize Earth Day events, globally, in 141 nations. Eight years after the publication of Carson's book, and amid widespread citizen concerns about pesticides and other sources of pollution, the federal government established the U.S. Environmental Protection Agency (EPA) on July 9, 1970. The agency began operating later that year on December 2 and would inaugurate a sweeping regulatory agenda throughout the next decade, including an expansion of the Clean Air Act, and new legislative initiatives, such as the Clean Water Act, Safe Drinking Water Act, Toxic Substances Control Act, and the Resource Conservation and Recovery Act.

An unexpected event occurred on December 7, 1972, when the Apollo 17 crew took the first complete photograph of the earth from space, "The Blue Marble" as the spacecraft was 28,000 miles from earth and traveling toward the moon. The photo represented the interplay between NASA's ambitious mechanized space exploration program and an aesthetic moment captured by its astronauts. This widely reproduced iconic image denotes an essential tension between technological rationality and a neoromantic dream of understanding humanity's place in the enigmatic universe during the 1970s.[46] Each position prevailed in the environmentalism that took a turn toward climate change at the time. Climate research involved government funding networks of academic natural scientists and engineers that engaged in technology development and commercial applications and seized the mantle of disseminating climate change knowledge in global public debates. Climate justice activists coalesced, as Andrew Jamison states, in "groups and alliances to save rainforests, preserve biodiversity, defend the rights of indigenous peoples, and develop sustainable forms of agriculture and industry,"[47] within a broader global justice movement.

GATES OF EDEN

The transitional period that followed the Second World War was arguably one of profound social and cultural transformation. A devastating new weapon, the atomic bomb, not only demonstrated a destructive power that was as horrifying as it was brutally effective in warfare but also brought with it the advent of "the nuclear age." The atomic era had an immeasurable impact on humanity and its place on the earth and on the human condition itself, with political, psychological, and existential effects. All of this would begin to play out internationally in power relations, domestically spiraling into a national consciousness based upon survival both real and imagined. There was increasing desperation over the aftermath of a global war witnessed first-hand by a generation that fought for democracy and its ideals against tyranny and abuses of power. That generational cohort observed as the world settled into a silent competitive drama between class-stratified political economies of capitalism urging a push toward prosperity, liberty, and "freedom," and the more poverty-depleted nations being targeted by centrally organized politically "classless" societies dominated by state socialism. Both sides sought expansion and domination in the postwar years, dividing the world into spheres of influence that would define and determine the premises of war and peace for the next half-century. Domestically, the nuclear age generated existential conditions influenced by a world better connected by communication technologies. The country was about to grow exponentially along with a flourishing educational arena that initially targeted the expanding middle class, with an intensified set of technological demands and a business sector that was market-centered and labor short. It was, indeed, a brave new world.

The postwar generation's coming of age occurred during extreme events, from Vietnam, political assassinations, and Watergate to an energy crisis, environmental degradation, and eventual industrial decline.[48] The postwar moment, characterized, by some, as a victory culture, was bolstered by hope and optimism, albeit with a sense of fear and ambiguity about the future. These were times of ferment when the authority of church, state, and empire was being questioned. In the face of changing lifeways, some held a steadfast insistence on tradition; while a rising generation welcomed the transition, many others remained in a state of social and cultural amnesia. The decision to follow in well-defined occupational and social trajectories made sense to postwar youth when the growing economy appeared to stabilize the work and family lives of countless middle-class Americans. By the 1970s, changes occurred on the socioeconomic surface of the United States; many of the time-honored ways just did not make sense anymore and were clearly not adaptive under the newer conditions. Circumstances altered dramatically in workplaces and families, with globalization and neoliberalism profoundly affecting the former, and the changed status of women influencing the latter. Beyond the "dark times" that Arendt, inspired by Bertolt Brecht's poem, "To Posterity,"[49] employs to describe the tragedies of the earlier decades of the century, a darkened world picture emerged, giving way to a cynical nihilism in ensuing decades.

Amid the growing darkness, a rich mix of musical styles could be heard on the radio and television in the postwar years: rock and roll, soul, the folk scene, British rock, jazz, and American roots music, including blues and bluegrass. Their melodies provided an aesthetic soundscape and their lyrics an imaginary landscape for growing up and accepting the tasks of an emerging adulthood, including intimacy and affiliation. Generationally, music was part of the soundscape that influenced youths' ways of speaking and their reflections upon their lives and futures. Greil Marcus understands rock-and-roll music as a "reaction against a programmed technological culture."[50] In this light, Louis Menand sees the "northern songs" of the Cold War era as being transformed "from an age-specific entertainment to a cultural form with political content."[51] The folk revival of the 1960s introduced plenty of country and blues lyrics about rambling, staying for a while, and then moving on.[52] Bob Dylan's album *Highway 61 Revisited* told of life along a seemingly never-ending highway. Dylan's Highway 61 stretches from northeast Minnesota down the Mississippi Delta to New Orleans, running through his hometown of Duluth.[53] Bruce Springsteen's album *Born to Run* restated the theme a decade later. Springsteen's Highway 9 stretches from upstate New York near the Canadian border, through the New Jersey blue-collar communities of his youth and ends in Delaware.[54] Their messages are similar: set out on the highway, discover other places, unfamiliar characters and voices, and along the way seek love and remake oneself. Eyerman and Jamison explain how folk revival, itself, was a social movement that linked personal and cultural transformation through a spirit of protest:

> In this period, popular music was one of the main mediating forces, forms of translation, between the movement's more obvious expressions—demonstrations, organizations, books and journals—and the wider population. Through the media of popular culture, the ideas, values and attitudes expressed in the movement reached a wider segment of people, and (perhaps) to a more long-lasting effect.[55]

Without these vernacular representations of emotional life and the historical moment, characterized by the production and distribution of news and commercial images, the lives of postwar youth would have been shaped only through tightly controlled television and print media.

There were other voices during those years, especially poetic ones, when it appeared as if the world turned upside down with political assassinations and the deaths of thousands of young men in an unpopular war. Postwar American poets were shaping a new conversation about personal survival in a world upended by war and motivated by the drive for success, offering an escape from a saturated media environment. In 1969, poet Gary Snyder wrote:

> No one today can afford to be innocent, or indulge themselves in ignorance of the nature of contemporary governments, politics and social orders. The national politics of the modern world maintain their existence by deliberately fostered craving and fear: monstrous protection rackets. The "free world" has become economically dependent on a fantastic system of stimulation of greed which cannot be fulfilled, sexual desire that cannot be satiated, and hatred that has no outlet except against oneself, the persons one is supposed to love or the revolutionary aspirations of pitiful, poverty-stricken marginal societies like Cuba or Vietnam. The conditions of the Cold War have turned all modern societies—Communist included—into vicious distorters of man's true potential. They create populations of "preta"—hungry ghosts, with giant appetites and throats no bigger than needles. The soil, the forests and all animal life are being consumed by these cancerous collectivities; the air and water of the planet is being fouled by them.[56]

Beyond Snyder's contemporary Buddhist influences, many postwar poets drew inspiration even farther back to rediscover a fully embodied poetics through the voices of shamans and storytellers, what David Abrams called "the spell of the sensuous."[57] Ethnopoetics redefines poetry by focusing on alternative traditions in language and performance. The "recovery of the oral," in Jerome Rothenberg's words, "involves a poetics deeply rooted in the powers of song and speech, breath and body . . . with or without the existence of a visible/literal text."[58] Ethnopoetics surfaced at this time—the end of European colonialism, with increasing risks of nuclear and environmental hazards, and a concern for the fate of wilderness and traditional societies—providing an entirely new reading of American poetics that included American Indian chants and African American blues. In Snyder's words, "the politics of ethnopoetics"

> is fundamentally the question of what occidental and industrial technological civilization is doing to the earth. . . . What we are witnessing in the world today is an unparalleled waterfall of destruction of the diversity of human culture:

> plant species; animal species; of the richness of the biosphere and the millions of years of organic evolution that has gone into it. In a sense ethnopolitics is like some field of zoology, which is studying disappearing species.[59]

Together with their subjective experiences, postwar poets and many others discovered historical, ethnic, and religious roots for inspiration. Perhaps this search for "a usable past"[60] to reshape cultural memory helped move them beyond the chilling Cold War present, one that left them, shivering and seeking truth, "outside the Gates of Eden," where Bob Dylan claims "there are no truths."[61] As Greil Marcus reminds us, artists helped to provide younger generations with a sense of continuity and even inspired reinvention of both a personal and national mythos, amid the unsettled postwar decades:

> It's easy to forget what remains to be settled. Since roots are sought out and seized as well as simply accepted, cultural history is never a straight line; along with the artists we care about we fill in the gaps ourselves. When we do, we reclaim, rework, or invent America, or a piece of it, all over again. We make choices (or are caught by the choices others have made) about what is worth keeping and what isn't, trying to create a world where we feel alive, risky, ambitious, and free (or merely safe), dispensing with the rest of the American reality if we can.[62]

The poets, together with the era's rock, jazz, and blues musicians, located in sound and rhythmical patterns a sense of time and truths perhaps only found inside Eden's gates, which William James termed "the specious present" or "the short duration of which we are immediately and incessantly sensible."[63] Again, in James' words, "the specious present, the intuited duration, stands permanent, like the rainbow on the waterfall, with its own quality unchanged by the events that stream through it."[64] It makes sense, then, that poets, musicians, and many others in the postwar generation sought this immediate state through aesthetic, meditative, and psychedelic means. This appeared necessary for many youth given the range of events that streamed through their subjective experiences and collectively through the media: an overseas war in Southeast Asia, political assassinations, extremist bombings, police violence against demonstrators, and the personal and social memories these events evoked.

TWILIGHT OF POSTWAR ACTIVISM

Postwar social movements brought about a new critical discourse for a culture in transition, especially to oppose the country's increasing militarism and both foreign and domestic surveillance, along with domestic social and economic inequalities. These were times when any criticism of America's wars and intelligence operations was silenced in the name of patriotism amid anticommunist fervor. Discontent also surfaced over the apparent contradictory ends of a national state that privileged corporate management and control of technology and information in manufacturing and services while neglecting the limited social mobility and widening economic disparities in the society. These grievances were activated through large-scale demonstrations organized by the Students for a Democratic Society, the National Mobilization Committee to End the War in Vietnam, and the Redstockings, New York Radical Women, and later the National Organization for Women, together with those of civil rights movements during this time, including CORE, SCLC, SNCC, and the Black Panther Party.

The various forms of organized resistance to economic and social conditions were viewed as "revolutionary" by the New Right and provoked a "counter-revolution." The movements thus prompted an initial fracture in the hardened postwar consensus-based political culture. With the death toll in the Vietnam War rising, and the draft hanging over the heads of millions of young men, youth exhibited a growing cynicism toward both political parties: the Republicans' for their long-standing jingoism and the Democrats who had escalated the Southeast Asian conflict. At the same time, many working people shifted to support Richard Nixon's 1968 presidential campaign to express their patriotism and their displeasure with the youthful antiwar dissenters and women's rights protesters, together with civil rights activists and urban rioters. Movements organized by black civil rights protesters, discontented white college youth, and vigilant feminists upended the status quo and threatened a business civilization that had grown prosperous during and after the war.

Conservative Chamber of Commerce-affiliated businesspeople witnessed civil disturbances on their main streets between black protesters and the police. They also saw white college students shut down their big state university campuses. As the war dead mounted, pacifist, socialist, and Trotskyite strategists mobilized antiwar rallies and marches that overtook the main thoroughfares of the big cities, followed by demonstrations in medium-sized cities and college towns. Ethnic violence, the campuses shutting down for teach-ins, and peace marches down Fifth Avenue in Manhattan and Michigan Avenue in Chicago and down the main streets in Far Western cities caused widespread alarm within conservative business circles. All were viewed as disruptive and potentially posing serious threats to property and profit.

The mass marches and rallies in downtown commercial districts tarnished the image of shopping as a safe and pleasant experience, for many protesters broke windows and set afire local stores and banks. Suburban families became apprehensive about going downtown to shop in department stores, frequent movie houses, eat in old-school establishments, and stay the weekend in downtown hotels. Angry veterans at local American Legion posts and Veterans of Foreign Wars lodges viewed strident antiwar demonstrators as unpatriotic, especially during an ongoing war in Southeast Asia. Then there were the unions threatening strikes on behalf of wages and benefits—shutdowns of newspapers, public transit, and school districts, and even a brief strike by National Football League players. All were troubling for conservatives seeking an orderly society with limited centralizing forms of power, but especially unsettling to them was an "activist" federal government. Lyndon Johnson's liberal agenda was disconcerting for conservatives. There were civil rights and voting rights legislations, together with the passage of Medicare and Medicaid bills. Then came the antipoverty programs, including the Head Start program for preschool children of low-income families and the centerpiece domestic public service corps named Volunteers in Service to America (VISTA). Johnson's Great Society campaign was influenced in part by Michael Harrington's writings. The intent was to break the cycle of poverty through federally funded community action programs that typically, though not exclusively, targeted low-income communities of color. The increased federal spending to ameliorate social problems and the new regulatory bureaucracies together with the growing progressive political constituencies inside the Beltway only increased dissatisfaction across business communities.

Out of a growing discontent among conservatives over the progressive course of national affairs and troublesome left-leaning youth, a corporate attorney from Virginia who served on the board of Philip Morris and represented various tobacco companies in law cases called upon corporate America to reverse these trends. The 1971 *Attack on American Free Enterprise System*[65] is a founding document of the "New Right" that coalesced after Barry Goldwater's failed presidential campaign in 1964. The confidential memo, known as "The Powell Memorandum," was written to the United States Chamber of Commerce by the soon-to-be appointed Supreme Court Associate Justice Lewis Powell, who served on the court from 1972 to 1987. Powell called on conservatives, specifically business circles, to defend against two trends in a postwar regulatory state that grew in size and power. Powell first took on federal governmental overreach, notably the liberal regulatory agenda passed by a Democratic majority in Congress. The increasing power of regulatory agencies, such as the Environmental Protection Agency, the Occupational Safety and Health Administration, and the National Traffic Safety Commission, each established in 1970 during Richard Nixon's presidency, were especially troubling to Powell and other free-market adherents. Powell then cited the threat to American business of the rising New Left counter-hegemonic movements that set out to confront antidemocratic tendencies in the country. Powell cited certain individuals in the memo, alleging their role as critics of the free enterprise system, notably University of California, San Diego professor Herbert Marcuse, Yale professor Charles Reich, public interest attorney Ralph Nader, civil rights attorney William Kunstler, and civil rights activist Eldridge Cleaver.

Powell's polemic conflated widespread dissatisfaction with Cold War foreign policies and failed military strategies, notably in Southeast Asia, with an attack on the business community. He understood these counter-hegemonic movements as threats to the "American system" of free enterprise, in his view, hence "economic freedom" itself. Together, these movement cultures were viewed as threats to business interests, as Powell's memo indicated. More than a rallying call, the memorandum became a catalyst for subsequent neoliberal ideological countering of progressive politics over the next five decades, through libertarian foundations, media outlets, textbook publishers, think tanks, and university programs. Powell's vision as set forth in the memo was the prospect of reforming governance at the state and federal levels through a pervasive neoliberalism, notably privatizing public sector functions. This would entail overturning Roosevelt's New Deal policies sustained through subsequent administrations, a goal that would take considerable funding to accomplish. Moreover, there would need to be a well-defined approach to communicate this goal to Americans to influence them to abandon a policy and regulatory agenda crafted since the 1930s. These same policies designed to meet the dual challenges of a global depression and a subsequent world war brought greater prosperity and a better quality of life to millions of families across the nation.

A liberal agenda at the federal level, and myriad disruptive events driven by unruly, and in some cases contentious, youth ran counter to the free-market values that defined the nation's business ethos. A new cultural debate then took shape within the political sphere. The clash of values focused squarely on the attitudes and behavioral styles of a massive postwar generational cohort. However, questions remain as to whether the short-lived 1960s counterculture was actually responsible for a disparate class system, a discontented populace, and a permanent urban poor dependent upon public entitlement programs. It was Lyndon Johnson's administration that proposed antipoverty, fair employment, and civil rights legislation; a progressive Congress that passed these bills, including legislation to create new federal regulatory agencies; and a liberal-tilting Supreme Court that upheld their efforts. All were in support of increasing middle-class opportunities for poor and minority populations. The business civilization reacted to the power politics inside the Capital Beltway, and in liberal state houses and city halls; each was clearly outside the purview of countercultural youth and their concerns. In the end, the youth that demonstrated for civil rights and against the war, smoked weed, and experimented sexually, may have been mere objects of their projections. Even so, David Chalmers indicates that the varieties of postwar activism framed the life trajectories of the 1960s generation and provided a model for subsequent generations of youth:

> The sixties left their mark not only on America's laws and politics, but also on lives and commitments, and on the way in which people related to society and the political system. Although the following decades came to look in a different direction, an augmented ethic of voluntarism and service remained. For the most part, it was to be found among college-educated, middle-class men and women in journalism, law, teaching, medicine, and the health care and social service professions. Its politics led through the feminist, environmental, antinuclear, consumer, and community service movements, whose causes continue to attract young high-school and college students.[66]

Moving toward the third millennium, the interests of groups adhering to strictly market-based precepts forwarded by the business civilization continued to clash with those of a counterculture embodied in the lifeways of activist African Americans and disaffiliated white youth. These tensions gave rise to the neoconservative movement that sought the country's return to "normalcy," in ways similar to the ascent of a pro-business agenda amid the waning of Progressivism in the 1920s. Nevertheless, postwar social movements would influence national political life into the first decades of the new millennium.

NOTES

1. Elaine Tyler May, *Homeward Bound: American Families in the Cold War Era*, fourth edition (New York: Basic Books, 2017), 1.
2. Lizabeth Cohen, *A Consumer's Republic: The Politics of Mass Consumption in Postwar America* (New York: Random House, 2003), 407.
3. Lewis Mumford, *The City in History* (New York: Harcourt, Brace and World, 1961), 558–559.
4. Sherry B. Ortner, *New Jersey Dreaming: Capital, Culture, and the Class of '58* (Raleigh, NC: Duke University Press, 2002), 91
5. Ortner, *New Jersey Dreaming*, 93.
6. Ortner, *New Jersey Dreaming*, 93.
7. Ortner, *New Jersey Dreaming*, 93.
8. Ortner, *New Jersey Dreaming*, 92.
9. Thomas Frank, *The Conquest of Cool: Business Culture, Counterculture, and the Rise of Hip Consumerism* (Chicago: University of Chicago Press, 1997), 118–119.
10. Maurice Isserman, "50 Years Later: Poverty and The Other America," *Dissent* 59:1 (2012): 85.

11. Michael Harrington, *The Other America: Poverty in the United States* (New York: Macmillan, 1962), 11.

12. Benjamin Aaron, "The Labor-Management Reporting and Disclosure Act of 1959," *Harvard Law Review* 73:5 (1960): 851–907.

13. Richard A. Hughes, "Boston University School of Theology and the Civil Rights Movement," *Methodist History* 47:3 (2009): 146.

14. Reinhold Niebuhr, "The Nature and Destiny of Man," *Theology Today* 1:2 (1944): 240.

15. Elisabeth Sifton, *The Serenity Prayer: Faith and Politics in Times of Peace and War* (New York: Norton, 2003).

16. Bernard M. Loomer, "Tillich's Theory of Correlation," *The Journal of Religion* 36:3 (1956): 150–156.

17. Paul Tillich, *The Courage To Be* (New Haven: Yale University Press, 1952).

18. Stanford University, The Martin Luther King Jr. Research and Education Institute, https://kinginstitute.stanford.edu/king-papers/documents/reinhold-niebuhr-1.

19. Stanford University, The Martin Luther King Jr. Research and Education Institute https://kinginstitute.stanford.edu/encyclopedia/tillich-paul.

20. Laura Visser-Maessen, *Robert Parris Moses: A Life in Civil Rights and Leadership at the Grassroots* (Chapel Hill: University of North Carolina Press, 2016), 3.

21. Eyerman and Jamison, *Social Movements*, 131–132.

22. Albert Camus, *The Rebel: An Essay of Man in Revolt*, translated by Anthony Bower (New York: Knopf, 1956), 294.

23. George Cotkin, *Existential America* (Baltimore: Johns Hopkins University Press, 2003), 235.

24. Tom Engelhardt, *The End of Victory Culture: Cold War America and the Disillusioning of a Generation* (New York: Basic Books, 1995).

25. C. Wright Mills, *Letters and Autobiographical Writings*, edited by Kathryn Mills with Pamela Mills (Berkeley: University of California Press, 2000), 4.

26. C. Wright Mills, *White Collar: The American Middle Classes* (New York: Oxford University Press, 1951).

27. C. Wright Mills, *The Power Elite* (New York: Oxford University Press, 1956).

28. C. Wright Mills, *The Sociological Imagination* (New York: Oxford University Press, 1959).

29. C. Wright Mills, "Letter to the New Left," *New Left Review* 5 (1960): 21.

30. Eric Foner, *The Story of American Freedom* (New York: Norton, 1999), 290.

31. Robert Cohen, *Freedom's Orator: Mario Savio and the Radical Legacy of the 1960s* (New York: Oxford University Press, 2009).

32. Lillian B. Rubin, *Worlds of Pain: Life in the Working Class Family* (New York: Basic Books, 1976), Quoted in: https://lillianrubin.com/worlds-of-pain/.

33. Mark Greif, *The Age of the Crisis of Man: Thought and Fiction in America, 1933–1973* (Princeton: Princeton University Press, 2015), 272.

34. Simone de Beauvoir, *The Second Sex*, translated by Howard Madison Parshley (New York: Knopf, 1953).

35. Betty Friedan, *The Feminine Mystique* (New York: Norton, 1963).

36. Cotkin, *Existential America*, 256.

37. Tilly and Tarrow, *Contentious Politics*, 156–157.

38. Christine Stansell, *American Moderns: Bohemian New York and the Creation of a New Century* (New York: Metropolitan Books, 2000).

39. Alain Touraine, "History, Modernity and Global Identities," *Glocalism: Journal of Culture, Politics and Innovation* 2 (2020): 3–4.

40. Touraine, "History, Modernity and Global Identities," 5.

41. Ralph Nader, *Unsafe at Any Speed: The Designed-In Dangers of the American Automobile* (New York: Grossman Publishers, 1965).

42. Rachel Carson, *Silent Spring* (New York: Houghton Mifflin, 1962).

43. Richard A. Liroff, A *National Policy for the Environment: NEPA* and *Its Aftermath* (Bloomington: Indiana University Press, 1976).

44. Denis Cosgrove, "Contested Global Visions: One-World, Whole-Earth, and the Apollo Space Photographs," *Annals of the Association of American Geographers* 84:2 (1994): 270–294.

45. Adam Rome, *The Genius of Earth Day: How a 1970 Teach-In Unexpectedly Made the First Green Generation* (New York: Macmillan, 2013).

46. Holly Henry and Amanda Taylor, "Re-Thinking Apollo: Envisioning Environmentalism in Space," *The Sociological Review* 57:1 Supplement (2009): 190–203.

47. Andrew Jamison, "Climate Change Knowledge and Social Movement Theory," *Wiley Interdisciplinary Reviews: Climate Change* 1.6 (2010): 818.

48. Morris Dickstein, *Gates of Eden: American Culture in the Sixties* (New York: Basic Books, 1977).

49. Bertolt Brecht, "To Posterity," *Selected Poems*, translated by H. R. Hays (New York: Grove Press, 1947), 172–177.

50. Greil Marcus, "Who Put the Bomp in the Bomp De-Bomp De-Bomp?" in *Rock 'n' Roll Will Stand*, edited by Greil Marcus (Boston: Beacon Press, 1969), 8.

51. Louis Menand, *The Free World: Art and Thought of the Cold War* (New York: Farrar, Straus and Giroux, 2021), 332.

52. Robert Cantwell, *When We Were Good: The Fold Revival* (Cambridge, MA: Harvard University Press, 1996); Ronald D. Cohen, *Rainbow Quest: The Folk Music Revival & American Society, 1940–1970* (Amherst: University of Massachusetts Press, 2002); Stephen Petrus and Ronald D. Cohen, *Folk City: New York and the American Folk Music Revival* (New York: Oxford University Press, 2015).

53. Bob Dylan, *Chronicles* (New York: Simon & Schuster, 2004), 240.

54. Bruce Springsteen, *Born to Run* (New York: Simon & Schuster, 2016), 77–79.

55. Ron Eyerman and Andrew Jamison, *Music and Social Movements: Mobilizing Traditions in the Twentieth Century* (Cambridge: Cambridge University Press, 1998), 119.

56. Gary Snyder, *Earth House Hold* (New York: New Directions, 1969), 90–91.

57. David Abram, *The Spell of the Sensuous: Perception and Language in a More Than Human World* (New York: Pantheon, 1996).

58. Jerome Rothenberg and Diane Rothenberg, *Symposium of the Whole: A Range of Discourse Toward an Ethnopoetics* (Berkeley: University of California Press, 1983), xiii.

59. Gary Snyder, "The Politics of Ethnopoetics," *Alcheringa: Ethnopoetics: A First International Symposium* 2 (1976): 13.

60. Warren I. Susman, "History and the American Intellectual: Uses of a Usable Past," *American Quarterly* 16:2 Part 2: Supplement (1964): 243–263.

61. Peter Bacon Hales, *Outside the Gates of Eden: The Dream of America from Hiroshima to Now* (Chicago: University of Chicago Press, 2014), 272–273; Bob Dylan, *The Lyrics 1961–2012* (New York: Simon & Schuster, 2016), 154–155.

62. Greil Marcus, *Mystery Train: Image of America in Rock 'N' Roll Music*, fourth edition (New York: Penguin Books, 1997), 5.

63. William James, "The Perception of Time," *The Journal of Speculative Philosophy* 20:4 (1886): 397.

64. William James, *The Principles of Psychology*, Volume 2 (New York: Henry Holt & Company, 1890), 630.

65. Lewis Franklin Powell Jr, "The Memo" (1971), *Powell Memorandum: Attack On American Free Enterprise System.* 1 https://scholarlycommons.law.wlu.edu/powellmemo/1.

66. David Chalmers, *And the Crooked Places Made Straight: The Struggle for Social Change in the 1960s*, second edition (Baltimore: Johns Hopkins University Press, 1996), 178.

Chapter 11

Millennium

MILLENNIAL TIME

The millennium, a calendrical moment denoting a period of 1,000 years, gave rise to any number of cultural tropes. While having theological, psychological, and political significance, the millennial moment represents a thousand years after a specific starting point. A millennium is calculated by the "powers of ten," the integer powers of the number ten in mathematical reasoning.[1] Contemporary thought, by comparison, conceptualizes temporal moments in terms of experience and memory, viewing time as a cognitive mapping of human experiences. Ludwig Wittgenstein views such mapping as a learned capacity that, through abstract linguistic means, permits our differentiating "memory-time,"[2] an arrangement or ordering of events relying upon subjective memory, from "information time,"[3] a public chronology based upon clocks and calendars. Edmund Husserl views time, kinesthetically, as discrete moments when we recognize embodied others through our intersubjective experiences forming consciousness as a result.[4] Maurice Merleau-Ponty then understands that our work grounds our bodies, the world, and others through a "field of presence" that is temporal in nature.[5] For Jean-Paul Sartre, "the lived moment"[6] within this field may be one of recognition, an essential moment of existence where time and timelessness intersect on behalf of individual freedom. Framing time as a conceptual map in these ways supports "meaningful learning." Such learning "constructs the cognitive structure—stored in long-term memory—which is enriched through the activation of metacognitive instruments (such as conceptual maps),"[7] permitting rational understanding of life history and historical development according to Monica Sorrentino.

Beyond a mere interval of time, the millennium has come to have cultural significance for some individuals. At such a moment, certain long-held cognitive and behavioral patterns, namely the world experienced spatiotemporally that forms a person's conceptual map, are confronted by a projected world envisioned or conceived imaginatively. The millennium can then be understood as a breaking point, a time of unrest when polarities, such as harmony and discord, stability and instability, security and insecurity, are met with both individual and collective anxieties and fantasies reflected in various cultural narratives about the place of humanity in the world. For some, the millennium is a transcendent marker with eschatological significance. The moment is denoted with uncharacteristic forms of collective behavior deriving from a utopianism or magical realism with uncertain consequences when projections become productions in real time. For others, the millennial moment is a personal benchmark, provoking reflection on the objectivity of the transition as an historic phenomenon and perhaps even the prospect of a subjective experience of freedom from previous entanglements and worn-out habits and modes of thought. Past millennial moments also brought about nefarious collective responses, notably millenarianism, a belief that surfaced in the second millennium.

Millennial movements began to take shape in the medieval world, drawing adherents seeking apocalyptic and political validation as a response to certain abrupt changes, globally, during the first centuries of the second millennium. The year 1000, a calendrical milestone, provided a gateway to the "modern revolution,"[8] when humanity crossed a threshold leading to a more complex society, according to David Christian. New global pathways were formed that interconnected agrarian civilizations and frontier regions, and there were significant impacts of humans on the biosphere at this time. The global population reached 250 million, giving rise to migration and travel that along with complex trade networks led to a cultural exchange of ideas and faiths. A nascent, modern rational society with changing patterns of demographic mobility emerging at this time provoked centuries-long chaos especially in Europe. Norman Cohn explored how medieval revolutionary millenarians and mystical anarchists, reacting to a changing world picture, fantasized a new paradise on earth and prepared for a war between the forces of good and evil.[9] In so doing, millennial movements eventually led to religious wars, inquisitions, and genocidal massacres of Jews and Roma. Throughout medieval times, these fanatical movements with utopian goals unleashed social unrest and terror on behalf of "the urge to purify the world through the annihilation of some category of human beings imagined as agents of corruption and incarnations of evil,"[10] in Cohn's words. In the third millennium, similar dynamics would pervade everyday life across the globe.

THE MONOPOLIZED MOMENT

At the dawn of the third millennium, as the Cold War receded, there was hope that global peace was at hand after a century of wars and violent revolutions.

The year 2000 saw the promise of prosperity and the spread of democratic ideals following an era of perpetual war. All was shattered on September 11, 2001, when the United States was attacked for the first time since Pearl Harbor, sixty years earlier, at the onset of the Second World War. The attack on the World Trade Center towers in New York City was a devastating pivotal event in global affairs. The "Twin Towers" collapsed, setting into motion protracted global military interventions, increased domestic surveillance and militarization, clandestine politics, and new media platforms. A renewed patriotism grew amid the shadows of the fallen towers.

The Patriot Act, passed by Congress to prevent future terrorist attacks, arose, phoenix-like, out of the ashes. An ensuing diminishment of civil liberties and press self-censorship came soon after the invasion of Afghanistan. Americans were willing to trade their freedoms for a perceived sense of security in an increasingly dangerous world. After the terrorist attacks, more people placed an overriding trust in governmental agencies, including the Pentagon, the Department of Homeland Security, the Federal Bureau of Investigation, and the Central Intelligence Agency. National security concerns extended into state and local police departments, school and college curricula, and the content of television programs and commercial films. New York was swept up in a foggy haze after 9/11, initially from the toxic air around Ground Zero. There was the fog from ongoing collective trauma as well. New York had become a traumatized city, a city in mourning. On the spirit level, New York City was reeling. Traumatic memories of the 9/11 attacks at Ground Zero were embodied in the generations that experienced them, and many people's lives were upended, geographically and emotionally. It would take considerable time for the city and the nation to recover from the shock.

After the Towers fell, civic, philanthropic, clinical, and spiritual efforts helped to clear the psychic fog. In *The Empire of Trauma*, Didier Fassin provided an ethnographic account of how the 9/11 tragedy created "a new language of the event"[11] where the media and the emerging field of psychiatric victimology, together, cast the psychic trauma of victims and their suffering as a public spectacle. It became permissible to air grievances in the broadcast media and to cast blame, which routinely occurs after a tragic disaster, in wider public arenas. As a result, the victims and survivors, themselves, achieved celebrity or heroic status, notably the "9/11 Widows," and the "9/11 First Responders," many of whom were exposed to toxic chemicals after the attack. There were also the 9/11 Commission hearings, with governmental officials called to testify. Over time, the city and the world came to understand the crisis in these celebrities' faces and, in their voices and testimonies, through a global media lens. Recurrent images of the event, the survivors, the public officials, and the commentators flowing across countless media screens intertwined with their narrative recounting of memories and recriminations. Together, they monopolized the moment. Global news media outlets located in Manhattan rearranged themselves. Network and cable news programming now reflected the new exigencies. In the immediate aftermath of the attacks, these outlets restricted reporting critical of the government, ostensibly for national security reasons. There was an effort to return to regular programming over the next eighteen months, although news reporting took on a jingoist tone again in the lead up to the 2003 invasion of Iraq, offering considerable time on air to pro-war advocates. Following the occupation of Iraq, editors and producers largely presented an uncritical view of the war and buried certain stories, notably about brutal counterterrorist actions. Network executives appeared reluctant to face the wrath of the White House and the Pentagon, and more so to lose the support of their sponsors. Viewers' preferences also shifted toward reality television, cable news, and news-based late-night and weekly comedy shows.

The nature of newscasting changed from headshot reporting to moderated panels of commentators, leaving audiences enveloped in a manufactured mediascape of conflicting positions. Media organizations fed content to one another so that a story moved through the news cycle in print, online, cable, and network television, and then quickly disappeared. Produced as objective representations of social reality, media stories triggered anxieties, at the same time limiting any sustained discord that could actually challenge overarching institutional structures. Whether covering the threat of a terrorist bombing or an anthrax attack, reporting was limited to official stories derived from government sources. Designated experts, often former government employees, were hastily called upon to articulate officially sanctioned narratives. While expert commentaries were clearly meant to reduce anxieties, these left aside any explanation that could support systemic change during this time of heightened fear and uncertainty. Such restricted focus on iterative alterations were then presented across media screens as satisfactory solutions, thereby limiting further public attention. As the media lens moved on to subsequent events, anxieties and concerns about a preceding threat or hazard were refocused on the next source of alarm, while the system remained unchallenged.

At the same time, most Americans felt a need to keep in constant touch with loved ones and close friends through telecommunications and social media platforms. There was an increasing dependence on personalized screen culture, with obsessive checking in on smart phones and tablets, and ongoing messaging across social media platforms. A sea change thus occurred in the flow of information as the power of social media firms gained increasing control over commercial marketing and political campaigns. Only after considerable damage was done to the public trust in the electronic commons, specifically the social media-based disinformation that compromised electoral integrity, did certain professional and governing elites initiate steps toward regulatory oversight of these firms' platforms. Nevertheless, there is an unfettered anarchy of largely unseen content below the surface of the social media, in what is known as the "dark web," where lawless transactions proliferate. The enveloping digital infrastructure had nevertheless spanned both the public and private domains of the twenty-first-century life, capturing personal attention and considerable profits through domination over communicative behavior. Big Tech platforms and social media were developed and then proliferated in domains of personal, work, and public life as the third industrial revolution supplanted Fordism, the manufacturing technology that defined the very way of economic life for nearly a century and that would wane with the coming of a third millennium. However, in the new century, there remained troubling consequences for the many families and communities in deindustrialized regions where Fordist production and consumption regimes sustained distinct patterns of living and working for well over a century.

DEINDUSTRIALIZATION

The entangled story of the American automobile industry and the UAW, its core industrial union, is instructive for an understanding of deindustrialization and its crises. In the industrial postwar decades, the UAW was the nation's premier union. In 1948, it led the way in collective bargaining through an agreement with General Motors. The company agreed to provide autoworkers with an automatic cost-of-living adjustment and a 2 percent "annual improvement factor" so that workers shared in the company's gains. Throughout the postwar period, the UAW negotiated wage increases that set the standard for workers in other industries. The union also greatly reduced wage differences between skilled and unskilled work and bargained for greater welfare benefits. By the early 1970s, there were two automobiles for every three American adults. Auto manufacturing and its affluent workers were the pride of mass production unionism. At the same time, there was a shift to flexibly specialized enterprises, increased reliance on information and communication technologies, weakening state regulation, and the rise of globalization.

Postwar Fordism ended in 1973 with the Arab oil producers' embargo on shipments to the United States and Western Europe. The rise in oil prices and federally mandated gasoline rationing changed how the nation consumed manufactured goods. Americans drastically reduced their consumption of full-sized cars and other mass-produced commodities. Factory output fell, doubling both unemployment and inflation, and this trend continued over the next ten years. America's dominance in the global economy was also challenged by Japan and West Germany. These countries' lower-priced and often higher-quality products flooded American markets, further eroding the nation's manufacturing base. By the decade's end, the UAW led the way to post-Fordism. In 1979, union officials convinced workers to accept wage rollbacks in a federally orchestrated bailout of the nearly bankrupt Chrysler Corporation. Since that time, the UAW lost 500,000 members as a result of the Big Three automakers' strategy to slash payrolls in the face of foreign competition. American workers, in all industries threatened by foreign imports, lost about $500 billion as their wages, and other concessions were bargained away in the economic recession of the 1970s and during corporate restructuring in the early 1980s.

The world economic recession of the early 1970s and the subsequent radical restructuring of capitalism along global lines signaled hard times for working-class and working-poor families. The new capitalist regimes of flexible accumulation, or post-Fordism, brought an end to many manufacturing jobs, lower-skilled service jobs within the unionized blue-collar workforce, and vendor and commercial sectors symbiotic with larger plants in industrial regions. Millions of blue-collar workers in aging industrial cities found their jobs abolished or plants shut down because of changing technologies and corporate investment strategies. With the near elimination of manual industrial labor in this technological revolution, the federal government became involved in moving the economy toward a postindustrial economy through displaced worker retraining programs.[12] A dual labor market of segmented work framed the new employment opportunities, characterized by a primary labor market of stable, high-paying jobs with good working conditions and chances for advancement, and a secondary labor market of less stable, low-paying jobs with poor-working conditions and limited mobility. The more highly regulated jobs requiring specialized training and education were found in the primary labor market; by contrast, the low-skilled occupations in the secondary market required little education and had relatively few barriers to entry. Between 1963 and 1989, average weekly wages for the least skilled male workers declined by 5 percent, while wages for the most skilled men increased by 40 percent.[13] The earnings gap between high school and college graduates has since grown dramatically, an indicator of the dual labor market effect on working lives. A College Board study[14] found American workers with a bachelor's degree earn more than 60 percent more than those with only a high-school diploma. Those with advanced degrees earn two to three times as much as high-school graduates. Between 1950 and 1980, the flow of American capital overseas increased sixteen fold. American companies were investing profits in Asian countries like Singapore and in Mexico's industrial corridor along the U.S. border. By the 1970s, America's steel, rubber, and automotive industries were hard hit by the loss of sales to imported goods. A decade later, the newer telecommunications, computer, and electronics industries were also facing strong international competition. In response, companies in older production regions in the Northeast and the Midwest began relocating their manufacturing operations to *right-to-work* southern states like North Carolina and Texas, and also to factories abroad. Plant closings and industrial layoffs in basic industries like automobiles, aerospace, rubber, steel, and consumer electrical goods became commonplace in post-Fordist America. Small shops, restaurants, and other services were disappearing from working-class communities in mass production regions throughout the nation in the early 1980s as a result of major plant closures. What followed was the remaking of blue-collar neighborhoods into new immigrant enclaves, which had a significant impact on state economies in California, Florida, Illinois, New Jersey, New York, and Texas.

The long transition and meandering economic turns attributed to a declining postwar boom closely linked with a defense economy finally came to a global collapse in and after 1989, the year the Berlin Wall fell. For the next two years, as the unstable command economic systems collapsed in Eastern Europe and the Soviet Union, popular movements would demonstrate against and eventually oust their communist governments. These events had profound effects on families a world away in the United States, where arms production meant prosperity for millions of defense industry workers, both union and nonunion alike. The conclusion of the forty-five-year global conflict, which had structured American life, brought a halt to their dreams. With the waning of the Cold War, the aerospace industry downsized as companies converted from defense to civilian production. In 1991, the defense budget declined 13 percent from its peak in 1985. Job losses from defense cutbacks averaged an estimated 275,000 per year between 1991 and 1995. Postwar workers, especially those in aerospace and defense industries, shared a belief with their fellow Americans in the power of science and technology to provide a secure world and secure lives for themselves and their families. Although they acknowledged that the nation's global economic power had been compromised by Japan and Western Europe, their sense of America's political superiority remained unchallenged. Since the hiatus in global confrontation, Cold War families, in particular, endured greater anxieties. They were ambivalent and had feelings of self-doubt, having experienced the loss of the political consensus and sense of conformity, features of the postwar suburban family ideal.[15]

SAVAGE INEQUALITIES

The massive layoffs as the Cold War receded devastated the household economies of many families accustomed to the abundance that accompanied their rising standard of living in the postwar decades. Unemployment and displacement resulted in profound economic and emotional consequences for blue-collar workers and their families in deindustrialized cities throughout the country.[16] There was a concurrent narrowing of life worlds associated with long-standing middle-class occupations, which had held the promise of a successful adult life course for those adhering closely to a traditional career path directed by trade union and corporate personnel systems. Hence, many were preoccupied with the specter of uselessness or the prospect of skills extinction and the fear of being made redundant. Some managers, labor union officers, and political leaders stigmatized employees, union members, or constituents for inadequate early career planning in the face of their deskilling by computers or robotic technology or for their fates during economic downturns, often including job loss, bankruptcy, and foreclosure. In not providing a viable alternative, areas undergoing economic crises sustained a population of adults burdened by their pasts and limited in their ability to move forward in life. Tales of discontentment were heard in communities where substantial numbers of residents experienced downward mobility. Some reflected upon consequences caused by an earlier distraction from secondary schooling. Their stories often told of how poor academic performance and a lack of motivation led to dropping out of high school for a job that failed to provide lifetime financial security. Others voiced their dissatisfaction with structural conditions that limited their prospects, notably housing and especially health care. When hourly, temporary, and part-time workers seek health care, many are forced to confront how structural factors that impede access to biomedicine are rooted in their limited educational attainment. Frequently, those lacking a high-school diploma will experience greater challenges accessing and navigating care, leading to consequential adverse health outcomes.

In deindustrialized regions, cultural identity drove many white working-class voters toward conservative politics. These voters held grievances wrought by their diminished economic power and status in a demographically changing country. These grievances began to shape contemporary politics with the political realignment of country, beginning in the 1970s. A Fourth Great Awakening took place as many disenchanted, displaced workers in the Midwest, Southern, and West, including those among the quarter million who lost their farms in the 1980s Farm Crisis, embraced an evangelical Christian faith as solace. Since that time, the country became polarized around diverse concerns, including race, encompassing diversity, economic opportunity, and immigration rights; gender, ranging from abortion rights to the increasing presence of women in the workplace; and geography, notably the rural–urban divide and sectional tensions. Underlying the divide is an "affective polarization,"[17] where partisanship induces animosity, lending an emotional component to the division that influences attitudes and behaviors beyond politics.

Many experiencing emotional consequences of polarizing events sought an ideology of outdated and dangerous beliefs about race. Racialist beliefs had worn thin after decades-long legal decisions and progressive legislation, causing many to become unmoored after such myths were shattered. Then cognitive dissonance and confusion set in about their place in a pluralist nation. Many chose to resolve the uncertainty stemming from an inconsistency in their thinking by embracing a new civil religion. Some were also drawn to far-right political rallies and extremist groups to enact their desperations publicly and to seek fellowship among others who felt estranged from the familiar world of their youth. This idealized collective memory included imagined religiously and racially homogeneous local communities, even as regional life was becoming ethnically and culturally diverse. A national mythos of exceptionalism also receded after defeat in the Vietnam War. Then came an economic transition that permanently altered life in both the city and the countryside.

Caught up in the changes that swept away traditional images of family, work, and community worlds, many eschewed a rational resolution. This could have taken the form of a social therapeutic in the educational arena. Small group and classroom work typically provide a means for people in transition to unburden themselves of shibboleths and presuppositions and explore options, especially regarding what to hold on to and let go of, forming a new mazeway for their lives going forward. Instead, a disenchanted and desperate people holding extreme political beliefs became dangerous to themselves and others within the national political arena. When modernizing, secular forces prevailed, these political communities created their own "social matrix" based upon various markers of their identities—either on regional, ethnic, class, faith, or an amalgam of one or more markers—that defined their place in the world. According to S.H. Foulkes: "The Matrix is the hypothetical web of communication and relationship in a given group. It is the common ground, which ultimately determines the meaning and significance of all events and upon which all communications and interpretations, verbal and non-verbal, rest."[18] This matrix then provided a narrative for group behavior and a set of shared meaning and interpretations. The emergent pattern of ongoing interaction defined by the matrix forms a "social unconscious," which Earl Hoppe and Haim Weinberg theorize as

> the social, cultural, and communicational constraints and restraints of which people are to varying degrees unconscious. The social unconscious emphasizes shared anxieties, fantasies, defences, myths, and memories of the members of a particular social system. Its most important building bricks are chosen traumas and chosen glories.[19]

There may be a deeper story, one that combines trauma, despair, belief, propaganda, and in the end another reminder of existential defeat for "expectant peoples."[20] This term originally designated postwar indigenous populations living under untenable social conditions and colonial inequalities that sought a path to national integration. With the promise of social modernity were certain limiting conditions, notably primordial ties sustained through affective kin- and faith-based loyalties, together with ethnocultural affiliations remaining from an older societal framework. Along the route toward political development, charismatic leaders emerged to offer their constituencies new ideologies, at times using traditional religious symbols to promote promises of personal and communal liberation from social and economic suffering. In a similar way, Southern mountaineers and Mid-Southerners in the Mississippi Valley embraced an aspirational politics, expecting altered life situations by participating in state and national elections. Their aspirations emboldened far-right partisans and media, faith-based and big money supporters to sustain a fragmented political landscape.

Neoliberal America brought about widespread social upheaval beneath the surface of mega-corporate capture, global militarism, and structural change. For two decades, millions underwent downward economic mobility characterized by income insecurities and attendant inequality. The suffering accompanying this transition disrupted white working-class life, most severely in the Mid-South and Appalachian regions. These conditions brought "diseases of despair"[21] that caused many deaths at midlife from drug overdoses, suicide, and alcoholic liver disease. Anne Case and Angus Deaton also found distressing trends related to social inequality, noting that for the first time in decades life expectancy in the United States began falling in 2015, caused by suicides, alcohol-related deaths, and drug overdoses. They attributed such "deaths of despair"[22] to growing economic insecurities consequent to globalization, automation, and the downsizing of the manufacturing workforce over four decades. Those most vulnerable are non-Hispanic whites without a four-year college degree, ages 45–54, with people born in 1980 at much higher risk. Case and Deaton suggest that working-class life in the United States is more fraught than in other high-income countries, largely due to inequality. They also found that many lower-skilled employees rarely work in jobs that provide structure, status, and meaning to their lives. An accompanying politics of cultural despair emerged from severe socioeconomic disruptions, notably job and community loss, and early death that occurred in the years before the Great Recession. Together, these circumstances induced cultural trauma from conditions of neglect and displacement in rural and semi-urban locales.

Similar epidemiological consequences occurred in the 1970s, when companies in western nations began laying off large numbers of industrial workers at midlife. Whole communities and production regions devolved into drug abuse, familial chaos, and personal suffering on a scale that could not have been imagined during the confident years of the postwar decades in the advanced capitalist countries. Harvey Brenner analyzed the data and found increased first incarcerations, first visits to mental hospitals, homicides and suicides, and first heart attacks in the aftermath of wide-scale downsizing and massive layoffs, and these numbers held in Brenner's analyses of the United States, the United Kingdom, and Sweden, and then for the rest of the world.[23] In analyzing historical data for the United States from 1940 to 1973, Brenner found that

> A 1% increase in the unemployment rate sustained over a period of six years has been associated (during the past three decades) with increases of 36,887 total deaths, including 20,240 cardiovascular deaths, 920 suicides, 648 homicides, 495 deaths from cirrhosis of the liver, 4,227 state mental hospital admissions, and 3,340 state prison admissions.[24]

Brenner also found that economic inequality was strongly related to increased mental illness, criminal aggression, and alcohol abuse, all deeply related to conditions of alienation that surface during periods of economic instability.[25]

Thomas Piketty analyzed evidence for the inequality of capital, especially that of capital accumulation and exploitation during recent decades.[26] Although a longer trend, Piketty states that in its origins "capitalism can be seen as a historical movement that seeks constantly to expand the limits of private property and asset accumulation beyond traditional forms of ownership and existing state boundaries."[27] Increasing inequality is accompanied by the co-option of political power through regulatory capture, similarly a long-term trend. Piketty's explanations of inequality may indeed be cogent at a time when a neoliberal politics has set out since the 1980s to remake the world. He views neoliberalism as a reversal of the trend toward lower inequality, characterized by progressive income taxation and egalitarian social democratic policies in the decades after the Second World War. Piketty states that in the 1980s "the conservative revolution in England and the United States, financial deregulation, and the fall of the Soviet Union," together, "changed the atmosphere and ambient ideology."[28] Throughout, these two societies continue to hold strong neo-propertarian policies. He argues that an economic crisis, as the 2008 recession, did not necessarily reduce inequality. Rather, any change in inequality regimes depends upon mobilization across intellectual, political, and institutional domains, preferably based upon the various solutions emergent from the crisis, hence an ideological sea change.

A third industrial revolution in the late twentieth century brought innovative communications technologies and energy regimes, and also a neoliberal economy. Though, as Robert Gordon hypothesized, future changes in the standard of living cannot compete with that of the 1870–1970 economic revolution. During that "special century,"[29] in Gordon's words, a concurrent revolution in technological improvements transformed life through electricity, telephones, and automobiles and increased the quality of life in the United States. Gordon expects any number of headwinds to reduce current living standards in the United States, notably increasing inequality, poor-quality education, an aging population, and growing government debt.[30] With the ending of the long economic wave of the special century, there was a hollowing out of middle-income jobs and stagnated living standards in American society. Nowhere is the suffering that accompanies inequality more apparent than in the experiences of youth over the past two generations.

TESTAMENTS OF YOUTH

Accompanying the economic transition is a widespread culture of fear especially suffered by the young who, in Henry Giroux's words, "live in an era of foreclosed hope," and are thus rendered unable to imagine a life beyond "the fragility and uncertainty of the present."[31] For millennial youth, then, there is a "politics of despair,"[32] following Giroux, characterized by indifference, as they do not see a future for themselves. An indicator of this constant feeling of precarity is the violence the young experience in their lives at school and in their neighborhoods, compounded by incidents of foreign and domestic terrorism, which has set apart recent generations from earlier ones. Beginning in the late 1980s, the incidence of shootings and other acts of violence increased on school campuses throughout the United States. Between 1993 and 2000, the percentage of American high-school students who were threatened or injured with a weapon on school property remained constant, between 7 and 9 percent.[33] The Centers for Disease Control and Prevention, the U.S. Departments of Education and Justice, and the National School Safety Center collaborated on the first nationwide investigation of violent deaths associated with schools.[34] The study identified common features of 105 school-related violent deaths, of which 85 were murders of children and youth aged five to nineteen in communities of all sizes in 25 states between 1992 and 1994. The results indicated that a majority of the deaths occurred in urban communities or were associated with secondary schools; they also revealed that 65 percent of school-associated violent deaths were students, 11 percent were teachers or other staff members, and 23 percent were community members who were killed on school property. Both victims and offenders tended to be young and male; firearms were responsible for a majority of the deaths.

A domestic terrorist act occurred in the American heartland—namely, the bombing on April 19, 1995, of the Alfred P. Murrah Federal Building in

Oklahoma City, which killed 168 persons, including government employees, other adults, and babies and young children who were in the American Kids Day Care Center located in the Murrah Building.[35] On April 20, 1999, two boys engaged in a school shooting rampage that killed twelve students and a teacher at Columbine, a suburban Denver high school.[36] During the 1998–1999 school year, when the shooting at Columbine occurred, there were 47 school-associated violent deaths in the United States, including 38 homicides, 33 of which involved school-aged children.[37] Between 1992 and 2001, there were 35 incidents in which students started shooting at fellow students and teachers at school or at a school-sponsored event, leaving 53 dead and 144 injured.[38] In the aftermath of these deadly school shootings, psychologists sought to link such aggressive behavior with exposure to violent video games that required players to identify with the aggressor. These games prime the player with aggressive thoughts and provide a way for a player to learn and practice new aggression-related scripts that can be accessed in real-life conflicts.[39]

In the wake of the acts of school violence and their aftermath came a "culture of fear" and perhaps even a "war against youth," according to Giroux,[40] with an attendant increased suspicion of the motives of young people and, consequently, calls for increased surveillance and greater administrative control over students' curricular and even extracurricular activities. Giroux views the public school as a contested terrain. There was a gradual move away from the view of education as a social investment in the young on behalf of democratic values toward deep distrust of the young. This has led to increased standardized testing in the schools and widespread drug testing for entry into many workplaces. Suburban communities became the locales for the unprecedented wave of lethal school violence, which began to erupt in the 1980s together with neighborhood violence. In the postwar decades, middle-class parents moved their families to the suburbs. They hoped the suburbs would be a place where they could raise their children and develop friendships. For many adolescents marginalized by their peers and the changing world around them, suburbs ceased to provide opportunities for social development and the "civilizing" propensities promised by Frederick Law Olmsted over a century ago.

Unlike the urban school violence fueled by poverty, interethnic conflict, and illicit drug subcultures, the violent rampages in the suburbs reflected a youth culture that was virtually unknown to the adults in these communities. In many cases, parents and teachers had no idea of how their children were responding to changed community conditions in the new suburbs emerging out of farmlands or to the sources of personal and interpersonal conflict, including bullying, that marginalized the more violent youth. The suburban youth violence of recent decades points out how poorly some adolescents identify with their communities, seeking out, instead, small cliques or global chat rooms for their friends. The safety net that the family, school, and community used to provide became unraveled; consequently, the world now appears to be as unsafe for children and youth as it is for adults.

Twenty-first-century youth are tragically overwhelmed with ongoing reports of school violence, natural disasters, and war horrors. Graphic images of death and suffering are commonplace in global media. Television and computer screens are filled with images of death, contorted bodies, and suffering people, especially children. The reporting on terrorist bombings, school shooting massacres, and the widespread carnage of warfare has left children and adolescents insecure and frightened. With repeated exposures, young people's minds become numb, as they attempt to accommodate these assaults on the idea of safety with their own need for security and a sane world. When we listen closely as they play or rap, we hear that they are worried, frightened, anxious, and insecure. They talk about the dangers of school violence, and they are apprehensive about natural disasters and pandemics that might occur. They ask questions about wars and death and are curious about weapons used in warfare. They disclose fears of potential bad things happening to them and their family. In the past, the family was an effective buffer against the horrors of daily events. But families have changed, ties have weakened, and conflicting interests have pulled members apart. The family is no longer the secure base it was in the past nor is the residential suburb able to maintain the promise of a secure domestic refuge from urban fear. In place of the family, contemporary youth engage in screen-based and consumer-based lifestyles to direct their lives and preserve their identities in the face of dangerous ecologies.

Private security personnel and electronic surveillance disrupt the lives of the young, together with constant watch from the police, either on the ground or in helicopters, especially within marginalized communities. The impact of any community-scale event is multiplied when friends and loved ones are killed or hurt, as in the case of school and neighborhood violence. Human catastrophes such as school shootings and neighborhood violence, pandemic disease, natural and technological disasters contribute to the young's ecology of fear. Manufactured risks brought about by technological modernization also render immediate surroundings unpredictable, exposing youth to the risk society. Youth are especially place-oriented, as locales convey a concrete sense of security and familiarity to their lives, and a sense of loss is common among those youth who have experienced community-scale crises.

Consequently, their ability to trust the adult world to meet their social, economic, and psychological needs—the very basis of the idea of sustainability—becomes greatly diminished. Rather than seeking long-standing forms of support in a dangerous world, youth represent their mistrust of an adult world that seemingly failed them through atomizing social media platforms. Within these digital subcultures, marginalized adolescents can express grievances and may even attribute blame for their chaotic lives to a particular social class or ethnocultural group. Reliable stereotypes that maintain ethnic, generational, or class boundaries thus displace personal resolutions to institutional and systemic concerns. Despite these attributions, in the face of extreme situations such as school and community violence that affect their sense of security and well-being, many other youth became engaged in proactive behaviors. Emergent groups that typically form to support survivors after a violent event are frequently heralded through mainstream and social media. These local responses may become broad-based movements, as evidenced by widespread social activism by youth for gun control measures and in mass protests of shootings in schools and violent policing in minority communities.

Max Weber viewed lifestyle as the interplay of life conduct and "life chances."[41] Life conduct refers to "choice and self-direction in a person's behavior."[42] Life chances indicate "the crystallized probability of finding satisfaction for interests, wants, and needs, thus the probability of events which bring about satisfaction."[43] Life chances are determined by a person's life situation, or structural conditions, such as social class, ethnicity, gender, and age, which shape lifestyle choice.[44] For Weber, the choice of lifestyle is related to socioeconomic status, and ultimately to consumer behavior. The global marketplace targets distinct consumer cohorts for products and services, but also for the lifestyles associated with their consumer behavior. Employing digital technologies, a millennial generation encompassing 92 million Americans born between 1980 and 2000 cultivated new media habits for accessing information and entertainment, retail shopping, dating, gaming, and activism. The young are digital natives who came of age during a third industrial revolution and neoliberal globalization. Online activities and generational experiences influenced their attitudes, behaviors, civic orientation, and voting preferences, setting them apart from previous generations. In the wake of threats to the public sphere, millennial youth, as a cohort, adopted the more privatized, individualized, and miniaturized mass technologies and their software programs, such as e-mail messaging and social networking sites, to define their lifestyles.

Personalized portable media players, such as the *Apple iPod* introduced in 2001 together with *iTunes*, came to supplant the social interactions that typically occurred around music in earlier youth cultures, including singing, playing, dancing, and listening together. Now, youth carry their music with them and live within a personal musical universe of their own making, downloaded from a streaming music website or mixed on a home computer. These cohort-focused buying decisions have increased demand for "next generation" hardware and software and up-to-the-minute digital content, in many cases to maintain their socially networked, computer-screen identities. Digital devices and communication platforms are the means for tracking peers' progress through the world and communicating with them. These are also platforms for purchasing varied commodities and for online gaming that supply distraction and support a consumer-focused personal identity and an atomized individualism.

MINIMAL SELVES

Christopher Lasch examined the move toward a postwar survival mentality that normalized political, military, and ecological crises and reaffirmed everyday life in the wake of these extreme situations. The "minimal self" is one that is ungrounded, threatened with disintegration, and preoccupied with self-concerns amid disruptive systemic crises.[45] While a theme within American writing in the postwar decades,[46] in sociological analyses of life in total institutions,[47] and in the works of psychologist Bruno Bettelheim,[48] the image of the survivor was latent in mass culture from 1945 through 1975. Exceptions were media coverage of basement and backyard fallout shelters and reminders, in Arendt's words, of the "radical evil" of Nazi genocide,[49] notably through the publication of Anne Frank's diary, which was dramatized on Broadway and in a Hollywood film.

For Terrence Des Pres, who draws from accounts of life in Nazi and Stalinist death camps, the struggle to wake up in the face of camp existence and the resolve to turn away from the horror and to take action toward life, together, characterize the survivor, who is "anyone who manages to stay alive in body and in spirit, enduring dread and hopelessness without the loss of will to carry on in human ways. That is all."[50] For Des Pres, extreme situations help point us away from nihilism toward a biological wisdom, both life affirming and life sustaining, derived from happiness in our work and sociality, and from our physical joy as natural beings.

Despite attempts by Arendt and Des Pres to explain the nature of "existence in extremity," in Des Pres' words, and its ethical lessons, what appeared to be most prevalent in the mass culture of the early postwar era that celebrated America's triumph over its external enemies as the dominant narrative. Growing up within the mass culture of 1950s and 1960s America meant witnessing a celebration of victory in war through toys, comics, films, news reports, and wartime stories told within families. Paul Goodman saw through these consumerist and media-informed images of American power targeted to postwar youth, and also the oppressive schooling for conformity to an organized society in *Growing Up Absurd*[51] and identified early on a youth subculture fueled by alienation from the very institutions such socialization was meant to legitimize.

The end of the Vietnam War, Watergate, and the 1973 oil crisis, each helped to erode the national self-image as victorious before all enemies and to move the mass society toward a survivor narrative, one that was based in the disillusionment of the adult Second World War generation and the bewilderment of its youth. The talk of being a survivor—of wars, disasters, and traumas experienced, and of psychic wounds received in the family, neighborhood, and workplace—emerged as a cultural narrative, an amalgam of media, consumer, and therapeutic elements.

After 9/11, individuals feared being caught or trapped in new public spaces, including streetscapes and city centers that were designed by expert architects and planners and under continuous monitoring and surveillance by expert systems.[52] The young men who mounted the terror trained in south Florida; abducted planes and passengers in Boston, the home of the Freedom Trail of American Revolutionary historic sites; and then used American technology to commit mass murder at two strategic locales: lower Manhattan, at the heart of the nation's global financial markets, and the Pentagon, the center of a global military–industrial complex. At the same time, they trusted safety experts and law enforcement officials to make the skies safe again, to keep anthrax out of the mails, to profile potential criminals, to detain potential shooters from entering schools and other public buildings, and to police the world to prevent terrorist attacks upon the homeland. With the terrorist attacks that took place on September 11, 2001, the narrative moved toward a convergence of personal and collective survival, including governmental strategies of survival in a dangerous and uncertain world where office buildings, airports, shopping malls, gas stations, and even schools became the targets of potential terrorist attacks. This more recent iteration of the "survivor narrative" was built upon the earlier iterations, but with invasive aliens penetrating American communities, whose actions and terrorist affiliations then darkened certain groups' views of refugees and immigrants.

With this emphasis on personal survival came a new individualism and "pathologies of the achieving self."[53] This took the form of sadness, anxiety, and obsessiveness in meeting one's self-expectations, but also as fear and terror when one falters or even momentarily slips off the ladder of ascent. These recent pathologies of high-achieving youth place them at greater risk for substance abuse, depression, eating disorders, suicidality, and other emotional states that fuel detachment from the self. For youths between fifteen and twenty-four, suicide is the third leading cause of death and the second leading cause of death for college students.[54] The tendency to conceal suicidal thoughts or intent from others may also reflect norms underlying the new individualism on the high school and college campus, which only heighten the sense of isolation experienced by troubled students.

The "Ground Zero" decade of contemporaneous influences imposed upon youth mixed messaging, which interfered in developmental cycles that previously were normal maturing stages of growth. The media ecology encompassed complex historic and social benchmarks to underscore a transitional context of dynamic societal changes that only compromised youth development. Conflicting demands were placed upon youth that essentially made them dependent, and even passive, subjects of institutional uncertainty. At the same time, there were pressures to achieve autonomous individuality as a form of self-preservation. Then a new rhetoric of self-improvement emerged as yet another contradiction of societal-induced cognitive dissonance. This brought about stigmatic levels of contractual conditioning within the school and society ranging from fear of failure to self-determination under managed behavioral expectations. These many levels of necessity share a common denominator of anticipatory anxiety and personal withdrawal into social media and other solitary pursuits. Accompanying the new "necessities" were stress-reduction techniques to cope within various "post-traumatized" social milieus in a world of unrelenting expectations and critical misgivings. Freudian psychoanalysis faded during this time in favor of more action-based cognitive formulations geared toward creating a present momentum of succession in real time, modeling reality based upon individual and situational awareness. The advent of cognitive therapy matches computational programing precisely, or at least coincidentally.

A form of "monitored behavior,"[55] originally the domain of bureaucratic supervisors or agencies, that documents every conceivable aspect of personal, school, or work life has moved into the domestic worlds of high-achieving families, with parents carrying out monitoring tasks. Even more intrusive on the inner lives of adolescents and emerging adults is the self-documentation of time and performance using personal organizers, now fully electronic, and even of reflection in the form of journaling or diary keeping, all to support personal effectiveness and life management. By the 1990s, a cognitive-behavioral therapeutic orientation came to dominate the clinical psychological universe to treat a host of problems within the person, under the direction of the therapist, who guides the patient through a standard, manualized treatment protocol. Cognitive-behavioral therapy is meant to help the patient to identify the cognitive structures and interpersonal behaviors that attend to consequent maladaptive behavior, with the goal of facilitating a new conceptual framework and inducing behavioral change.[56] Self-monitoring techniques, such as "mood monitoring" of daily changes in expression of affect in relation to everyday events, and "thought records," where patients practice writing down their thoughts about certain situations that bring about anxiety, are integral to cognitive-behavioral therapy. These tracking tools are meant to help patients cope with their distressing thoughts through a focused self-evaluation of their environment.

Contemporary technologies of body governance[57] have increased patient demand for pharmaceuticals and have shaped consumer behavior in medical self-care and even self-experimentation in the interest of "maximizing our corporeal existence."[58] These hybrid initiatives in biomedicine include notions of the active patient, consumer choice, individual responsibility, self-regulation, and frequently, self-surveillance. Each involves some degree of engagement with medical professionals and adherence to their advice. Attendant with these trends is "biocommunicability," as understood by Charles Briggs and Daniel Hallin, or the manner through which informational flows, communicative ideologies, and discourses about biomedical knowledge together produce a sense of freedom for some and feelings of exclusion and even constraints for others.[59] Pharmaceutical marketing and media messages for biopsychiatric medications, together with health news about new drug therapies, typify this trend toward communicating to patients on how best to choose from available treatment options.

Affluent families do not await such marketing reminder messaging or for a crisis of survival in the physical, psychological, or social health domains, to prepare the next generation for the prospect of precariousness. Wealthier parents set about to protect their children from birth, through high-quality housing, together with privatized childcare, health care and education. In these ways, affluent parents provide their progeny with the ability to withstand the economic consequences of the business cycle through transgenerational transfer of wealth. With economic security, children of elites are assured immensely better life chances than others in their age cohort, as their educational, health care, and housing needs will be met throughout their lives, should personal catastrophes not befall them.

Poorer families, by contrast, rely on public schools and safety net programs for their children's preparation for life. To support early childhood in lower-income families, publicly funded Head Start programs promise a good beginning, educationally, medically, and psychologically, for children in these developmental settings. However, what occurs in their public schools and neighborhoods will need to be factored into children's developmental trajectories. As a consequence, poorer children's life chances are frequently limited by their growing up in disruptive and often dangerous social environments and by governmental budget austerities that affect their education and health care. Disparities then occur among these children in social indicators, such as educational attainment and quality of life.

Social-class disparities thus govern children's life chances and opportunities in neoliberal cities where per capita spending on schooling has diminished in low-income neighborhoods. The decreased tax base accompanied white flight and suburbanization, as wealthier families continue a century-long process of abandoning the urban core. The more affluent millennials also moved to the suburbs for educational opportunities they view as far better and safer than city schools. There they sought the security of suburban life over the risks and dangers to their children and themselves, should they remain living near the dense urban core. The suburban areas devised neoliberal solutions to their problems through rearranged class relations, retooled educational systems for class reproduction, and an entrepreneurial governance style, each adaptive to the new economy. These same arrangements affected urban-rooted families remaining traditional in both lifestyle and mindset, often holding outdated beliefs about their children's educational achievement that made it difficult to adapt to changing conditions.

A global economic crisis would touch the disparate worlds of city and suburb, each experiencing precarity in a neoliberal era characterized by privatization, monetization, and fiscal austerity. The global crisis would in time further unsettle and upend the lives of the young—those caught between two cultural worlds in the cities and others facing risk and insecurity in the suburbs—in a country that for three decades had undergone a transition from industrial to postindustrial life marked by decreased social spending, increased military budgets, and market deregulation.

NEOLIBERAL MONETIZATION

With the monetizing of communication and data, the intensification of capital, itself, was converted by computerized systems. This third industrial transition encompassed land domination, commercial asset coordination, and control of information, thus not solely based on the distribution of goods or real property transactions. While technological advances appeared to drive the transition, the actual change was not mechanical but that of capital intensification. Capitalization itself was the dominant driver, and not the move from manual to automated labor, or the conversion of natural and synthetic materials into manufactured products. Transformation of natural, human, and information resources through capitalization cuts across diverse global geographies and populations. A favorable environment for capitalizing is characterized by wide-scale diffusion of information, consistency of markets and the regulatory environment, and flexible movement of people, the very basis of globalization.

The financial sector—banks, investment firms, insurance companies, real estate firms— generated substantial wealth for global elites over the decades. The digitization of the financial services industry and the exponential growth in information technology brought about a "new financial order."[60] So many people entrust these firms with managing their mortgages, pension funds, and retirement savings. Though with the Great Recession, banking instututions became unstable when the real estate bubble burst causing borrowers to default on their mortgage payments upending markets worldwide. The economic crisis that began in December 2007 affected millions of families. Americans were shocked by the financial crisis and the economic near collapse from internal banking fraud and the subprime mortgage crisis. The Great Recession eventually abated at the end of the decade and the nation's economic health was restored for families at the top of the income distribution. Most other families holding significant debt faced downward mobility, and for the poorest families, there was the prospect of homelessness.

On Monday, September 15, 2008, or "Black Monday," the Dow and the S&P fell precipitously. After that day, Lehman Brothers filed for bankruptcy, and Bank of America said it would buy Merrill Lynch. In 2009, 1,306,315 bankruptcy cases were filed in federal court, with homes in foreclosure increasing to 4.58 percent that year. Between 2007 and 2016, 7.8 million foreclosures occurred. The federal Emergency Economic Stabilization Act of 2008 authorized $700 billion for the Treasury Department to selectively bail out certain institutions to prevent a collapse of the financial system.[61] The ailing automotive industry, an early adopter of robotics, artificial intelligence (AI), and machine learning in its operations notably in production and logistical management of the supply chain, also required a federal bailout. Should the Big Three automakers have declared bankruptcy, countless hundreds of thousands of workers in manufacturing, and many hundreds of thousands more in secondary and service industries in workers' home communities, would have lost their jobs. This would have severely limited a means of livelihood that supported an urban modernity in the Midwest, a key production region for a century.

The American Recovery and Reinvestment Act of 2009 (ARRA), a $787 billion stimulus legislation appropriated $267 billion in direct spending for state fiscal relief, including unemployment benefits, food stamps, family assistance, and Medicaid. The ARRA appropriated another $308 billion in discretionary funding for "scientific and medical research, clean energy, transportation and water infrastructure, public housing, education, modernization of medical records and the electric grid."[62] Intended as a quick economic response, funding decisions were based upon a distributive politics and not necessarily where economic precarity was predominant. Federal stimulus funds bolstered certain regional and economic sectoral interests, thus creating an uneven geography of recovery that privileged locations that supported the federal policy agenda, specifically energy, infrastructure, and research. The ARRA provided National Institutes of Health with $10 billion in new funding and $3 billion for the National Science Foundation. This reinforced a national research and development infrastructure of large universities and industrial laboratories and in so doing their regional economies, typically comprised of advanced technological enterprises.

Over a decade later, there remain local, national, and global consequences, both political and personal, as a result of economic precarity wrought by the 2008 crash. According to Peter Cappelli,

> One in five employees lost their jobs at the beginning of the Great Recession. Many of those people never recovered; they never got real work again. . . . A generation of young people entering the job market had their careers disrupted by it. The fact that this age group continues to delay buying houses, having children, and other markers of stable, adult life is largely attributed to this.[63]

The country eventually recovered, more racially divided and with greater social inequality and the near elimination of repetitive middle-skill work that required more than a high-school diploma and less than a four-year degree. These include routine-cognitive occupations, such as clerical, administrative and sales, and routine-manual jobs, including production and operatives. Christopher Foote and Richard Ryan studied job losses in middle-skill occupations, the group that suffered the greatest from automation and international trade and found "Unemployed middle-skill workers also appear to have few attractive or feasible employment alternatives outside of their skill class, and the drop in male participation rates during the past several decades can be explained in part by an erosion of middle-skill job opportunities."[64]

In the financial downturn, Richard Wilkinson and Kate Pickett[65] hypothesized that social inequality leads to considerable deleterious outcomes, including lowered educational quality, unequal opportunities and limited social mobility, together with health effects, such as obesity and chronic diseases. A decade after the recession, Wilkinson and Pickett found that people get sick and worse during economic downturns and when in a chronic state of economic distress there are widespread psychological costs, which affect a deeper and more intimate level, typically in relationships with friends and family. Feelings of self-doubt, social anxiety, stress and fear, wrought by stagnant wages and economic austerity measures and attendant cutbacks in the social safety net of public benefits and services result in surging mental health problems.

Federal bailouts and the stimulus reoriented the country's economic geography and reflected the consequences of a neoliberal economy that had developed since the 1980s. Entering with the twenty-first century was a two-tier economy and society and a dual labor market with many people living at the high and low ends and with fewer in the middle. There continued to be booming areas, where energy production, information technology, and biotechnology reign. However, beyond the decades-long impoverished rural areas, hard times remained for many living in new urban wilderness areas such as Camden and Detroit that emerged bankrupted by the economic crisis. These cities appeared to have been bypassed by a federal government fixated on bolstering the local economies in regions that enhance elite power and control over innovative technologies, rendering other areas the domain of forgotten men and women. To be sure, these deindustrialized zones are the result of a longer postindustrial transition, and a privatized, consumption-driven, individualism that dominated the socioeconomic landscape.

Many came to realize that the United States had moved toward becoming a nation with far too many cars on the freeways; increasing numbers of television channels; a craze for purchasing personal goods online; and skewed salaries of corporate executives, professional athletes, and college coaches. Young people across the country, feeling disenfranchised by the austere socioeconomic relations and the atomized digital worlds that condition life under neoliberalism, participated in short-lived localized movements in the Great Recession's wake.

OCCUPY WALL STREET

Activists coalesced in the Occupy Wall Street (OWS) movement that began on September 17, 2011, in Zuccotti Park, or Liberty Plaza, in Manhattan's financial district. The movement designed an organizational repertoire of contention that included an online presence, notably Facebook. These protest-related tools and actions allowed OWS actors to enter a public sphere typically dominated by elite discourse, and eventually capture global attention. An essentially leaderless movement, OWS inspired an alternative community and a communal discourse on economic inequality centrally located in a place that represented corporate financial control. David Graeber understood that the movement would operate on the basis of direct democracy:

> . . . organizers made the audacious decision to operate not only by direct democracy, without leaders, but by consensus. The first decision ensured that there would be no formal leadership structure that could be co-opted or coerced: the second, that no majority could bend a minority to its will, but that all crucial decisions had to be made by general consent.[66]

The collective nature of leadership in the "occupied" spaces[67] also made it difficult for law enforcement to identify "the head" so as to crush it. How the movement communicated key information indicated the nature of dispersed leadership. Throughout the day, a call of "Mic check!" signaled participants to break away from private conversations and unite in communal discourse. Participants in the large OWS general assembly responded to a speaker as a "human microphone" analogous to repetition of religious or party dogma.

The group in the front repeated the speaker's words in waves of repetition, so those behind them could understand what is being said.

Young people in parks and public squares clearly resonated to a feeling of *communitas,* or an unstructured, somewhat liminal, or temporary state where many people feel equal. This community spirit was generated and reinforced by the secular rituals spontaneously created in the occupied spaces, where many lived and slept for the duration of the occupation. For the occupiers, collective passions and experiences ran counter to their experience of growing up in digital-oriented homes and atomized communities. A group voice radiating out from the front of the assembled throngs through crowd repetition of speakers' comments conveyed a sense of collective identity. This communication mode reduced the digital loneliness typically experienced when using smart phones, *iPads,* and other tablet PCs. Though with online activity on Occupy-linked websites, organizers were able to "turn the structural potential for contention into confrontations with authorities and elites," as Tilly and Tarrow point out, "through electronic mobilization empowered by the recent innovations in digital technology."[68] In this manner, participants built "an electronic repertoire of contention" for diffusion of movement activities.[69]

Wall Street workers, for the most part, watched from their windows with smug indifference. Though as younger workers looked on, some may have understood, albeit momentarily and from afar, that the OWS protests externalized their apprehensions. The constant fear of getting fired or being laid off, and the endless hours spent working, may have led a few to fantasize about living in the alternative world being enacted just below them. For many younger financial sector employees, their workplaces were a source of psychic stress, where they are always on their phones and available for text messages throughout the day and evening. First-year analysts worked roughly 105 hours per week, leaving little time for much else. Socialization into a hard-driving workplace culture typically involves the newly recruited to compete with their peers, beat out competitors, and often mistreat subordinates to please their superiors. Analogous to an aggressive fraternity culture characterized by excessive drinking and bullying, new recruits suffer physical pain and mental anguish. A sign of mental captivity is sleep deprivation and other forms of hazing with the goal of maintaining control and loyalty amid suboptimal working conditions and attendant mental stresses. Gareth Morgan refers to such work organizations as "psychic prisons,"[70] to suggest that people bound in situations beyond their control. In these workplaces, aggressive and often sadistic employers haze new hires, demanding compliance with the deadening routines attendant with the cultural logic of late capitalism. Emotional consequences for these overly stressed young workers include prescriptions for antidepressants and antianxiety medications, and more desperate actions when these no longer work.

For New York City elites, the OWS events moved from slow institutional denial, aided by initial media neglect, to attempts by authorities to forcefully manage the situation with barricades and police patrols on the street. As the early autumn chill enveloped lower Manhattan, fire marshals monitored the encampment, directing police to confiscate generators and remove gas cans from the park, ostensibly for public safety reasons. Expropriation of power-generating sources was a pretext for seizing control of a place that flourished for weeks as a "hypercity," an essentially liquid spatial formation with an assemblage of workstations, working groups, and any number of political and artistic events, or "scenes."[71] The Federal Bureau of Investigation and Department of Homeland Security, together with the local police and private security firms, are affiliated as the Domestic Security Alliance Council. The Council then coordinated to take back control on behalf of safety concerns, seizing the moment away from the spectacle of resistance and insurgency of movement activists. In the end, those involved understood implicitly that all would eventually return to "normal" routine security, with assurance of safe streets and city sidewalks for an unimpeded journey to and from work for financial district employees.

Contemporary social movements protesting conditions of inequality and their psychic stressors are typically constrained by elite attempts to co-opt them. Weber observed one such process, namely, "the routinization of charisma"[72] in both religious and political movements, whereby the followers' early enthusiasm over a charismatic leader who spearheaded the movement eventually becomes consolidated into an easily co-opted rationalized organization. The once extraordinary becomes the everyday, and the collective behavior within a movement shifts from early adherents' oppositional stance toward traditional authority to more institutionalized and routinized organizational styles of later followers. The short-lived OWS activities challenged Weber's thesis. Rather than establishing an entrenched hierarchy, OWS participants dispersed together with their organizational repertoire. They diffused the movement's central messages, provoking global conversations about corporate power, systemic inequality, and a counterpolitics.

Sensing a collective narcissism in the protesters, philosopher Slavoj Žižek warned that, should these warm, loving feelings associated with *communitas* continue beyond the initial social euphoria, the OWS movement will have lost sight of its initial, formidable political impulses.[73] Žižek predicted that these sentiments conveyed a false sense of *gemeinschaft*, or close social relations based upon personal ties, among OWS adherents, and perhaps even contributed to a political amnesia, rendering them incapable of remembering the conditions and the passions that initially attracted them to the movement. Beyond Žižek's pessimistic assessment of the authenticity of this anticapitalist movement, others observed that OWS insurgents activated a movement culture that sought a change in consciousness. Their protests served to call attention to the seeming inauthenticity of neoliberal discourse surrounding fiscal austerity in the wake of the Great Recession. OWS promoted increased awareness of neoliberal property and regulatory regimes and their computational logics based upon what Mark Fisher designates as "capitalist realism."[74] The term implies the limited viable political and economic alternatives to the current system regulating work, education, and the production of culture, one that relies on a fossil fuel-based carbon economy. Despite the abrupt end of the two-month occupation and a scattering of the activists, a public conversation continues on the current modes of systemic capitalism's control of individual's lifeworlds, a system that OWS protests attempted to resist in their miniaturized hypercity formation. As a movement, OWS elucidated certain features of contemporary life that had become increasingly abstract and controlled through dual forces of advanced digital technologies and the monetization of everyday life.

A COUNTRY OF THE YOUNG

Younger generations enter an adult world that continues to support a fossil fuel economy and global militarization but neglects to deliver enough well-paying jobs and low-cost higher education. Elected representatives persist in funding protracted wars and homeland security emergency and surveillance measures while appropriating only limiting funding of public goods. Giroux explains how governmental disinvestment in public goods and the curtailment of civil liberties only heightened older generations' ontological insecurities, decreasing their motivation to focus on younger generations' needs:

> Limiting civil liberties, cutting back social programs, defining democracy as expendable as part of the discourse of emergency time, and appealing to the culture of fear prevents adults from focusing on young people as a symbol of the future and creating the symbolic and material conditions for increasing the scope of those values and freedoms necessary for the young to become active and critical citizens willing to fight for a vibrant democracy.[75]

With increasing discontent over lives rendered abstract by digital technics and monetized norms, many among the young were eager to participate in mass social movements. However, activism was dimmed momentarily by the two early millennial global crises. Following the 9/11 attacks, executive crisis managers and bipartisan legislators achieved consensus toward military solutions, which limited movement activity to only brief demonstrations against waging war in Iraq. After the Great Recession, while there was the short-lived Occupy movement, top-down executive action and legislative consensus prevailed in resolving the financial crisis. Protracted wars and economic recovery efforts thus expended considerable political capital, pushing aside many domestic and global issues. However, the changed conditions that surfaced during the two crises eventually provided organizing opportunities for civic engagement for social justice, climate action, and school safety in the educational arena.

Christopher Thomas explores the strong tradition of student and teacher civic activism that are intertwined and rooted in larger societal civic movements.[76] He points to connections between student protests and teacher strikes, especially ways that the efforts of student and teacher activists inform future educational policy. Activist movements in education date back to the common school movement that increased public schools in the mid-to-late 1800s. In the postwar era, teacher civic activism emerged, including unions and strikes in the 1940s through the early 1960s, and on to contemporary labor activism. Thomas then writes about the history of student protests in the United States through to the protests following the school shooting at Marjory Stoneman Douglas High School in Parkland, Florida, in the spring 2018, which were a catalyst for the National School Walkout, teachers' strikes, and women's marches at the same time. The activism of both students and teachers extended into these larger movements.

Throughout the twentieth century, teachers and students mutually influenced one another beyond the classroom. In the postwar era, they protested alongside one another against segregation and the Vietnam War, and also for the expansion of student rights. Thomas locates recent protest formations through what he calls "liminal shadow spaces"[77] away from the mainstream policy-informed sites, where people frame counter-hegemonic initiatives and thereby exercise their civic agency to coalesce on behalf of systemic and societal change. These shadow spaces, within local teachers' unions, and in students' delocalized social media, gave voice to the larger protest cultures that proliferated nationwide.

Thomas views counter-hegemonic activism over the past decades as a response to the 1983 federal report, *A Nation at Risk*, which held that the state of American public schools threatened the country's economic and political dominance. This policy report provided a master narrative that emphasized economic development as the primary goal of education. The central premise of the report is that of schooling for economic competitiveness and the need to create federally supported centralization of educational decision-making, together with market logic, to impose tighter controls on schools, teachers, and students. These included tighter accountability measures and school disciplinary procedures, a narrowing of the curriculum, and the abandonment of community control over educational decisions.

Students' and teachers' activism thus constitutes a rejection of the "A Nation at Risk" paradigm that oversees the modernization of public education in a neoliberal framework. Activist movements provide a counter-hegemonic context for the civic agency of both students and teachers. Students no longer wished to be the "products of education" or mere "passive consumers of educational policy."[78] According to Thomas, students rejected mainstream criticisms that they protested solely as victims of gun violence, or too young to participate in political discourse and activism on behalf of their position. Through their activism, notably through strikes as a form of collective action, teachers likewise rejected "passive conceptions of citizenship."[79] In this way, teachers and students set out to restructure education policy from the ground up within the public sphere. They were disillusioned when partisan dynamics crushed progressive legislative initiatives and became eager for an abrupt and immediate change in political direction that would affect their lives. They spoke up to shape an informed constituency representing their political hopes in massive public demonstrations on behalf of social justice with the media looking on. They also took the lead on climate change with the promise of a new political movement. Fridays for Future, also known as the "School Strike for Climate," is a global climate action movement of school students led by Swedish activist Greta Thunberg, which organized thousands of protests, notably skipping school on Fridays. Beginning when fifteen-year-old Thunberg led a strike outside of the Swedish parliament in August 2018, the initiative grew to include mobilizations by students from over 150 countries.

Despite mass movements, at election time, politicians flush with cash from big money interests campaigned with standard messaging. When younger generations turned to social media for coherent answers, they found little consequence. Regardless of these constraints, students who took up activism and engaged in political campaigns and single-issue coalitions began to learn how power works. Conceivably, this political and social knowledge may transfer toward critically examining diverse public issues. Many live and work in regions where social and political power based on maintaining a high-carbon economy is entrenched and elite-controlled. While working and paying high rents and inflated living costs, finding time for advocacy may be difficult at best. Countless others remain burdened with student loan debt for a college degree that failed to yield the jobs they envisioned. Instead, many wound up working for low wages that hardly make a dent in their debt obligations. The young are divided between those who have found a landing place and others who are locked in a persistent struggle to survive. However, they are unified by their pursuit of personal agency and a place in the world. The young now comprehend their dependence on a system where daily survival necessitates skillful navigation through often-enigmatic social pathways. They are also coming to understand the structural dimensions of large-scale processes, including dynamics of population, resource governance, and political power that influence their lives and the fate of humanity.

Education can move in this direction to enlarge the public sphere by connecting project-based learning and critical pedagogies to common concerns, including greater accountability. The discourse of accountability recently emerged through relationships between and among those involved in collective action, notably organized efforts in human rights, independent media, and environmental concerns. These efforts promote practices of deliberative democracy, involving consensual decision-making to create diverse styles of autonomous associational life. Such practices are important since national elites are slow to forward transparency across the society. The pursuit of accountability then emerges through social formations where people associate through collective action for concerns that are critical to their lives. Deliberative democratic practices, including schooling for democracy, are thus key in cultivating a sense of social responsibility for openness and change within the varied arenas of public life. For this to occur, it will be necessary to adopt public pedagogies, especially in later adolescence, as a preparation for life and engagement in the public sphere.

NOTES

1. Philip Morrison, Phylis Morrison and The Office of Charles and Ray Eames, *Powers of Ten* (New York: Scientific American Books, 1982), 114.
2. Jaakko Hintikka, "Wittgenstein's Times (And Ours)," *in Time and History*, edited by Friedrich Stadler and Michael Stöltzner, Publications of the Austrian Ludwig Wittgenstein Society-New Series (Frankfurt: ontos verlag, 2006), 540–541.
3. Joachim Schulte, "Wittgenstein on Time (1929–1933)," *in Time and History*, 557–559.
4. Joona Taipale, *Phenomenology and Embodiment: Husserl and the Constitution of Subjectivity* (Evanston: Northwestern University Press, 2014), 24–25.
5. Maurice Merleau-Ponty, *Phenomenology of Perception*, translated by Donald A. Landes (New York: Routledge, 2012), 438–439.
6. Armando Mascolo, "L'évasion de l'être. Jean-Paul Sartre and the Phenomenology of Temporality," in *The Concept of Time in Early Twentieth Century Philosophy*, edited by Flavia Santoianni (Cham: Springer, 2016), 78; Jean-Paul Sartre, *Being and Nothingness*, translated by Sarah Richmond (London: Routledge, 2018), 180–184.
7. Monica Sorrentino, "Space in Education," in *The Concept of Time in Early Twentieth Century Philosophy*, 32.
8. David Christian, *Maps of Time: An Introduction to Big History* (Berkeley: University of California Press, 2004), 364–367.
9. Norman Cohn, *The Pursuit of the Millennium: Revolutionary Millenarians and Mystical Anarchists of the Middle Ages* (New York: Oxford University Press, 1970).
10. Norman Cohn, *Europe's Inner Demons: The Demonization of Christians in Medieval Christendom* (Chicago: University of Chicago Press, 1993), xi.

11. Didier Fassin and Richard Rechtman, *The Empire of Trauma: An Inquiry into the Condition of Victimhood*, translated by Rachel Gomme (Princeton: Princeton University Press, 2009), 6.

12. Eric Mann, *Taking On General Motors: A Case Study of the UAW Campaign to Keep GM Van Nuys Open* (Los Angeles: UCLA Institute of Industrial Relations, 1987); Barry Bluestone and Bennett Harrison, *The Deindustrialization of America: Plant Closings, Community Abandonment and the Dismantling of Basic Industry* (New York: Basic Books, 1982); Gregory Pappas, *The Magic City: Unemployment in a Working-Class Community* (Ithaca, NY: Cornell University Press, 1989).

13. Chinhui Juhn, Kevin M. Murphy, and Brooks Pierce, "Wage Inequality and the Rise in Returns to Skill," *The Journal of Political Economy* 101:3 (1993): 410–411.

14. Jennifer Ma and Matea Pender, *Education Pays 2023: The Benefits of Higher Education for Individuals and Society* (New York College Board, 2023).

15. Elaine Tyler May, *Homeward Bound*, 25.

16. Carl A. Maida, Norma S. Gordon, and Norman L. Farberow, *The Crisis of Competence: Transitional Stress and the Displaced Workers* (New York: Routledge, 2015).

17. Shanto Iyengar, Yphtach Lelkes, Matthew Levendusky, Neil Malhotra, and Sean J. Westwood, "The Origins and Consequences of Affective Polarization in the United States," *Annual Review of Political Science* 22 (2019): 129–146.

18. S. H. Foulkes, *Therapeutic Group Analysis* (London: Allen and Unwin, 1964), 292.

19. Earl Hopper and Haim Weinberg, editors, *The Social Unconscious in Persons, Groups, and Societies. Vol. 3: The Foundation Matrix Extended and Re-Configured* (London: Karnac, 2017), xxii; cited in Bill Roller, "Review: *The Social Unconscious in Persons, Groups, and Societies, Volume 3: The Foundation Matrix Extended and Re-Configured*," *International Journal of Group Psychotherapy* 69:3 (2019): 374.

20. Kalman H. Silvert, *Expectant Peoples: Nationalism and Development* (New York: Random House, 1963).

21. Emily Brignone, Daniel R. George, Lawrence Sinoway, Curren Katz, Charity Sauder, Andrea Murray, Robert Gladden, and Jennifer L. Kraschnewski, "Trends in the Diagnosis of Diseases of Despair in the United States, 2009–2018: A Retrospective Cohort Study," *BMJ Open* 10:10 (2020): e037679.

22. Anne Case and Angus Deaton, "Rising Morbidity and Mortality in Midlife among White Non-Hispanic Americans in the 21st Century," *Proceedings of the National Academy of Sciences* 112: 49 (2015): 15078–15083; Anne Case and Angus Deaton, "Mortality and Morbidity in the 21st Century," *Brookings Papers on Economic Activity* (2017): 397–476; Anne Case and Angus A. Deaton, *Deaths of Despair and the Future of Capitalism* (Princeton: Princeton University Press, 2020).

23. M. Harvey Brenner, "Economic Change, Alcohol Consumption and Heart Disease Mortality in Nine Industrialized Countries," *Social Science and Medicine* 25:2 (1987): 119–132.

24. M. Harvey Brenner, *Estimating the Social Costs of Economic Policy: Implications for Mental and Physical Health and Criminal Aggression*, Report

to the Congressional Research Service of the Library of Congress and Joint Economic Committee of Congress (Washington, DC: US Government Printing Office, 1976), 5–6.

25. M. Harvey Brenner, *Mental Illness and the Economy* (Cambridge, MA: Harvard University Press, 1973).

26. Thomas Piketty, *Capital in the Twenty-First Century*, translated by Arthur Goldhammer (Cambridge, MA: Harvard University Press, 2014).

27. Thomas Piketty, *Capital and Ideology*, translated by Arthur Goldhammer (Cambridge, MA: Harvard University Press, 2020), 154.

28. Thomas Piketty, "Thomas Piketty Looks Back at the Success of *Capital in the 21st Century*," *The Tocqueville Review/La revue Tocqueville* 37:2 (2016): 60.

29. Robert J. Gordon, "Perspectives on *The Rise and Fall of American Growth*," *The American Economic Review* 106:5 (2016): 72.

30. Robert J. Gordon, *Is U.S. Economic Growth Over? Faltering Innovation Confronts the Six Headwinds*, Working Paper 18135 (Cambridge, MA: National Bureau of Economic Research, 2012).

31. Henry A. Giroux, "Domestic Terrorism, Youth, and the Politics of Disposability," *Knowledge Cultures* 3:5 (2015): 118.

32. Henry A. Giroux, "Doing Cultural Studies: Youth and the Challenge of Pedagogy," *Harvard Educational Review* 64:3 (1994): 289.

33. Jill F. DeVoe, Katharin Peter, Phillip Kaufman, Sally A. Ruddy, Amanda K. Miller, Mike Planty, Thomas D. Snyder, Detis T. Duhart, and Michael R. Rand, *Indicators of School Crime and Safety: 2002*, NCES 2003–009/NCJ 196753 (Washington, DC: US Departments of Education and Justice, 2002).

34. S. Patrick Kachur, Gail M. Stennies, Kenneth E. Powell, William Modzeleski, Ronald Stephens, Rosemary Murphy, Marcie-jo Kresnow, David Sleet, and Richard Lowry, "School-Associated Violent Deaths in the United States, 1992 to 1994," *Journal of the American Medical Association* 275:22 (1996): 1729–1733.

35. Edward T. Linenthal, *The Unfinished Bombing: Oklahoma City in American Memory* (New York: Oxford University Press, 2001).

36. Dave Cullen, *Columbine* (New York: Twelve, 2009).

37. DeVoe et al., *Indicators of School Crime and Safety*, 2002.

38. National Research Council and Institute of Medicine, *Deadly Lessons: Understanding Lethal School Violence, Case Studies of School Violence Committee*, edited by Mark H. Moore, Carol V. Petrie, Anthony A. Braga, and Brenda L. McLaughlin. Division of Behavioral and Social Sciences and Education (Washington, DC: National Academies Press, 2003).

39. Craig A. Anderson and Karen E. Dill, "Video Games and Aggressive Thoughts, Feelings, and Behavior in the Laboratory and in Life," *Journal of Personality and Social Psychology* 78:4 (2000): 772–790.

40. Henry A. Giroux, *The Abandoned Generation: Democracy Beyond the Culture of Fear* (New York: Palgrave, 2003).

41. Max Weber, *Economy and Society: An Outline of Interpretive Sociology*, edited by Guenther Roth and Claus Wittich (Berkeley: University of California Press, 1978), 375.

42. William C. Cockerham, "Medical Sociology and Sociological Theory," in *The Blackwell Companion to Medical Sociology*, edited by William C. Cockerham (Malden, MA: Blackwell, 2005), 417.

43. Ralf Dahrendorf, *Life Chances* (Chicago: University of Chicago Press, 1979), 73.

44. William C. Cockerham, ed., *The Blackwell Companion to Medical Sociology*, 12.

45. Christopher Lasch, *The Minimal Self: Psychic Survival in Troubled Times* (New York: Norton, 1984).

46. Warner Berthoff, *A Literature without Qualities: American Writing since 1945* (Berkeley: University of California Press, 1979).

47. Erving Goffman, *Asylums: Essays on the Social Situation of Mental Patients and Other Inmates* (Garden City, NY: Doubleday, 1961).

48. Bruno Bettelheim, *The Informed Heart: Autonomy in a Mass Age* (Glencoe, IL: Free Press, 1960).

49. Hannah Arendt, *The Origins of Totalitarianism* (New York: Harcourt Brace Jovanovich, 1973), 459.

50. Terrence Des Pres, *The Survivor: An Anatomy of Life in the Death Camps* (New York: Oxford University Press, 1976), 6.

51. Paul Goodman, *Growing Up Absurd: Problems of Youth in the Organized Society* (New York: Random House, 1960).

52. Richard Maxwell, "Surveillance: Work, Myth, and Policy," *Social Text* 23:2 (2005): 1–19; Michael J. Shapiro, "Every Move You Make: Bodies, Surveillance, and Media," *Social Text* 23:2 (2005): 22–34.

53. Joseph E. Davis, "Adolescents and the Pathologies of the Achieving Self," *Hedgehog Review* 11:1 (2009): 40.

54. Centers for Disease Control and Prevention, "Suicide Trends among Youths and Young Adults Aged 10–24 Years—United States, 1990–2004," *Morbidity and Mortality Weekly Report* (MMWR) 56:35 (2007): 905–908.

55. Daniel A. Foss and Ralph Larkin, *Beyond Revolution: A Theory of Social Movements* (South Hadley, MA: Bergin & Garvey, 1986), 154.

56. Philip C. Kendall, ed., *Child and Adolescent Therapy: Cognitive-Behavioral Procedures* (New York: Guilford Press, 1991).

57. Michel Foucault, *The History of Sexuality, Vol. 1: An Introduction* (New York: Vintage Books, 1990).

58. Charles L. Briggs and Daniel C. Hallin, "Biocomunicability: The Neoliberal Subject and Its Contradictions in News Coverage of Health Issues," *Social Text* 25:4 (2007): 44.

59. Briggs and Hallin, "Biocomunicability," 46.

60. Robert J. Shiller, *The New Financial Order: Risk in the 21st Century* (Princeton: Princeton University Press, 2004).

61. Archit Shah, "Emergency Economic Stabilization Act of 2008," *Harvard Journal on Legislation* 46 (2009): 569–584.

62. James G. Gimpel, Frances E. Lee, and Rebecca U. Thorpe, "Geographic Distribution of the Federal Stimulus of 2009," *Political Science Quarterly* 127:4 (2012): 575.

63. Peter Cappelli, Iwan Barankay, and David Lewin, "How the Great Recession Changed American Workers," *Wharton Business Daily* (September 10, 1918), https://knowledge.wharton.upenn.edu/podcast/knowledge-at-wharton-podcast/great-recession-american-dream/.

64. Christopher L. Foote and Richard W. Ryan, "Labor-Market Polarization over the Business Cycle," *NBER Macroeconomics Annual* 29.1 (2015): 371.

65. Richard Wilkinson and Kate Pickett, *The Inner Level: How More Equal Societies Reduce Stress, Restore Sanity and Improve Everyone's Well-Being* (New York: Penguin, 2019).

66. David Graeber, "Enacting the Impossible: On Consensus Decision-Making," *Occupied Wall Street Journal* 3:4 (2011) quoted in Zeynep Gunel and Lemi Baruh, "Social Networking Technologies and Social Movements," in *Social Networking: Redefining Communication in the Digital Age*, edited by Anastacia Kurylo and Tatyana Dumova (Lanham, MD: Rowman and Littlefield, 2016), 138.

67. Russell P. Skelchy and Jeremy E. Taylor, "Introduction: Sonic Histories of Occupation," in *Sonic Histories of Occupation: Experiencing Sound and Empire in a Global Context*, edited by Russell P. Skelchy and Jeremy E. Taylor (London: Bloomsbury Academic, 2022), 8.

68. Tilly and Tarrow, *Contentious Politics*, 220.

69. Brett Rolfe, "Building an Electronic Repertoire of Contention," *Social Movement Studies* 4:1 (2005): 65–74.

70. Gareth Morgan, *Images of Organization* (Thousand Oaks, CA: Sage Publications, 2006), 207.

71. Jonathan Massey and Brett Snyder, "Occupying Wall Street: Places and Spaces of Political Action," *Places Journal* (2012), https://placesjournal.org/article/occupying-wall-street-places-and-spaces-of-political-action/.

72. Max Weber, *Economy and Society*, 246.

73. Slavoj Žižek, "Occupy Wall Street: What is to Be Done Next?" *The Guardian* (April 24, 2012), https://www.theguardian.com/commentisfree/cifamerica/2012/apr/24/occupy-wall-street-what-is-to-be-done-next.

74. Mark Fisher, *Capitalist Realism: Is There No Alternative?* (London: John Hunt Publishing, 2009).

75. Giroux, *The Abandoned Generation*, 15.

76. Christopher D. Thomas, *Reclaiming Democratic Education: Student and Teacher Activism and the Future of Education Policy* (New York: Teachers College Press, 2022).

77. Thomas, *Reclaiming Democratic Education*, 71.

78. Thomas, *Reclaiming Democratic Education*, 81.

79. Thomas, *Reclaiming Democratic Education*, 81.

Chapter 12

A Place in the World

TECHNOLOGY AND THE LIFEWORLD

Hannah Arendt distinguishes the human condition from the earth and the natural world based upon relational and interdependent living. Arendt understands the necessity of three fundamental human activities that constitute the *vita activa*, namely, labor, work, and action. Together, these activities enable humans to create their own world and share it with others. Arendt sees the human domain as a common world—a shared reality constructed by work:

> This world, however, is not identical with the earth or with nature, as the limited space for the movement of men and the general condition of organic life. It is related, rather, to the human artifact, the fabrication of human hands, as well as to affairs, which go on among those who inhabit the man-made world together. To live together in the world means essentially that a world of things is between those who have it in common, as a table is located between those who sit around it; the world, like every in-between, relates and separates men at the same time.[1]

The public spaces devised by humans as they go about their work give rise to a community's collective life. Humans articulate this common world within a semantic domain, a "specific area of cultural emphasis"[2] where people share a set of meanings to describe certain things. Husserl understood that such embodied experiences within "a communal nexus of meaningful situations, expressive gestures, and practical activities"[3] constitute a lifeworld. Within any domain are clusters of words indicating meaningful relations between discrete concepts that share overlapping asymmetrical boundaries. Related is dominion, which in sovereign rule or law refers to boundaries of territorial power. Dominion came to indicate authority over things, specifically to subdue under some command, or alternatively place under some cultural stewardship. In computing, a domain refers to a network of users, workstations, computers, devices, and databases controlled by an authority that governs network functions.

Arendt also considers the relational nature of the artifact as uniquely generating a world of fabricated things. The human domain, constituting a world apart from nature, inheres in the archaeological category of an artifact, through which, as V. Gordon Childe noted, humans make themselves.[4] The artifact, a human-made object, is derived from the Latin *arte*, by skill, and *facere*, to make. From an existential perspective, Sartre understands that it is within continuous experiences where we find ourselves enmeshed in the world through our projects.[5] The original sense of humanity being constituted by artifacts has, over time, moved to things as being produced. A factory (from Latin *factorium* or a "place of doers, makers") was the common name in preindustrial times for an entrepôt or depot where local artisans could interact with foreign merchants, referred to as factors or agents. Only after the industrial revolution does the term represent a facility for mechanized production and manufacture. Technologies are artifacts by virtue of their materiality and can therefore be examined as any other human-made objects in relation to praxis or action. Because humans develop, use, and relate to these technologies in unique ways, Don Ihde explores the structures that mediate humans and technological artifacts.[6] For Ihde, the technological lifeworld is understood on the levels of both experience and culture, as technologies are always "technologies-in-use."[7]

Facts and the terms that fall under the province of *facere*, doing or making, are foundational to a modern emphasis on the material conditions of human reality and gave rise to debates between facts and things. Wittgenstein's early proposition, "The world is the totality of facts, not of things,"[8] asserts that things are created by facts that can be represented by thought. A proposition remains meaningful if it can be defined and pictured logically in the real world. However, any proposition as to whether facts or things are paramount may lose its meaning, notably its pictorial representation, within digital media ecologies. Wittgenstein later viewed definitions and meanings as emerging from forms of life, that is, within the cultural and social contexts where words are used. Even this view of the social aspects of cognition may be confounded by contemporary techno-cultural mediations of the real by simulations. The world of things is now embedded in information flows that constitute cyberspace: an environment interconnected by digital technologies. Algorithms and computing accelerate a global economy based upon information and technological advances that move things faster and more efficiently and, it seems, more remotely and invisibly.

The binary oppositions between nature and culture that endured since classical times emphasizing humans' separation from nature become even more complex within the media environment. According to Sy Taffel, human entanglements within techno-cultures impose similar "dualistic ontologies"[9] as exemplified by Neil Postman who sought

> [t]o make people more conscious of the fact that human beings live in two different kinds of environments. One is the natural environment and consists of things like air, trees, rivers, and caterpillars. The other is the media environment, which consists of language, numbers, images, holograms, and all of the other symbols, techniques, and machinery that make us what we are.[10]

Further alienation then ensues through engagement in seemingly fragmented media ecologies that impede personal control over immediate and remote environs. As a counterpoint, Taffel's multiscalar approach is based upon an assessment of media assemblages that reconnect the binaries provoking dualism and disaffection. Borrowing from cultural geography, Taffel seeks to overcome such dualisms by reconnecting humans and technology through scale, defined as "a relational quality between elements of ecological systems occurring within a single ecological register but requiring different perspectival approaches to be adequately addressed."[11] One such perspective, Gregory Bateson's ecology of mind, posits three ecologies: mind, society, and environment, "a triadic schemata of entangled assemblages designed to reconnect ethical and political issues pertaining to mental, social and environmental issues."[12] For Taffel, Bateson's distributed cognition model encompasses mental pathways beyond internal dynamics between an individual's body and brain to include connections between bodies and larger assemblages, including social groups, technology, and ecosystems:

> The ecology of mind then does not refer merely to an individual human brain or nervous system but also encompasses flows which pass through synapses, software, semiotics, and silicon—through various elements of the social and environmental ecologies with which it is thoroughly entangled and so cannot be functionally cleft. The three ecologies do not imply ontological separation but an epistemological tool presenting overlapping lenses through which to view particular modes of relation within entangled assemblages.[13]

Whether endorsing Postman's humanistic critique of media ecologies or Taffel's multiscalar approach, any effective method to navigate a human future would need to be built upon personal agency and autonomous human development. Education must then help the young to think critically and creatively about the human world in a unified manner. Being prepared for life entails a collective effort, for children and youth require educational approaches that connect them, as makers, to the human province. This necessitates providing them tools to engage the world and a sense of efficacy that permits them to forge their own vocational and life paths. Education is constantly reorganizing, although the semantic domains surrounding the young person's experience of observable practical consequences, such as "circumstances producing a result" through calculation or experimentation, have withstood the test of time. The promise of a sustainable future will occur in the hands of the adolescent. Moreover, that adolescent's life chances remain in the hands of education, where an enduring aim is one of shaping informed outlooks toward the human prospect.

The culture of humanism in Renaissance Europe brought about a fully modern critical rationality and a new historical perspective. Renaissance culture held learning as a distinctively human activity. This perspective built upon the idea of *paideia*, a Greek word indicating the shaping of character through education of the body and mind to produce a mature, enlightened outlook, which took hold in classical Mediterranean societies. A central aspect of classical pedagogy included moral development to cultivate a civilized conscience, initially toward a public morality consistent with the law of the community or the *polis*. Renaissance humanists claimed that *studia humanitatis,* or the humanities, involved learning virtuous conduct through the study of rhetoric, ethics, and history. This perspective is instructive for reorienting contemporary learning and identity formation, as Charles G. Naubert states:

> More broadly, Renaissance humanism has permanently affected the way in which modern people conceive their own identity, for humanist historical thought first taught us that we, too, are the products of an ever-changing flow of events, and that within the bounds set by human physiology, all human values, ideas, and customs are contingent products of time and place.[14]

Education was thus central to the humanist project of instilling civic and moral virtue through instruction in these core disciplinary practices. This project inheres in the habits of mind, the thinking dispositions developed through active inquiry, including evidence, perspective, connection, supposition, and relevance. These dispositions framed the rational post-Renaissance print-based crafts, notably politics, law, medicine, clergy, science, the fine and applied arts derived from humanistic education, which were frequently at odds with intuitive-based, a-rational traditions. The dominance of rational practices in early modern capitalist economy was accompanied by organized state and ecclesiastical stigmatization, condemnation, persecution, expulsion, and cultural annihilation of groups practicing predominately a-rational and intuitive disciplines. After the scientific revolution, Enlightenment *philosophes* questioned everything as human constructs, from sociality to the body and the mind, and most notably religion and a-rational and intuitive practices. Alongside this critique, Europe's population exploded, together with the full flowering of post-Renaissance crafts in later capitalist economies.

This humanistic outlook was mediated by social class, cultural, and political polarities within and across societies. Commerce and long-standing trade routes also assured a network of interests that crossed cultural boundaries and accommodated tolerance. The advent of mercantilism was thus central to humanism, though acceptance of many cultural differences was clearly out of self-interest. Nevertheless, this outlook supported an exchange network of ideas and intercultural influences that mediated closed regimentation and accommodated a more worldly sensibility. Rather than viewing humanity universally, many interpreted identity, ethnicity, and nation through binary frames, differentiating other cultures from their own civilization as they defined them. Binary distinctions have a long history, dating back to the ancient world and were then revived in Renaissance thought alongside the era's contrasting cosmopolitanism.

A public sphere opened in the Enlightenment as expanding cities and transformed hinterlands became sites of contested terrains for heated debates over binaries of gender, class, and conscience, which then gave rise to social movements that forged deep channels beneath the surface of modern life. The eighteenth century extended the polarities between the two ways of knowing, notably rational externality and intuitive internality. This dichotomy contributed to the rise of this public sphere, following Jürgen Habermas, characterized by professional strata, citizen engagement, and state structures based on rational externality and a private sphere that enclosed intuitive internality within family and personal life.[15] There were hybrid formations that combined the two in certain arts communities, where intuitive and rational practices coalesced. This conflict between two ways of knowing continues today and manifests in any number of rational and intuitive social formations, with membership and practices based upon education, training, income, and class. Going forward, an intertwining of the two epistemologies will be necessary for psychic survival in an increasingly precarious world.

Over the centuries, the humanities provided perspectives and rational dispositions for living, learning, and working amid diverse crises, from nationalist and religious wars, famines and economic downturns to pandemic outbreaks and their consequences. As in early modern Europe and later in postcolonial North America, humanistic knowledge can again inform critical reflection on contemporary systemic crises, including advanced warfare, technological change, economic depressions, pandemic diseases, climate change, and most notably fire. In William Cronon's words, "the fate of humanity, like the fate of the Earth, is tied to the fires that made the world as we know it."[16] Humanistic analyses of combustion and inferno thus convey the many instances, past and present, of cultural resilience amid fiery disruption.

ELEMENTAL FIRE

Earlier civilizations understood the centrality of the hearth to the well-being of families and communities. From this seemingly domestic concern came the primacy of caring about broader economic and ecological matters. Through deliberation on such matters, anticipatory behaviors advanced as strategic elements of culture, for humans spend considerable time reorganizing the world to resist countless random events. Fire continues to empower the imagination, provoking a deeper historical consciousness of the human condition. Fire thus remains central to our being along with our fate, entrancing, mystifying, supporting, and destroying life as an elemental foundation of our natural and cultural experiences through time.

The early use of fire transformed our ability to survive, nourish, adapt to new habitats, create light against darkness or threat, and ultimately change the face of the earth. Internally, fire evokes spirit and living movement and inspires phantasms of the mind. Since antiquity, fire shaped the symbolic modalities of community life through artistic and architectural representations. Externally, fire came to dominate human survival through everyday modifications of its physical forces and the harnessing of those forces for social and economic power. While acknowledging a profound and deep mental and spiritual historicity to its influences, fire has both created and destroyed human ambitions to control nature.

This duality begins with the archaeological record and interpretations of the evolving central significance of fire in human settlements and the deep history of the hearth in the evolution of human lifeways.[17] In the ability to transform landscapes and mindscapes, war and peace, the dual nature of this elemental fire has both served and provoked human survival through to the present. This includes the magnitudes and dimensions of a changing face of the earth, but also climate change and combustive materialism of our dependence on burning fossil fuels. So much of modern life remains dominated by the origins of elemental fire and its consequence for survival in diverse habitats in the early human world. Archaeologists exploring the Neolithic found the domestic hearth and household as central to domestication of "the savage mind" initially forged outside when hunters sat around the fire in a now vanished world, telling and listening to life-nourishing stories.[18] Paleolithic peoples were in many ways embedded in the elements; although their control of fire provided tool-making skills that would later assist Neolithic peoples to control nature and accelerate the shift to food production. In the Neolithic, fire as "the mark of settlement,"[19] in Joseph Rykwert's words, was brought inside as the hearth, or heart of the home—a place in the world. More than a place for food storage and meal preparation, the home became a vehicle for symbolic storage of information that frames identity and relational values.

With domestication, we began a long process of thinking about the world outside from the perspective of domestic life within.

The building of homes and the internal furnishings and symbolic objects within were part of a "deep history" of humanity, according to Clive Gamble.[20] The domestic sphere became an arena for memory construction, notably intergenerational social memory, where the daily activities within houses built to withstand the elements helped to form and reform the social world, creating stronger temporal depth within the society. In these Neolithic settlements, domestic architecture created "theatres of memory"[21] in Trevor Watkins' words, providing a cognitive web of knowledge and social memory in an external form. As settled humans, we began to give up the idea of ourselves as part of nature. Under civilized conditions, we tended to identify more with our cultural selves, as our individual and collective memories and social landmarks become more domesticated, together with our plants and animals. In earlier millennia, house-based societies represented in their materiality the symbolic forms and values of a community and promoted a "deep enculturation" of human learning through the socialization of infants and children within homes and villages, as stated by Merlin Donald.[22] *Hestia*, the Greek goddess of fire, presided over the domestic hearth and the sanctity of family life. Aristotle dichotomized the household (*oikos*) and the state (*polis*) as separate communities, with the former built upon elemental human relationships based upon kinship and the later as the basis of relationships among free citizens. Together, these permit the realization of happiness, although Aristotle distinguishes between an economic sphere, that of the household organized around reproducing life and managing the necessities to sustain it, notably care and socialization, and a political sphere that exists "for the good life" (*eudaimonia).* Although the public and private realms appear to exist separately, Arendt articulates the necessary intertwining of the two spheres on behalf of the good life:

> The "good life," as Aristotle called the life of the citizen, therefore was not merely better, more carefree or nobler than ordinary life, but of an altogether different quality. It was "good" to the extent that by having mastered the necessities of sheer life, by being freed from labor and work, and by overcoming the innate urge of all living creatures for their own survival, it was no longer bound to the biological life process. . . . Without mastering the necessities of life in the household, neither life nor the "good life" is possible, but politics is never for the sake of life. As far as the members of the *polis* are concerned, household life exists for the sake of the "good life" in the *polis*.[23]

The term, economy, is derived from the word for home, *oikos*. Economics, *oikonomia* in Greek, meant the management of the hearth implying stewardship and control over both fire and the domestic environment.

In the recent millennium, fire stands as an apocalyptic symbol and realistic reminder of its ferocious power over human landscapes, households, communities, and our collective memory. The fear of catastrophic wildfire is often accompanied by a sense of impermanence, and an anxiety over loss of control of the home as a shelter and locus of family life, and of the fire within. The two fire realms—burning landscapes and burning fuels—complement each other according to Stephen J. Pyne.[24] Facing the prospect of displacement and dispossession from home and community, hundreds of thousands of residents flee wildfires annually, as the size, intensity, and burn area of these fires increase each year in the earth's fire-prone forests and dry grasslands. Downed or sparking power lines often start the most destructive and dangerous wildland fires. This interaction of transmission lines with the natural landscape in precipitating fires reveals linkages between the two fire realms, as the electrical power passing through these lines relies on the combustion of coal and gas. Previously utilities only repaired wildfire-related damage to transmission and distribution lines. Recently, the power lines themselves are triggering fires when trees or branches come into contact with them. When high winds threaten to produce electric line damage, causing sparks that ignite dry brush and other vegetation, hundreds of thousands of households lose power in preemptive shutoffs on behalf of wildfire prevention.

Wildland fires are inevitable and are usually unpredictable. These fires affect diverse populations in a region: suburban, urban, and rural, wealthy and poor, culturally assimilated and recently immigrated. They disrupt the lives of individuals from every social class and ethnic group. When there is no loss of life, loss of residence with its attendant displacement is the central crisis for survivors. Those displaced from their homes often feel suddenly rootless, an emotion marked by depression, despair, and longing for a secure base. The circumstances that accompany a major wildfire are generally unfamiliar to most people. These include the physical sensations when experiencing the terrifying forces of strong winds and firestorms; the ruthlessness of wide-scale damage to property; the frustrating disruption of transportation and communication systems; the numbing witnessing of death and injuries, and, in some instances, looting and violence. Wildfires disrupt families and social networks that often share the same physical environment after the event. A community of survivors typically remains in place and continues on as part of the recovery environment. Amid a social atmosphere of attributed blame and causality, survivors frequently become involved in collective endeavors out of a sense of communal loss and a common need that typify the shared experience of catastrophe. From these experiences come ways to narrate a disaster, for the most part collective interpretations of cause and consequence that may, according to Rolf Lidskog, affect material, political, and social worlds. Lidskog characterizes this form of *communitas*, or intense feelings of belonging, as an "*invented* community in the sense that people talk about it in new ways because of the wildfire."[252]

Invented communities can take many forms, according to Lidskog. A "therapeutic community"[26] mobilizes resources for survivors, often with outside help and support. A "contaminated community"[27] emerges when a toxic exposure generates a boundary around a polluted area, together with a culture of fear, apprehension, and anxiety. A "corrosive community"[28] comes about when survivors create a counternarrative to the ones forwarded by the media, public agencies, and companies, viewing these outsiders' efforts as unresponsive and forging a culture of distrust and divisiveness that shapes survivors' identities far beyond the precipitating event.

As a socially interpreted event, a wildfire can also affect community identity and promote social trust among residents who may, together, embrace recovery efforts on behalf of resilience in a distinctly local manner. Resident engagement and community prevention behavior, such as risk assessment and defensible space practices, have minimized risk and thereby limited resource loss after a disaster event. Despite these measures, challenges remain, notably that of increasing resilience and decreasing the vulnerability of communities-at-risk in advance of extreme events in an era of anthropogenic climate change.

Dual threats of burning landscapes and burning fuels exemplify what Beck's risk society,[29] characterized by a preoccupation with catastrophes in the form of external risks, such as fires, and manufactured risks, brought about by technology, as with carbon emissions and their consequences. The second industrial revolution based upon industrial chemicals and the internal combustion engine eventually brought about conditions that prevail in global capitalism today. The current global level of expanded capital also has deep roots in the rational administrative and monetization practices of an earlier industrial capitalism. Under recent globalization, humanity is integrated through transnational production, finance, and accumulation. A high-carbon society based on fossil fuels and other nonrenewable sources underlies the current variant of capitalism. William Robinson sees the interplay of systemic global capitalist crisis and a "crisis of humanity" threatening the survival of billions of people in an overheating world, stating:

> Our world is burning. We face a global crisis that is unprecedented in terms of its magnitude, its global reach, the extent of ecological degradation and social deterioration, and the scale of the means of violence. This is a time of great upheavals, momentous changes, and uncertain outcomes; fraught with dangers, including the very real possibility of collapse as well as the growing threat of repressive social control systems that serve to contain the explosive contradictions of a global capitalism in crisis.[30]

With the transitions from a carbon-based global capitalism, exiting the entrenched arrangements will therefore entail major political and economic readjustments. These could result in widespread social conflicts, both state and nonstate violence, and accompanying suffering similar to previous industrial shifts and energy transitions. The debate centering on energy resources thus reimagines powering global industrial civilization in a warming world. With growing concern about an overheating planet, an "energopower," or "political power through the twin analytics of electricity and fuel,"[31] arose to control a resource-scarce world, in Dominic Boyer's words. Arguments surround the prospect of an energy transition, from an age of extreme energy based upon hydropower and fossil fuels, notably coal, oil and natural gas, to a low-carbon economy relying on renewable energy technologies. Although decarbonizing the economy, in itself, cannot solve the problem, natural solutions will also be needed.

Carbon-rich natural ecosystems, such as forests, can also offer climate solutions through natural forest restoration practices. Though climate change, itself, threatens forest stability, as William Anderegg asserts, through "fire, drought, biotic agents, and other disturbances."[32] Climate-driven disturbances from rising global temperatures can reverse forests from being carbon sinks that store carbon in their biomass and soils, to carbon sources emitting carbon dioxide into the atmosphere. As a response to deforestation and massive infrastructure projects disruptive of local ecologies, indigenous activists and nongovernmental organizations in the global South set out to challenge global capitalist impositions as threats to their moral economies within a global justice framework. These include consequences of global change caused by population pressure and resource scarcity in a warming world, notably environmental devastation in war zones, famine, pandemic diseases, natural and industrial disasters. Such apocalyptic scenes are portrayed by global media "in a visual landscape filled with the spectacle of destruction and decay,"[33] as Giroux suggests. Still, neoliberal policies are flexible enough to limit the successes of activist movements and the prospect of any local and regional efforts to move toward a post-carbon society. Despite such obstacles, there are challenges to the global dominance of systemic energy regimes, both within debates surrounding symmetrical and asymmetrical access to, and utilization of, the electrical grid and in the actions taken by localities.

A NETWORK SOCIETY

Across the globe, high-voltage electrical systems maintain large-scale electricity grids that have altered corporate industrial operations, personal life and interpersonal behavior, further separating people from natural cycles and their landscapes. The new technologies powered by electrification also brought with them environmental consequences, provoking those declaring a "resistance of grid and its cultures,"[34] in Dominic Boyer's words, to seek local-level alternatives to energy resources. As central plants became unreliable in delivering electricity, and power outages became a norm, remote customers and industries severely hampered by power failures sought alternatives to conventional power stations. Customers then began to rely more on power generation through a system of distributed energy resources, such as fossil fuel generators, fuel cells, solar, rooftop wind, or other energy sources. One example is the microgrid, which is a digitally controlled grouping of energy sources, renewable and nonrenewable, that may include energy storage capacity. The microgrid can function more autonomously than centralized grids, which deliver electricity from power plants through an interconnected system of high-voltage and low-voltage power lines, and also distribution transformers.

Many now even choose to conduct their lives "off the grid," severing dependence on conventional supplies of electricity through systems of wire and power stations. Off-grid living moves away from utilities that generate, transmit, and distribute power, toward alternative power options using onsite renewable energy sources, such as wind, solar, and geothermal energy. Certain countercultural activists even eschew digital technologies, moving away from digital-based lifestyles, altogether. Their retreatist stance toward digital ecologies is analogous to "Back-to-the-Land" movement of the 1960s and 1970s that promoted communal values and self-sufficiency. Many back-to-the-land adherents sought "the good life" away from cities and out of reach of state control as Scott Nearing and Helen Nearing promised in their writings on rural homesteading in Vermont during the 1930s and 1940s.[35] Pursuing self-reliant lifestyles, contemporary intentional communities formed in later decades as experiments in alternative living and care of the earth. One such community, The Farm in Summertown, Tennessee, established in 1971 as a countercultural escape from the San Francisco Bay Area, continues to support an ecovillage model.[36] Off-grid ecovillage approaches tend to support alternative technologies on behalf of sustainability and common pool resources, such as water, solar, and wind energy. An ecovillage will typically use ideas, tools, and other forms of *bricolage,* a patchwork of well-worn practices borrowed from older institutional traditions, which are then adapted to new conditions. These include investments in renewable energy sources and sustainable farming practices that involve incremental tweaking of older techniques, from composting and dryland farming to wind power, to carry out the necessary adaptations.[37] As small-scale social experiments formed in the face of centralizing governmental power and corporate capture of public resource allocation, a micro-electronics-based network society grew around them. In contrast with older notions of impersonal machines, newer technologies enmesh users in a web of sociability, hence their social life worlds.

Electric environments became key to accessing mass media and networked communication throughout the twentieth century. Electronic computational engineering before the Second World War provided a pathway to subsequent developments in electronic media and new information environments. The immediate postwar generations were conditioned to relate to an analog world in their childhood and youth, notably where binary modes of thought were the touchstones for shared realities. As adults, they accommodated to a world that was becoming digital through the Internet, WiFi, search engines, personal computation, and communication devices, accelerating the move toward a "network society,"[38] in Manuel Castells' words, with consequent transformations of the material world. Those born after 1980 may have limited knowledge of earlier generations' mass communicative arrangements and their cognitive effects, and how their lives may differ from them as a result of experiences with newer technologies, creating intergenerational tensions and conflicts. Such technological change notwithstanding, the family requires sustained dialogue to survive as a unit, although as an institution the family morphed into a consumer-based household at least since the mid-twentieth century when it became a cost center within late capitalist economies. Despite newer technologies' disruptive attempts to move families away from a conduct of life that emerged centuries ago, the domestic group's survival still requires a relational approach based upon cooperation and dialogue. Beyond the domestic group, private power is increasingly based upon wealth sustained by work using digital technologies or investment in digitally operated enterprises. Dependence on these technologies influence a person's life chances, from early education and advanced training, including access to good schools, colleges, apprenticeships, and internships to reliable marriage partnerships, neighborhood choice, and ways of living. In the wider society, public enterprises that adopted digital technologies as early adopters were positioned to gain greater control over their populations than later adopters.

The digital divide created new forms of economic dependency, with digital-limited households and workplaces, locales and regions dependent upon corporate and public institutional arrangements with widespread digital capacity, including technology manufacture, for the sale of digital technology and the digital distribution of information. At the national and global levels, military dependencies changed with digital technologies, as clientism among national states became increasingly based upon access to these new technologies and then ongoing tribute to digital-rich states. With multilevel transactions came greater dependence on the authority of energopower, making possible new instruments of control.

Computer science, cognitive and behavioral sciences, and ecology would subsequently embrace "digitality" as a common condition, supplanting previously accepted conceptions of being human, the mind, and knowledge, itself. Computer scientists produced a set of models to create a new epistemology for an emergent digital culture, as Claus Pias states:

> Using new terms such as "information", "feedback", and "analogical/digital" as starting point, the participants tried to develop a universal theory of regulation and control, that would be applicable to living beings as well as to machines, to economic as well as to mental processes, and to sociological as well as to aesthetical phenomena.[39]

Complexity and chaos theories emphasizing nonlinear functions of dynamical systems originating in the physical sciences became widely used in the social sciences, unseating rational models of human motivation and behavior. Evidence-based biomedical and biopharmaceutical research altered clinical understanding of the body and also the mind, as psychiatry moved beyond analytic and dynamic theories. Neuroscience and its handmaiden cognitive science swept away earlier paradigms of the self to make way for modeling the neural mechanisms underlying cognition and the mental processes based upon computational analogies.

In the postwar decades, human ecology moved from an earlier natural history perspective that addressed the power of nature over humanity toward a global interest in habitat corrections to industrial and war time destruction, and then to a more elaborate concern for the role of human agency in complex ecological systems. The latter view, a complete reversal of perspective, repositioned humanity within the web of life, acknowledging the impact of humans on their natural and built environments. At the time, the international symposium, "Man's Role in Changing the Face of the Earth," was held in 1956.[40] Funded by the Wenner-Gren Foundation for Anthropological Research, the conference brought together scholars from a range of disciplines, from anthropology to zoology, to evaluate "man's capacity to transform his physical-biological environment and upon his cumulative and irreversible alterations of the earth." Participants discussed the limits of the earth, including population, energy, and resources, and also the unstable relationship between humans and nature under present conditions. In subsequent decades, the human sciences set out to study the energetics of subsistence and survival, leading to a more nuanced understanding of the relations between humanity and nature.

These twin developments refocused humanistic thought away from mid-century disciplinary conventions. Michel Foucault addressed the consequent "biopower"[41] unleashed with technologies of normalization that subjugated bodies and minds. Foucault maintained that "biopolitics"[42] developed a regime of "governmentality,"[43] in his words, to manage and control populations through security apparatus, risk regulation, genomic medicine, and public health. At the same time, Bruno Latour's concept of "hybridity,"[44] combining features of the natural and social realms, articulated diverse hybrid linkages, exemplified by pasteurization, climate change, and human–animal entanglements in pandemic diseases. Even these humanistic assumptions grounded in a philosophical biology based upon features of life, itself, notably metabolism, reproduction and death, eventually gave way to dominant mechanistic-reductive accounts of life.

This shift in perspectives came at the moment of a third industrial revolution with the ascent of the microchip, which impacted everything. This economic transformation brought the convergence of new communication technologies with new energy systems, mainly renewable electricity. An intensification of microchip technology was a systematic progression from vacuum tube and earlier silicon transistor technologies to the integrated circuit, leading to microprocessor design. With these efficiencies in communications systems, the chip revolutionized diverse industries, including learning practices and schooling regimes. Knowledge was likewise transformed as computers converted text and other data into binary code. The encoded data not only altered communication but brought into being a new logic in knowledge. In the Information Age, as Debra Borkovich and Philip Noah explain new hardware, software, and database dominance and control of grid-based "energic systems" over the household, workplace, community, and region proliferated.[45] A parallel digital revolution, relying on energopower, organizes the social lifeworlds of skilled adopters, while limiting opportunities for nonadopters, disorienting the lives of people caught up in the transition.

BEYOND THE TRANSITION

A century of dynamic economic growth that began in the second industrial revolution and included postwar expansion and affluence waned in the mid-1970s. Subsequent decades brought a new level of politics to the economy, and an economic polarity to the politics, itself. Modes of thought that came into being after the war and discussed only within expert circles gained wider recognition. General systems theory, based upon understanding the dynamic processes in natural systems, was one such concept. This paradigm influenced cybernetics, a postwar-era discipline rooted in mechanical and electronic engineering, which used applications of feedback in electronic circuits to study communication and control in living organisms.

A signal event in the immediate postwar years, the ten conferences entitled "Cybernetics. Circular, Causal, and Feedback Mechanisms in Biological and

Social Systems" held between 1946 and 1953, would change the course of science and technology, and consequently have a pervasive impact on society and the progress of history thereafter. The "Macy Conferences" were co-organized by neuropsychiatrist Warren McCulloch and Frank Fremont-Smith, a physician who headed the Josiah Macy Foundation. The conferences brought together scientists from mathematics, psychology, engineering, anthropology, physics, ecology, psychiatry, neurophysiology, linguistics, and sociology. According to Johan Fredrikzon:

> By combining the insights from several fields of knowledge, the driving forces of the conferences sought to establish cybernetics as a new, universal science, positioning machines and organisms alike as systems operating to reach goals by using negative feedback mechanisms as their main control architectures allow data to be combined, processed and organized into information; then analyzed and transformed into knowledge.[46]

Algorithmic models of mental and somatic phenomena, and sociality, together with machine learning, provided data analytics to automate analytical model building. Although in this process, access to knowledge has become a terrain of power and contestation, as Melucci states:

> Control over informational production, accumulation, and circulation depends on codes which make information understandable. In complex societies, power consists more and more of operational codes, formal rules, knowledge organizers. In the operational logic, information is not a shared resource accessible to everybody, but an empty sign, the key of which is controlled by only a few people.[47]

Disruptive innovations occurred in the multiple domains of technology, society, and knowledge production, each affected by digital communication. This shift involved novel complexities in the digital transformation of information that forced innovative procedures and patterns of communication behaviors, thereby disrupting earlier regularities. Global realignments came about as the Cold War receded, in part based upon technology transfer of digital products, services, and modalities, but mainly from broader structural processes of financialization, globalization, and neoliberalism that were changing the landscape of the economies of the world.

The conditions surrounding higher learning altered as well. Postwar universities provided a sanctuary for creative thinking following years of global war, genocide, atomic bombings, devastation, and displacement. During the transitional decades, these institutions changed dramatically, away from humanistic academic practices and modes of thought upon which previous generations based their teaching and scholarship, toward operating as corporate firms with profit-oriented motivations. Rearrangements in governance, financing and, recently, digital coursework brought considerable gains in contract, grant, and tuition based revenues.

Universities thus appropriated rational planning and operations management for competitive advantage in a privatizing educational marketplace. Autonomy and collegiality laid the basis for freedom of thought within a scholarly community before the transition. Thereafter, humanities-driven narratives were occluded in an ongoing, breathless present. These narratives often comprised dissent that produced teach-ins and critical pedagogies; the vital and at times prophetic power of the African American voice; judicious appraisals of the social and political consequences of technology; and cultural feminist models that challenged deep-rooted cultural constructions and representations of gender identities. With the digital technologies that transformed teaching and knowledge creation, there is much to reflect upon in the half-century trajectory of bureaucratization and commodification across the educational spectrum.

AUTONOMOUS TECHNOLOGY

Langdon Winner uses the term "autonomous technology"[48] to frame the view that when technological innovations move out of the control of designers or makers, these could then render humans permanently bound to their inventions. He critiques a line of thought, beginning with Mary Shelley's *Frankenstein*, suggesting the paradoxical way that machines can gain autonomy and come to govern human activity. Winner's concept of "Technics-out-of-Control" is especially cogent as the information revolution rendered highly skilled and middle-skill knowledge workers reliant on AI applications, including expert systems and machine learning. AI expands algorithmic computational decision-making, accelerating both financial transactions and information flows, while robotics manages information storage and retrieval. Integrating production, services, and financial arrangements within a global network intrudes on the spaces of everyday living, working, and learning. Many remain concerned about the prevalence of smart machines and expert systems that may diminish personal agency. The Pew Research Center convened an international expert panel of some 979 technology developers, innovators, policy leaders, researchers, and activists to understand networked artificial intelligence and the human prospect. While viewing the possibilities of algorithm-driven AI in augmenting effectiveness in health care and education systems, the majority of the experts expressed concerns about human agency, evolution, and survival within an "AI-infused future." Most frequently mentioned were

> the use of AI reduces individuals' control over their lives; surveillance and data systems designed primarily for efficiency, profit and control are inherently dangerous; displacement of human jobs by AI will widen economic and digital divides, possibly leading to social upheaval; individuals' cognitive, social and survival skills will be diminished as they become dependent on AI; and citizens will face increased vulnerabilities, such as exposure to cybercrime and cyberwarfare that spin out of control and the possibility that essential organizations are endangered by weaponized information.[49]

With increasing smart machine-based control, considerable apprehension surrounds the widespread adoption of digital technologies in multiple spheres of life, with many viewing these technologies as working against them, rather than on their behalf.

Increasing conflicts also develop between those who advanced with the technological shift to digitality and others left behind in its wake. With digitalization also come global electronic surveillance and the tools to manage the political consequences of global economic crises that disrupt marginalized communities in both cities and the countryside. Some have argued that newly designed surveillance and military technologies arising from AI programs have the capacity to neutralize oppositional elements and counter-hegemonic movements, even revolts, in the aftermath of emergent financial, food, and energy crises. According to Robinson, new technologies that result from digitalization, a so-called "fourth industrial revolution," make possible a global police state, for these new technologies have "revolutionized warfare, social control and the modalities of state and private violence in the new century, including the military application of these technologies and the further fusion of private accumulation with state militarization."[50] Hence, those marginalized and disenchanted by the digital regimes and their conditioning of both life and livelihood will be subject to dystopic "panoptical surveillance and social control technologies,"[51] as stated by Robinson. Anxieties thus result from the potential loss of autonomy under machine domination. There are worries as to whether or not smart machines will capture state functions, diminishing the public voice. Then there are concerns related to lifestyle, from those who choose lives tethered to newer digital technologies, or who found a virtual "homestead" on the new electronic frontier, to still others who live off-grid on the "wastelands." Whether at home or in the office, or in the public square, access to newer and more powerful technologies will be asymmetrical. Finally, machine learning and computational decision-making will clearly affect how knowledge is derived, analyzed, disseminated, and used.

Older generations have struggled with the fast-moving technological changes in the new digitally enhanced world of computers. Adolescents by comparison use digital communication platforms at a higher rate than other age brackets. Ninety-five percent of teenagers in the United States have access to smartphones and about 45 percent are online almost constantly, according to the Pew Research Center.[52] Across the board, younger adults frequent social media platforms at a higher rate than their older counterparts, potentially indicating that subsequent generations of adults will consume social media more than current ones. Digital devices and communication platforms became the key domains for tracking one's peers' progress through the world and communicating with them,[53] as well as for purchasing commodities that supply distraction and, with them, a consumer-focused personal identity. With the processing power of newer smart phones, the pace of data—video, music, online gaming, conferencing—streaming across digital pathways are excessive. Large amounts of information become obsolete almost immediately, and much of it is corrupted and untrustworthy. The attention economy, while profit driven, can be mindless, amoral, and without context. Hence, for many adolescents, feelings of loneliness and isolation accompany their increasing social connectivity, for cyberspace is a placeless universe.

The trajectory of growing up for many is one of conditioning to take a role within a media-controlled, consumer-oriented "one-dimensional society" as termed by Herbert Marcuse.[54] In a society defined by technological rationality, human development is commodified, with a person progressing from early schooling and recreational video gaming; through high schools and colleges that reinforce digital learning to the detriment of direct experience; moving on for many youth to the prospect of "just a job" or a stint in the service. Digital approaches thus dominate the contemporary learning landscape, one that prepares the young for the loneliness of the digital workplace. Military careers and trade apprenticeships depend upon digital technology as do the engineering and media industries. Laboratory science is digitally encoded and analyzed. This is the case for any workplace a young person enters, whether the law, accounting, finance, or a desk job, including teaching, which essentially involves considerable clerical work in the digitally reinvented school or college. The digital world will thus be their reality, as smarter digital technologies will accompany the younger generations throughout their lives. Facilitating dialogue and critical inquiry regarding these technologies' power and potential control over the "whole of life" is the best precautionary remedy, rather than encouraging students to take on these powerful energetic modalities. With the prospect of digitally shaped life trajectories, young people require the means to theorize these forms of schooling and work, to reflect upon and critique how digital technologies shape their lives, and to consider the ethical and everyday life dimensions of contemporary technologies.

Though imagined dystopic consequences are endless, essential concerns revolve around the nature of work, for newer technologies create controlling patterns in workers' lives, as Albert Borgmann suggests.[55] Central is the fragmentation of work and loss of craft mastery and the prospect of individual de-skilling in a new "smart machine" age. This condition is characterized by the economic devaluation of craft knowledge, loss of control over one's labor, and according to Michael Lerner attendant powerlessness and resigned helplessness.[56] What occurred with the ascendency of mass production technologies and then when the second industrial revolution waned appears to be happening again as advanced computerized operations and disruptive AI-based technologies control most work. These conditions provoke long-standing questions about the direction and limitations originating with the ascendency of earlier forms of industrial and service work. Weber warned about the onset of deadening, mechanized work in white-collar worlds beyond the factory, based as they are on an "iron cage" of technological efficiency, rational calculation, and bureaucratic control. Harry Braverman held that work "as purposive action, guided by the intelligence, is the special product of humankind."[57] However, an ongoing degradation of skilled work ensued due to managerial strategies of workplace control in industrial workplaces. It then follows that such deskilling of workers leads to consequent alienation from their intellectual capabilities. Stanley Aronowitz explored how this same alienation even shapes the free time of workers, seeing the weekend itself as a colonization of leisure.[58] Advancing Marshall McLuhan's media ecology perspectives, Neil Postman addressed the rise of trivialized leisure and how televisual media culture promotes passivity and, unlike print technology and its cultures, limits intellectual involvement.[59]

As a remedy, Richard Sclove, reflecting on the social dimensions of technology and on their compatibility of decisions about technologies with democracy, stated the following argument, "Insofar as (1) citizens ought to be empowered to participate in shaping their society's basic circumstances and (2) technologies profoundly affect and partly constitute those circumstances, it follows that (3) technological design and practice should be democratized."[60] The idea of democratizing technology supports efforts to democratize education. Together with understanding theoretical perspectives, discussing problem-based cases derived from practical ethics can support students' critical thinking. Central is humanity's "imperative of responsibility,"[61] in Hans Jonas' words, for the fate of the world in a technological age. Jonas asserts that our present actions especially in the scientific and technological spheres affect future generations, often with unforeseen consequences. He argues for the ethical necessity of those alive today to take responsibility for future generations.

On behalf of understanding technologies as social practices, teachers can engage students through film, television, and online media; through interactive role plays, peer-based teaching and learning, small group work, and through reflective exercises that integrate what students have learned in their experience-based practices and engagement with the world. For any such reflective inquiry to occur, instructors may need to "leave the lectern" for a time on behalf of active learning with students. In class circles, both teachers and students may then engage in critical reflection and discussion about digital technologies' dominance over their everyday lives. Creative involvement in the arts and design may also move students beyond atomizing pedagogies to better understand the constructed nature of digital worlds.

Countercurrents in many students' lives, notably long hours of work in the last two years of high school and then in college, may mitigate against their embracing more time-consuming instructional approaches that support active learning. Many other students holding a narrowed outlook, who disregard project-based learning, see the approach as a detraction from conventional schooling. A deadening process may have set in during later adolescence under demanding conventional schooling regimes where many lose interest in an intellectual path and narrow their career focus. Open-ended exploration and creativity may be stifled at this time, especially in families where grades and test scores are overemphasized. High-scoring students will often arrive at college with this mindset, frequently testing out of general education through Advanced Placement coursework and moving early into their majors. They go on to comply with academic routines focusing on grades and earning a high-grade point average. They then graduate as "excellent sheep,"[62] in William Deresiewicz's words, with a set of marketable skills derived from coursework and summer internships and with the networking capacity to move on to a job or professional degree program. Viewing college as an investment, many parents see themselves as customers paying for a service, and then pass these attitudes on to their children. Students are thus motivated toward completing the requisite courses on their study lists and making an exit. In the end, a baccalaureate diploma symbolizes, for them, yet another moment in a conditioned life. Throughout, digital technology will have supported the move toward their career goals, defining their educational trajectory from early schooling to the university.

DIGITAL SCHOOLHOUSE

Information Technology (IT) has proven to be a disruptive innovation in every occupation, and teaching is no exception. Sweeping technological developments in digital learning platforms gradually supplanted in-person teaching in many classroom settings. Managerial oversight of these technologies rearranged the relationship between administration and faculty, altering long-standing school and college cultures. Early adopters of classroom IT remain confident that digital technologies will potentially lead to greater "student success" or academic achievement. At the same time, there is the risk that, in Winner's words, "patterns of perceptive thinking that were entirely reliable in the past now lead us systematically away."[63] While seemingly assuring greater accountability in monitoring attendance, recording grades, and diminishing plagiarism, the trend toward standardization increases a machine's ability to appropriate the task of teaching. Standardized tests and syllabi can be readily put online with administrative support and minimal faculty oversight.

There are now fully online and asynchronous courses, largely dependent upon publishers' materials, where classes do not have a set meeting time and with no on-campus meetings nor for that matter any live online work. Hence, there is no student–teacher interaction in real time. Lectures can be prerecorded by the instructor, and then uploaded to Box, a portal where students listen on their own time as with a podcast. Students watch videos downloaded from a web-based learning management system, perhaps discussing them in online chat rooms. Midterm and final exams uploaded to the portal remain open for a certain number of hours so that students can take their tests online. Instructors increasingly use test banks and randomized test questions to help decrease dishonesty in digital learning.

Massive open online courses (MOOCs) are taught remotely on distance learning platforms. MOOCs allow for unlimited participation and open or free access via the Web. Using conventional course materials, filmed lectures, problem sets and online tests, and discussion forum, the intent is self-paced learning, with an optional course certificate. Based on connectivism, a learning theory was originated by George Siemens integrating chaos, network, complexity, and self-organization theories. Siemans states, "We can no longer personally experience and acquire learning that we need to act. We derive our competence from forming connections."[64] Students are then taught to navigate digital networks and to form social networks to help them find needed information and support one another, thus "collecting knowledge by collecting people,"[65] in Karen Stephenson's words as cited by Siemans. Learning is no longer under an individual's control; rather it resides in "non-human appliances," typically a database where students connect "specialized information sets" on behalf of continual learning.[66]

The pandemic brought pervasive use of online pedagogy, although a fast-moving eclipse of in-person instruction occurred earlier through the above-mentioned modalities. With classrooms doors shuttered, teachers worked from hastily refashioned home-based offices. All tried their best to learn how to use digital conferencing technologies, at times mandated practically overnight. Virtual classrooms surfaced disturbing gaps in student attention, participation, and low emotional commitment to learning. Many students did not log on because of limited access to computers and Internet connections, while others remained completely out of touch with their teachers, apparently embedded in a cascading situation of chronic absenteeism and unfinished learning. Within atomizing virtual classrooms, where each learner is a mere node in a digital network, youth experienced fatigue and burnout from hours on end of these classes. Faculty began to pare down their curricula to bare essentials, flattening the digital learning environment, rendering teaching a hollow enterprise for many.

Early in the pandemic, then, digital classrooms appeared to lack the necessary depth provided by face-to-face learning and peer interactions in the classroom. Widespread dependence upon digital pedagogies became the new normal. As the school doors reopened for in-class instruction, administrative decisions regarding these instructional modes determined learning going forward. When schools and colleges reinvented themselves as digital schoolhouses during the pandemic, the young found it difficult to adhere to the precautionary-based virtual status quo. The school, and for many the college, was always a transitional life space between family and work worlds where the young could negotiate both their hormonal turbulence and peer relationships. While at the beginning of the pandemic, students felt isolated as they tried to adapt to abrupt changes in their learning, they soon accommodated to virtual and remote learning, often preferring this to in-person coursework. Then there was a digital divide exposed by the shift to web-based and distance learning platforms, together with a widening social-class divide. Many students left before completing their coursework, hence forsaking their educational goals toward a diploma or degree.

Digital platforms hold certain advantages over other instructional modalities, notably cost-saving and other efficiencies, therefore the prospect of schools and colleges abandoning these technologies remains unlikely. Based on cost and outcome data, there will come decisions about the usefulness of autonomous technologies on retooled digital campuses. Often tethered to these technologies themselves, managers' work lives are typically governed by computer-generated data and algorithmic models. To remain employed, teachers and lecturers increasingly accommodate to the ongoing digital Taylorization of their own occupational lives. There are relatively few options, especially when budgetary austerities and more recent pandemic disease safety precautions limit their choices. Nonetheless, computerized learning platforms and smart digital classrooms have become the preferred ways to instruct low-income and middle-class students who rely upon public education for social mobility. For affluent families desiring more interpersonal pedagogies, private institutions offer the amenities of in-person learning opportunities at a higher cost. The outlook for most other students may be learning via machine.

Digital learning brought increased surveillance and work-flow analyses, from above, in education systems, although Taylorism has been a part of the human services since the 1970s. Mechanistic trends, such as operations research, were slowly emerging over the ensuing decades, initially as adaptations to neoliberal austerities. Schools and colleges are now managed more like public entitlement agencies, such as unemployment or social security offices where supervisory personnel monitor the flow of clients and assess routinized procedures and the time it takes to perform them. Teaching and support staff are expected to adhere to similar procedural regimes and perform work and render expected outcomes accordingly. Teachers now set aside increasing amounts of time for in-service trainings on managing digital classrooms, improving digital security, handling plagiarism, monitoring online incivility, and the like. Such mandates, together with managing unexpected situations and conflicts in digital schooling, make it difficult to keep up in their respective fields. In the end, the new arrangements take their toll on the physical and psychic health of teachers and lecturers, and ultimately their will to continue to work under these conditions.

Previously, PowerPoint presentations and "clickers," or classroom response system devices, competed with earlier pedagogies for students' attention. Textbook publishers found ways to allure faculty with a box of instructor resources delivered in advance of the academic year, including PowerPoint slides, manuals, test item files, software, and course design materials, so that a syllabus follows closely to the text. However, under smart machine dominance, schools and college moved beyond even these text-based materials and canned presentations to guide instruction in the virtual classroom. Learning management systems, such as Moodle, Blackboard, and Canvas, provide teaching staff with tools for online learning and assessment of student learning, together with instructor feedback about skill development and learning achievement.

Digital communication platforms, such as Slack, combined with Zoom video conferencing, support virtual lectures and office hours, with class- and even project-based channels. Sway, a presentation program, creates class materials, lectures and projects, combining photos, audio, video, charts, and even tweets into corporate-style layouts and presentation formats. Packback designed a platform for online instructor-led and peer-based discussions. Students are coached on how to choose appropriate questions based on the course material and to formulate open-ended questions for online discussion. The platform uses AI to provide students with "instant 1- to-1" personalized feedback while they write their questions, permitting more iterative work, and AI will even moderate the online discussion automatically. A "digital TA" reviews every post and sends instructors the top posts from each online discussion.

Tensions between creative and progressive types and those overseeing the digital schoolhouse will continue during this technological transition.

Teachers remain under significant pressure to adopt technological innovations and uphold the primacy of commercial textbooks, corporate-influenced standardized curricula, and the AI-based platforms. Multiple levels of administrative review assure that student learning objectives are explicitly detailed at the beginning of the syllabus and then embedded within course elements through oversight by outcomes assessment committees. As instructors follow administratively approved syllabi and textbooks, they eventually become recorded voices uploaded to a learning management system, and clerical workers recording grades, in part, based upon AI recommendations.

The pandemic thus cast light on rearranged working conditions, as digital technologies captured faculty members' seemingly secure worlds. For decades, career faculty were cavalier about the slow eclipse of a familiar school and college environment. With the pandemic's abrupt appearance, a digital canopy fully enveloped their work lives. Many avoided or even denied the growing control revolution in education, in part, based upon their satisfaction with a more in-person relational work life, which continued in most classrooms during the transition. The pandemic interlude brought abrupt changes in work conditions, constraining time-honored practices of creative freedom within education. In the aftermath, only certain instructional approaches are viewed as having value, just as only specific styles of advocacy are deemed appropriate in the school and society.

Well before the advent of virtual learning, a two-tiered education system existed in the United States. An elite tier provides students with a twenty-first-century variant of the progressive approach, based upon small face-to-face classes, project-based learning, mentoring, and reflective inquiry. A mass tier, with most campuses transformed into large schooling hubs more frequently operated through IT and AI technologies under administrative command, continues to privilege screen- and web-based approaches, with tighter controls on teaching and learning. Graduates of the elite tier flow into post-Renaissance skilled, craft-based trades practiced for the most part in-person or on bodies and objects mediated by technologies; most will be housed in large institutions; a few will remain fee-for service. Post-Renaissance professions and occupations employ expert knowledge and craft practices focused on working with people, ideas, and things; although all are being altered by digitally based approaches.

These pursuits surfaced with printing, the guilds, and university training, and include physicians, attorneys, scientists, engineers, and the clergy. Through guilds and apprenticeships, skilled tradespeople, or "makers," who work with materials to build and modify things also emerged at this time, together with creative practitioners, such as musicians, artists, performers, writers, and designers. Mass-tier graduates move into screen work or positions using screen-based technologies supportive of the aforementioned dominant professional and occupational groups. Their work is controlled from above by AI-based supervisory techniques using algorithms and machine-learning models to assure compliance.

During the recent pandemic, virtual communication became established, not just in schools and colleges, but in most knowledge-based work settings. As the immediate crisis abated, virtual work landscapes did not disappear. Routine interactions are handled through e-mail, on social networking sites, and on conference management platforms. Many question paying high fees to commute via transit or auto, and in the latter case purchase parking, to sit at a desk in front of a computer screen in these alienating spaces, empty of human interaction and void of meaning. Though executives prate about the need for their home-based workforces to return to the office, threatening to dismiss insubordinates, the question of employee compliance remains.

Younger office workers appeared unwilling to go to their former work routines on a daily basis, including two or more hours of commuting and eating poor-quality food on the run. Once in the office, they then sit at their digital workstations hounded by heartless supervisors, whose approval is required for time away from their desks to relieve themselves and freshen up in corporate restrooms. The Occupy movement exposed these conditions decades ago after a near meltdown of the financial system, and there have been legal challenges to the bullying of newly hired employees who work the unbearably long hours expected of those in high-pressure occupations. On the demand side, consumers experience recorded messages and other call center-style approaches to health care, prescription drugs, insurance, banking and financial services, real estate, auto purchases, travel, and other brokered commodities. There is waning interest in walking into a bank to make deposits or spending time in a department store or bookstore to browse the stock, when one can buy these goods online, and have them delivered. Younger generations are fast cultivated into living within virtual worlds, where they take classes on digital platforms, get their news from unvetted sources on the web, stream music, and monitor their food intake, exercise, and vitals on smart watches and smartphone applications.

As the trend toward virtual work on the campus and beyond continues in the pandemic's wake, the positions of Frankfurt School thinkers, such as Marcuse, are instructive, specifically their critique of a dominant instrumental rationality and its impact on consciousness and lived experience. Under these highly rationalized regimes, greater control will be levied on the knowledge-based professions and occupations. An amalgam of atomization and control from above, algorithms and machine-learning programs now regulate virtual processes, frequently supplanting real-time managers and supervisors. Digital management easily bypasses conflicts and dilemmas associated with face-to-face interactions by imposing new rules and governance structures in the emergent virtual or hybrid educational, work, and consumer worlds. The young and their families are perplexed and struggling in their efforts to adapt to digitally organized schooling, work, and consumption regimes. However, these trends are overarching, and a future regulated by smart machines will likely continue beyond the precautionary practices established during the pandemic. As humanity and global society undergo shocks, be they epidemiological, ecological, or economic, or the three together, their impacts will synergistically disrupt every element of life. Administrative and managerial reliance upon smart machines may then become even more prevalent in the future.

Despite digital schooling, some teachers still adhere to student-centered pedagogies derived from the work of John Dewey, and later Paulo Freire, in the face of administratively sanctioned IT and AI control. Yet the choice for both teachers and students upon returning to their digital schoolhouses, fully retooled as a consequence of the pandemic, may be to remain and comply with, or exit from, dominant virtual learning regimes. Those choosing to remain may accommodate to the new schooling regimes and work for a time within their institutions. Teachers and students thus seeking to survive this transition may find a way to move beyond "boxed" instructional practices within IT-saturated educational media ecologies. They can engage public intellectuals and nonacademic thought and action leaders in the wider society for support of their projects and through them to create something new in and for the world. Working side by side with them to reimagine learning, they may even come to experience the next culture of humanism. Short of cultivating habits of mind provided by the humanities, rising generations could lose their way as they navigate disorienting digital ecologies. Their horizons, or perspectives toward the world, will be clearly challenged in this reordering of life by scientific technologies and their applications, which clearly present ethical quandaries that require reflective inquiry.

A NETWORKED WORLD

An emergent economic model brought about by financial crisis, demographic change, and information technology impacts our everyday lives, according to Paul Mason, creating greater global inequalities. Lower-skilled groups in global Northern nations and entire countries in the global South appear to have been left behind due to an inability to access advanced technologies accompanying the next economy. Such disparities place those living at the margins of the recent economy at risk of further decline amid the consumption ethos associated with a globalized Internet. Even wealthier classes within industrialized nations likewise encounter specific risks associated with the pace of late capitalist development and its disruptive technologies. Certain craft-based practices, such as "the humanists, scientists, craftsmen, lawyers, radical preachers and bohemian playwrights,"[67] as Mason suggests, will likewise be disrupted by new technologies. These traditional crafts employ occupational techniques, habits of mind, attitudes, and social arrangements to protect intellectual private property similar to when they arose in the post-Renaissance world. Lowered marginal costs of the production of knowledge, or information goods, may upend these occupations as their craft knowledges become accessible through commons-based goods, notably free and open-source resources, including music, software, databases, and collaborative digital workspaces.

At a time when networked technology is viewed as the solution, Taffel foresees potential disenfranchisement for many traditional occupations. As a consequence of computational efficiencies, Taffel envisions a counter-Renaissance, a counter-Reformation, and even a counter-Scientific Revolution. These epochal transitions could occur as the social power associated with long-standing expert knowledges and practices gives way to peer-to-peer approaches brought on by digital technologies.[68] Jonathan Crary writes:

> All the seemingly altruistic fervor about overcoming "the digital divide" continues to be a unified campaign by corporate interests to require digital compliance everywhere, including the use of computer-based learning in schools for even the youngest of students. The suggestion has been that people without broadband access are living in a condition of deprivation, cut off from the possibility of upward mobility, career opportunities and cultural enrichment. However, the primary goal of the most powerful stakeholders is the eventual transformation of everyone into captive and obedient consumers of their products and services. The unspoken truth is that as internet access and use expands, economic inequality is heightened, not diminished.[69]

In the digitally based transformation, subjectivity and emotionally centered discourse and the fleeting nature of trending media forms are replacing the critical rationalism of early modern European public consciousness that came about with printing and public rhetoric. A triumphant Western rationality seemingly peaked in the Enlightenment, accompanied by a public sphere of courts, colleges, coffeehouses, labs, theaters, pulpits, and craft studios that flourished at the time. Within this enlarged public sphere, new social contracts were negotiated between craft-based occupations, the society, and the state. However, certain other craftspeople lamented contractual models that elevated rationally organized occupations, based as they are on individualist and mechanistic conception of human nature.

William Blake as a poet, craftsman, and political radical best exemplifies a multilayered and multi-intelligence-based character style of the era's creative types. An engraver-poet, Blake's perspective appeared to derive from fine-honed technical skills, an extraordinary attention to detail in his illustrations, matched by an explosive mythopoeic imagination in his poetry. At the time, political elites advocated solutions based upon scientific dogmas espoused by earlier rationalists Isaac Newton and Francis Bacon that limited consciousness to sense perception. Blake instead called for freedom of the imagination so as to create new forms of expression to confront existing political systems and the tragic consequences of the industrial revolution. Blake lamented the transition from hand production methods to the machine and condemned the attendant exploitation of child labor and the repression of workers' life energies by hard labor.

Not unlike Blake's earlier demand for a "politics of vision"[70] that would liberate consciousness from an imprisoning discourse of rationality, a group of digital utopians at the dawn of the Internet espoused a belief that computing could be personally liberating. Stewart Brand, Kevin Kelly, Harold Rheingold, and other techno-utopians, under the spell of Marshall McLuhan, advocated a collaborative vision that drew upon a countercultural ethos and technological breakthroughs of Silicon Valley to close what they viewed as a digital divide between the tech-savvy haves and techno-peasant have nots on the electronic frontier.[71] Rather than a liberating digital ecology of the techno-utopian variant, Taffel sees new modes of exploitation and hierarchical arrangements arising in the networked world, even as attempts were underway to enact commons-oriented peer-to-peer technology. These include the nonprofit P2P Foundation advocating that individuals, organizations, and governments adopt open design and open-source software through licenses, such as the Creative Commons, to change personal and societal consciousness. Beyond the imagined virtual communities of earlier techno-utopians, current efforts toward collaborative digital economies based upon participatory production and commons-based approaches engender new risks to personal freedom and individual consciousness. Those enmeshed in digital ecologies are frequently caught up in an attention economy that alters people's sense of time, with the fleeting nature of words and images that arise with information and communication technologies. These technologies provided financially elite consumers with considerable information-at-hand, often diminishing the value of expert knowledge, especially in a tech-saturated world where everyone has something to say.

Facing the prospect of marginalization, entrenched expert and craft-based post-Renaissance occupational groups have attempted to resist threats to their knowledge monopolies through efficiencies founded in these very technologies. However, austerity policies employ technological solutions and advanced metrics, which contribute to a calculative rationality. Machine-learning technologies in most cases surpass these professions' distinctly early modern attitudes and deliberative practices that accompanied their craft's ascendency. Countering the myriad technological solutions and the progressive impulse at the core of Renaissance and Enlightenment optimism that continues to define elite occupational groups is a recent sense of ecological catastrophism that has brought about an apocalyptic discourse of decline. A feature of this mode of thought is a critique of the "ecological costs of microelectronics,"[72] in Taffel's words, including the extraction and processing of raw materials to produce the hardware, industrial microelectronics production techniques and labor regimes, anthropogenic carbon emissions, energy costs, and the e-waste that accompanies planned obsolescence of devices. These consequences of energopower thus call into question the manner in which recent technologies altered the character of contemporary life and what it means to be human under these circumstances.

BOUNDARY SITUATIONS

The humanities inform recent considerations of the human condition under changed technological circumstances. The centuries-old dialogue between two ways of knowing, rational externality and intuitive internality, is reiterated in current debates surrounding the modern self. Charles Taylor's notion of a late modern self-identity, with features of both an enchanted Western deism and a disenchanted secular skepticism, frames one such communitarian approach.[737] Following Aristotle, Taylor understands humans as social beings that develop moral outlooks within "common spaces" where individuals engage in discursive interactions. Taylor argues that community and civic life are best realized through conversation and debate within a social framework, and through action, essentially in carrying out common projects in a local milieu. Along these lines, Taylor recently endorsed local community projects designed to remake deindustrialized communities from the ground up.[74] These bottom-up pursuits hold the promise of rebuilding new solidarities among demoralized deskilled individuals displaced through technological advancement so they may regain their self-esteem and self-worth. The projects contribute to reconstructing a community's capacity to self-organize by remaking institutions that provide constituents a voice in a system of representative democracy. The intent is to increase social capital by mobilizing the skills and resources of local societies and organizations to respond to crises that erode community in advanced economies, such as the introduction of industrial robotics and the shift from manufacture to human services.

Taylor locates our time as a pluralistic late modern "age of authenticity"[75] defined by disenchanted post-Reformation humanism and secularization. The ensuing culture provides modern people a social imaginary based upon mutuality. A secular age opened new spheres for social interaction, notably, "a space of fashion."[76] In this space, individuals define personal identities and fulfill themselves through displaying their stylistic choices among others in their circle. A moral order based on mutual benefit thus sustains a modern culture of "expressive individualism"[77] that contrasts with the unifying solidarity of an enchanted medieval cosmos with ritual and worship interwoven into the fabric of everyday life.

The relational construct of personhood as a socially defined conditional state is being challenged by robotics and strong artificial intelligence that, together, replace a neo-Darwinian DNA-based conception, according to Bruce R. Reichenbach.[78] As such, personal identity is thrust into a dystopic mechanical future, based on "replacement of DNA-based human evolutes with human-created, silica-based computational machines that can process information, without loss, millions of times faster and more accurately than the carbon-based neurons in our brain."[79] Knowledge acquisition thus transpires through "self-taught and self-communicating" computational machines,[80] following Kurzweil,[81] that may limit human freedom and moral agency and threaten existential personhood.

An emergent postmodern self-identity may form, one that requires neither a classical mind–body dualistic sense of embodiment nor a post-Darwinian acceptance of the varieties of religious experience, forwarded by William James, as part of the human condition. As computational intelligence-based expert systems and machine-learning approaches advance globally, communitarian-inspired moral pluralism may diffuse across the global North. This ethical dimension came about with the advent of social media personhood as a representation of an economic identity within a market-based cyber-communitarianism. Personal identity may now be constructed through virtual social relations within chat rooms, online gaming communities, and avatars in networked virtual environments.

The virtual self—a narrative identity that develops with ongoing story construction through online peer interactions—engages within a network of meanings and plural truths divergent from those found in embodied interactions. The political and ideological implications of a self that is constituted in curated digital environments remain unclear, especially since such socially constructed categories are not as yet fully defined, leaving questions about clarity and coherence. This curated self, though a categorical feature of an interdependent attention economy, frequently acts as a moral agent in the digital milieu, prompting questions concerning ethical responsibility in placeless communities. The abstracted nature of digital encounters and associated transactions has made it difficult to establish norms related to privacy and democratic deliberation in cyberspace.

Bernard Stiegler finds young people feeling trapped within the constraints of a "computational capitalism."[82] Stiegler locates this malaise in a held perspective that equates rationality with mathematical calculation. Calculative thinking often produces political powerlessness when facing technological consequences of a world fashioned by algorithms and automated systems. Such nihilistic thought diminishes any authentic thinking and gives way to machine thinking, which elevates the computer as an expert system that acts as the ultimate authority. With digital technology's ascendency, entropy sets in, as all machines eventually break down.[83] Hence, there is a two-fold disruption: initially, from the machine's expropriation of both the knowledge from inside the worker's brain and the dexterity accompanying hand production, and then from the disruption of thought, itself, as entropy, or chaos, triumphs over negative entropy. The capacity to reverse a disruptive, entropic turn resides in acquiring new perspectives through active thinking, a building site for intellectual construction. This requires regaining one's attention so as to decode and analyze information and to find new ways to relate to one another beyond those advocated by culture industries dominated by computing and telecommunications. Seeking alternatives to entropic powerlessness, Stiegler forwards a reframing based on the power of negentropy, where living things become more orderly, thus the opposite of randomness. Any reorganized outlook thus requires new forms of intellectual energy, moving beyond mere passive reflection and taking an active position involving a dialectical relation of thought and action, rather than one or the other. Stiegler then finds hope in the ability to employ active thinking while moving into an unseen and unknowable future.[84]

Given the scale and complexity of the earth's natural resources, and the human activities that have lasting impacts on the biosphere, sustainable and adaptive human–environment interactions will rely upon coherent thinking about the path forward. Anthropogenic climate change is taking its toll on dryland and coastal regions worldwide and that affects the lives of those living in global cities. The drought and fires in Western North America, Northern and Western Europe, East Asia, and Australia, together with sinking landscapes, water shortages, and rising energy costs are going to wreak havoc on the contented worlds of global elites, as class conflicts escalate. An Intergovernmental Panel on Climate Change report on climate change mitigation recommended an immediate move away from dependence on fossil fuels and a transition to renewable energy sources to limit the emission of greenhouse gases.[85]

A consequence of late modernity is the need for widespread conversations among experts and the lay public about the fate of the earth facing global change. However, given the magnitude of the climate crisis and in the wake of persistent disasters, various scenarios under consideration typically evoke fear and threats to personal existence, in many cases pointing toward inevitable community displacement and residential uprooting.[86] Humanistic perspectives may inform such necessary dialogues, with insights about the cultural forces accompanying, or even driving, climate change and the necessary energy transition. At mid-century, Karl Jaspers contrasted our era of global change with more sedate times, stating:

> There are tranquil ages, which seem to contain that which will last forever, and which feel themselves to be final. And there are ages of change, which see upheavals that, in extreme instances, appear to go to the roots of humanity itself.[87]

Boundary or limit situations are ones that are both physically and psychologically concrete which arise when persons confront frightening and anxiety-provoking encounters. In Jaspers' words, these are "situations such as those of which I am always in a situation, that I cannot live without struggle and without suffering, that I inevitably assume the guilt, that I have to die, are what I call limit situations."[88] For Jaspers, such "unconditioned moments" evoke an intense impulse to move beyond everyday thinking and employ rationality to reflect and communicate troubling concerns and then transcend "conditioned thinking" on behalf of resolution.[89] Paulo Freire appropriated this line of thinking to frame conditions of underdevelopment as a limiting situation of humanity in the global South. Jaspers' approach to crises is thus instructive in understanding the merit of sustained reflection as an appropriate response to climate change as an existential threat.

In the Anthropocene, contemporary ethical ideas, religious and secular, reconsider eschatology, and for evangelicals "end times," amid these tipping points. There is also a long-standing narrative of the need to diminish such despair and instill hope in the face of such finality, especially among the young in affluent global Northern countries. Viktor E. Frankl forwarded a post-Holocaust "survivor narrative" in the mid-twentieth century.[90] In post-Holocaust thought, psychological resilience is expressed as having a sense of hope for a living future based upon active thinking, which became paramount in helping many survivors remain alive through the camp experience. Geoffrey Hartman provides an instructive critique of the post-Holocaust narrative, viewing the "rebirth of language"[91] in survivor narratives: "After the camps, then, the survivors not only testify, that is describe terror undergone, but speak, they testify to speech itself, as an act of which they had been deprived and that enter once again into normal human intercourse."[92] Similar to Arendt and Habermas who forward "postwar imperatives of memory,"[93] Hartman understands the necessity of "aesthetic protest,"[94] in his words, against silence and on behalf of an enlarged public sphere.

Likewise, George Steiner understands that the events that have damaged humanity since 1914, owing to the magnitude of their violence, oppression, economic enslavement, and social irrationality, have led to the "darkened condition of grammar."[95] Steiner goes on to define grammar as "the articulate organization of perception, reflection and experience, the nerve structure of consciousness when it communicates with itself and with others."[96] Against darkening historical forces, and those of technology, which impede grammatical expression, Steiner forwards creativity so as to provoke a future tense or "grammars of creation" to sustain hope. Creative experimentation, including expansive grammatical expression and representation in writing, film, and musical composition, design and the visual arts, and innovations in the art of living, can serve as a forward-looking back loop strategy. This strategy would hold for both Western and postcolonial aesthetic practices. These practices may come to connect the seemingly unbridgeable divides among individuals and groups in the face of global collapse of the current socioeconomic system, although neoliberal austerity regimes remain persistent in global Northern countries.

Jem Bendell posits the idea of "deep adaptation"[97] as response to anthropogenic climate change, conveying an ambivalent sense of hope in the face of, in his view, a predicted near certain societal collapse. Bendell forwards the need to assume an ethical position that involves envisioning a post-collapse future while consciously anticipating societal breakdown. He critiques the progressive framing of resilience, forwarding instead a deep adaptation that encompasses resilience, relinquishment, and restoration. His position requires reconciliation with nonelectrical grid-based power and climate action, each of which is enmeshed in the governance of markets, finance, and banking. However, corporate firms, governmental agencies, scientific laboratories, school systems, higher education, and nonprofit organizations, since the last half-century, were reorganized and currently operate within long-standing market-based financing arrangements. Such deep adaptation will only occur, according to Bendell, when these entities change perspective and direction provoking questions as to the kinds of events that would initiate such a change in view and directionality. This will clearly require complex organizations, together with the ones charged with managing societal crises, to initiate change in the face of a "bounded rationality,"[98] which holds that in decision-making, people are limited to the information at hand, by their own cognitive limitations and by the time allotted to planning. Bendell advises the rest of us to face the prospect of societal collapse and the fracturing of local communities and to consider a decentralized life within intentional communities. At the same time, he questions the limits to self-sufficiency imposed on this form of community life. Bendell proposal evoked concerns about the capacity of various global constituencies to gain access to the requisite knowledge and techniques for deep adaptation and creative resilience in the transition from a globalist, fossil fuel-based, growth economy to a post-carbon one.

These dire predictions provoke a consideration of the viability of implementing state solutions to the impending ecological crisis, notably Giorgio Agamben's conception of "bare life," or biological life *(bios)* constrained by states through sovereign power and control of technology. Those consigned to bare life are stateless persons and others excluded from legal protections. The state holds the sole power to transform bare life into "the good life" through its politics, following Aristotle, and to exclude natural or reproductive life (*zoē*) by confining it to the domain of the home (*oikos*):

> In the "politicization" of bare life—the metaphysical task par excellence—the humanity of living man is decided. In assuming this task, modernity does nothing other than declare its faithfulness to the essential structure of the metaphysical tradition. The fundamental categorical pair of Western politics is not that of friend/enemy but that of bare life/political existent, *zoē/bios*, exclusion/inclusion. There is politics because man is the living being who, in language, separates and opposes himself in his own bare life, and at the same time, maintains himself in relation to that bare life in an inclusive exclusion.[99]

In Agamben's view, the modern state's executive power, especially in a state of emergency, is often enacted, beyond political crises, during precarious economic conditions. Agamben constitutes a politics that takes these polarities into consideration through its systems of laws, with their exclusionary and inclusionary boundaries, making it difficult for citizens to act against state sovereignty. Michael Hardt and Antonio Negri view Agamben's ontological "original either-or structure" as limiting and politically disempowering. Seeking the more "productive dimensions of '*bios*'," they state:

> When human power appears immediately as an autonomous cooperating collective force, capitalist prehistory comes to an end. In other words, capitalist prehistory comes to an end when social and subjective cooperation is no longer a product but a presupposition, when naked life is raised up to the dignity of productive power, or really when it appears as the wealth of virtuality.[100]

Hardt and Negri forward an alternative conception of politics as creative resistance to state power on behalf of preserving the planetary commons: "We might call this an ecology of the common——an ecology focused equally on nature and society, on humans and the non-human world in a dynamic interdependence, care, and mutual transformation."[101] This open-ended conception, based upon increased cultural and communicative capacities, creates a path for climate justice as a transnational movement, according to Emanuel Leonardi, to transform "the notion of planetary climate as a global common/s."[102]

Critical political and ethical analyses of prospective societal collapse, together with cognitive reframing of the climate crisis, may be necessary to cut through the current milieu of avoidance and denial. This could then permit reflection, and perhaps broader creative responses to the existential situation at hand, as environmental and epidemiological events, and their economic consequences, continue to move in the direction that climate scientists predict. Many inquire whether any political intervention can coalesce regions and nations to advance politics that will sustain the nine planetary boundaries. Classical liberalism posits market-based counter arguments to large-scale intervention into society, citing the underlying limits of human nature. Following Arendt's view of the human condition, our activities enacted within a *polis*, or a public "space of appearance,"[103] can give rise to our capacities and personal identities. These worldly actions, together with speech, take place within a "plurality," or the presence of others, providing the potential for people to act "in concert," representing multiple perspectives in a common, or shared, "public world."[104] In this manner, threats to personal and community well-being, which could seriously alter countless individual life trajectories, can be discussed and acted upon so as to evoke ethically grounded decisions and responses in an enlarged public sphere.

THE FATE OF THE COMMONS

A critical loss stemming from neoliberal privatization is that of the commons, beyond the traditional use of the word to denote shared pasturage to currently indicate shared responsibilities for maintaining equitable access to public goods. In many ways, the loss of the idea of the commons, broadly understood as an eclipse of an ethos of shared public resources, has led to considerable economic precarity in communities left behind in the recent socioeconomic transition. This position is reinforced in recast regional economic geographies based upon asymmetrical access to training and consultation in the transfer of ascendent technologies, although there may also be resistance to such change. An example is that of agriculture, where seeds are emblematic of advantageous human entanglements with the natural world, for as Ashley Dawson and A. Naomi Polk state, "seeds are the living embodiment of the commons."[105] As a result, traditional farming practices were an ongoing deterrent to capitalist enclosure until even these were captured by advances in agricultural biotechnology, upending long-standing agrarian patterns and giving way to "patented and restrictive corporate ownership."[106] The current era is thus marked by innovations in materials science, engineering, information, and biology-based technologies that drive global demand. A typical example is that of when biomedical and microchip-based technologies are introduced, early adopters who anticipate a place in the transition and change their skills accordingly fare better than the others who adhere to outmoded skill sets. Certain occupations and firms thus ascend while others are rendered obsolete by technological advances and their vicissitudes. This is true for technological changes in the recent transition as it was more than a century ago when advances in chemicals and mechanical engineering displaced various artisan groups.

Attendant with limited mobility and resource scarcity are the new measures to control those sickened, disenchanted, and even enraged over being left behind in the new economy. Giroux refers to disciplines of control embedded in the contemporary economy as a "zombie politics" of rationalized technocracies and neoliberal austerities.[107] Together, these promote untold personal suffering and cultural deadening in an "age of disposability" following Giroux: "The death-haunted politics of disposability evident in the wave of austerity measures at work in North America and Europe is a systematic outcome of neoliberal capitalism as it actively engages in forms of asset stripping and social control."[108] With hundreds of billions spent annually to support technologies of control, governments and corporate firms have remapped the economic landscape. Consequently, the life chances of youth in marginalized communities are narrowed, while those within communities held necessary to the new economy find increased opportunities in ascendent occupations.

The new economy depends upon diverse state and corporate efforts to maintain the fossil fuel energy regime, ranging from extraction to production and distribution. All guarantee unrestrained global logistics to support construction and flow of transnational pipelines, increase drilling capacity and productivity, and protect global extraction areas in Louisiana, Texas, California, western Canada, the Persian Gulf, Congo Basin, and circum-Arctic petroleum fields. Consequently, the grid will likely remain fired by electricity, supported by nonrenewables in certain regions. Global oligarchic power inheres in ownership of these production regions and global commodity chains, both under intensive military surveillance and protection, deter any move to a post-carbon world economy. Until the consequences of climate change are mitigated by binding agreements among nations of both the global North and South, the environmental movement as enacted by diplomats, advocates, and activists will be stalled. Any future quality of life, perhaps humanity's very survival, will depend upon transition to a post-carbon economy. In our "time of need," with a global civilization based on the complex organization of technology, William Barrett asserts:

> So long as we think that the solution of the problems technology has brought must be purely technological, then we are involved in coordinating and

> subordinating machines to machines; and so are already on our way to building The Machine, any break in whose complex network could mean disaster.[109]

There is an increasing awareness of the shortcomings of governing elites in the wake of human and natural disasters, ranging from poisoned water supplies to flood tides of historic proportion and wildfires of unprecedented destruction and duration. Despite countless disastrous consequences, governmental and corporate elites upholding normative decision-making continue to marginalize those who engage in ethical and political discourses on human-induced environmental impacts. Well before the dawn of the Anthropocene, there were movements to support the aesthetic improvement of landscapes, including preservation, conservation, and restoration efforts to counter external threats to their sustainability. Earlier, the commons was a central paradigm for understanding adaptation and resiliency of diverse landscapes under urbanization.

To counter current normative ecological bypassing of anthropogenic climate change compels a reoriented commons paradigm to account for exigencies of geological time. The Anthropocene, an earth-centered time frame, also requires alignment with a human-centered, historical temporal dimension and acknowledge the material conditions, social relations, and cultural practices that accompany environmental transformations. "The Anthropocene debate thus entails a constant conceptual traffic between earth history and world history," in Dipesh Chakrabarty's words.[110] Rapid and harmful development, the exploitation of natural and human environments for profit and growth of human populations, brought complex environmental challenges. With these conditions came changes in norms and ways of living for many concerned with their consequences, together with new modes of thinking through critical global pedagogies that "reassemble" the natural and social commons.[111]

A "commons imaginary" recently surfaced as a political argument "against commodification, privatization, or enclosure and in favor of egalitarian, grassroots approaches to resource management,"[112] according to John Wagner, who identifies this imaginary as "a response to more recent but equally fundamental changes in our social, economic, and political lives, particularly those associated with economic globalization, the accelerated pace of erasure of place-based communities and social identities, and global environmental and economic crises."[113] Wagner understands how commons-as-social-imaginary corresponds to an "ideoscape,"[114] following Arjun Appadurai, albeit seeing it as a counternarrative to the project of modernity. Constituents within a commons "ideoscape" seek alternatives to current governance of common resources, notably water and energy, bringing to their discussions cultural values, political arguments supportive of human rights, and personal concerns for the good life.

Across generations, people are engaged in climate action for renewable energy sources to reduce dependence on coal and oil. They advocate for measures leading to a post-carbon economy that markets and governments can advance on behalf of the earth system. The intent is not necessarily to fully correct the imbalance between humans and nature but to muddle through, nevertheless. Along with such action is a move to involve people in the construction of narratives around their understanding of current and future energy regimes. These "sociotechnical imaginaries"[115] advance from diverse understandings of what currently constitutes "the good life" and the various modifications to daily life and routines that are acceptable given any future energy system transitions. Sociotechnical storylines may come from the government, media, and popular narratives, the latter based upon people talking about their own experiences.[116] The prevalent scientific understanding of energy encompasses features of Western modernity, notably a mechanistic and instrumental view of the natural world, and the primacy of a sociotechnical world. Narratives imagining a move from a hegemonic global energy system to a sustainable alternative typically address three concerns: satisfying human needs, ensuring social justice, and respecting environmental limits.[117] At the same time, dominant political and economic elites move to control narratives about current sociotechnical regimes and to even shape the ones concerning sustainability transitions toward renewable energy futures.

Three overarching themes persist in such rhetorical initiatives, both hegemonic and counter-hegemonic, to navigate sociotechnical futures. According to Laurence Delina and Anthony Janetos, whether normative, aspirational, or desired, energy narratives demonstrate "an embrace of plurality, high regard for reflexivity, and understanding of the politics."[118] Such inclusivity necessitates an acceptance of others who express plural viewpoints in a different voice, as Sheila Jasanoff states:

> Coping effectively with the future of energy will require every human on the planet—scientists and lay people—to operate at new scales: accepting relationships with persons from very different cultures; putting up with political decisions reached outside the processes of our nation states; adapting to changes originating beyond our local control; and thinking in extended time spans that dwarf human imagination and experience.[119]

With this in mind, inclusive dialogues across diverse publics who hold different explanations and alternative solutions may require a dual emphasis. Discussions grounded in the materiality of energy, located in technologies and technical systems, would also need to consider the nonmaterial aspects of energy, most notably the social world.[120] Interlinking the two dimensions can help participants representing diverse and often conflicting positions to move beyond polarizing conversations and unproductive debates.

A global initiative based upon a narrative approach, "Biodiversity Revisited," set an agenda with an overarching goal "to contribute to sustaining diverse and just futures for life on earth."[121] A two-year dialogue ensued among 300 social and biophysical scientists, together with participants from across the humanities and law, and also from governmental and nongovernmental organizations, representing 46 nations. Their starting point was to revisit current biodiversity narratives that frame the ecological crisis. Participants discussed the various new narratives that could reframe nature and sought alternative narratives to replace worn-out biodiversity stories to incentivize action on behalf of sustainability. The group came up with three core areas and accompanying questions to focus research and action.[122] The first included empirical examinations of the narratives that underpin destructive systems to answer the question: Why do some narratives become authoritative, while others are silenced or deliberately ignored, and what are the results? The second was to bring together diverse approaches to narrative to enrich biodiversity research and address the question: How can biodiversity research listen to and learn from narratives that have been traditionally outside biodiversity research? The third involved exploring the role of narrative in imagining alternative futures and enabling transformative change to understand: How can narratives and narrative approaches be used to foster productive engagement with contested and uncertain futures? The intention was to create "equitable spaces for exchange"[123] to enable participants from diverse disciplines to listen, and then consider and critically reflect on the various positions and narratives forwarded in the dialogues. Through a deliberative process, scientists, practitioners, policy makers, and representatives from local communities across the globe came up with "more life-sustaining narratives"[124] to "reconfigure current practices."[125]

Interpersonal processes based upon group dynamics help participants to "frame" the reality of the energy transition, rapid loss of biodiversity, and living beyond planetary boundaries. Goffman's idea of frames as "schemata of interpretation"[126] that enable people to make sense of events and organize their experience can be applied to meaning construction within social movements. Robert D. Benford and David A. Snow hold that

> collective action frames also perform this function by simplifying and condensing aspects of the "world out there" . . . [they] are action-oriented sets of beliefs and meanings that inspire and legitimate the activities and campaigns of a social movement organization.[127]

All of this takes considerable political will and organizational acumen on behalf of collective action, not mere reliance on political gatherings and partisan campaigns. Mass social movements are necessary to craft life-affirming politics to promote participants' deeper connection to nature and to the well-being of their communities. For this to occur, environmental educational praxis, based upon meaning construction, will need to be included in movement repertoires. Thomas Beery and his colleagues view environmental education, broadly understood, as a means to strengthen the human relationship with nature "along a continuum from individual to the social":

> Connection to nature is not simply a measure of individual relationships, but is also a product of the social world of people, one that comes with social and cultural organization, power structures, and collective infrastructure. We must guide educational efforts to empower individual relationships between people and nature without losing sight of education's role in illuminating social and cultural ways to collectively support the idea of people as a part of nature.[128]

The various dialogues, movements, and educational efforts requisite at the global level on behalf of a "planetary just transition,[129] must also take place in communities where localized trust can be built. In this way, localities can come to understand how catastrophic global climate change adversely impacts their quality of life and the health of the local ecosystem. Diverse constituencies will thus need to coalesce and create place-based arenas for critical inquiry and reflection around biodiversity, energy, and sustainability concerns.

Education is therefore critical to the public sphere, which is polarized by contentious debates over class, ethnicity, culture, and more recently the fate of the planet. With this in mind, two provinces, notably the school and society, can join together to afford students greater freedom to produce future knowledge as humanity faces profound challenges to its existence. Accordingly, experiential instructional approaches are being designed and currently flourish in schools and their communities that vary in social and economic mobility, risk and security, affluence and poverty, subcultures and identity. These pedagogies are sustained amid prevalent institutional arrangements, including a neoliberal framework, administrative controls, and governmental regulation of credentialing, licensing, and financing. Efforts to reframe social and cognitive dimensions of learning and its enculturation through more engaged and meaningful approaches meet a range of challenges.

In light of this, Arendt reminds us that the public sphere, while remaining a common world, will always involve dialogues, at times contentious disputes, among people holding divergent outlooks and positions:

> The reality of the public realm relies on the simultaneous presence of innumerable perspectives and aspects in which the common world presents itself and for which no common measurement or denominator can ever be devised. For though the

> common world is the common meeting ground of all, those who are present have different locations in it, and the location of one can no more coincide with the location of another than the location of two objects. Being seen and being heard by others derive their significance from the fact that everybody sees and hears from a different position. This is the meaning of public life, compared to which even the richest and most satisfying family life can offer only the prolongation or multiplication of one's own position with its attending aspects and perspectives.[130]

Contested cultural terrains in school, at work, and in the home remain challenges that young people perennially confront on their paths to adulthood. Public pedagogies provide youth with the necessary preparation for life to meet a contentious world and to lead productive and critically conscious lives amid the ensuing dialogues and debates within the public sphere. Education is thus crucial to "life politics," Anthony Giddens' concept of a politics of life decisions, including decisions that connect the personal aspects of human activity to planetary problems.[131] A deeper politics, life politics takes place when people come to understand how globalizing processes influence their self-identities. Educational movements that combine Dewey's project-based learning with Freire's language of critique and liberating praxis, in a similar manner, promote a critical consciousness. With such in-depth understanding, diverse constituents—teachers, youth, parents, and other adults—can come together to critically reflect and then take critical action within their schools and communities, and in this way provide alternative narratives for change. Forwarding proposals based upon emancipatory politics begins with building local-level coalitions around matters affecting individual and community life and planetary life, as well.

NOTES

1. Hannah Arendt, *The Human Condition* (Chicago: University of Chicago Press, 1958), 52.
2. Harriet Joseph Ottenheimer, *The Anthropology of Language: An Introduction to Linguistic Anthropology* (Belmont, CA: Wadsworth, 2013), 23.
3. Elizabeth A. Behnke, "Edmund Husserl: Phenomenology of the Body," *The Internet Encyclopedia of Philosophy*, ISSN 2161-0002, https://iep.utm.edu/husspemb/; Elizabeth A. Behnke, "Edmund Husserl's Contribution to Phenomenology of the Body in *Ideas II*," *in Phenomenology: Critical Concepts in Philosophy*, Volume 2. *Phenomenology: Themes and Issues*, edited by Dermot Moran and Lester E. Embree with Tanja Staehler and Elizabeth A. Behnke (London: Routledge, 2004), 235–264.
4. David R. Harris, "Introduction," in *The Archaeology of V. Gordon Childe: Contemporary Perspectives*, edited by David R. Harris (Chicago: University of Chicago Press, 1994), 1–7.

5. Jean-Paul Sartre, *Existentialism is a Humanism*, edited by John Kulka, translated by Carol Macomber (New Haven, CT: Yale University Press, 2007).

6. Don Ihde, *Technology and the Lifeworld: From Garden to Earth* (Bloomington: Indiana University Press, 1990).

7. Peter-Paul Verbeek, "Don Ihde: The Technological Lifeworld," in *American Philosophy of Technology: The Empirical Turn*, edited by Herman John Achterhuis (Bloomington: Indiana University Press, 2001), 134.

8. Ludwig Wittgenstein, *Tractatus Logico-Philosophicus*, translated by David F. Pears and Brian F. McGuinness (London: Routledge & Kegan Paul, 1961), 7.

9. Sy Taffel, "Scalar Entanglement in Digital Media Ecologies," *European Journal of Media Studies* 2:1 (2013): 235.

10. Neil Postman, "The Humanism of Media Ecology," *Proceedings of the Media Ecology Association* 1:1 (2000): 10.

11. Taffel, "Scalar Entanglement in Digital Media Ecologies," 241–242.

12. Taffel, "Scalar Entanglement in Digital Media Ecologies," 240.

13. Taffel, "Scalar Entanglement in Digital Media Ecologies," 240–241.

14. Charles B. Nauert, *Humanism and the Culture of Renaissance Europe*, second edition (Cambridge: Cambridge University Press, 2006), 222.

15. Jürgen Habermas, *The Structural Transformation of the Public Sphere: An Inquiry into a Category of Bourgeois Society*, translated by Thomas Burger (Cambridge, MA: MIT Press, 1991).

16. William Cronon, "Introduction," in *Fire: A Brief History*, edited by Stephen J. Pyne (Seattle: University of Washington Press, 2001), xiii.

17. E. B. Banning and Michael Chazan, "Structuring Interactions, Structuring Ideas: Domestication of Space in the Prehistoric Near East," in *Domesticating Space: Construction, Community, and Cosmology in the Late Prehistoric Near East*, edited by E. B. Banning and Michael Chazan, Studies in Early Near Eastern Production, Subsistence, and Environment, 6 (Berlin: *ex oriente*, 2006), 5–14.

18. Trevor Watkins, "Architecture and the Symbolic Construction of New Worlds," in *Domesticating Space: Construction, Community, and Cosmology in the Late Prehistoric Near East*, 15–23.

19. Joseph Rykwert, "House and Home," *Social Research* 58:1 (1991): 51.

20. Clive Gamble, "The Anthropology of Deep History," *Journal of the Royal Anthropological Institute* 21:1 (2015): 147–164.

21. Trevor Watkins, "Architecture and 'Theatres of Memory' in the Neolithic of Southwest Asia," in *Rethinking Materiality: The Engagement of Mind with the Material World*, edited by Elizabeth DeMarrais, Chris Gosden, and Colin Renfrew (Cambridge: McDonald Institute of Archaeological Research, 2004), 97–106.

22. Merlin Donald, *Origins of the Modern Mind* (Cambridge, MA: Harvard University Press, 1991); Merlin Donald, "Human Cognitive Evolution: What We Were, What We Are Becoming," *Social Research* 60:1 (1993): 143–170.

23. Arendt, *The Human Condition*, 36–37.

24. Stephen J. Pyne, "From Pleistocene to Pyrocene: Fire Replaces Ice," *Earth's Future* 7 (2020): e2020EF001722, https://doi.org/10.1029/2020EF001722.

25. Rolf Lidskog, "Invented Communities and Social Vulnerability: The Local Post-Disaster Dynamics of Extreme Environmental Events," *Sustainability* 10:10 (2018): 4457, https://doi.org/10.3390/su10124457 (2018): 14.

26. Lidskog, "Invented Communities and Social Vulnerability," 4.

27. Lidskog, "Invented Communities and Social Vulnerability," 4.

28. Lidskog, "Invented Communities and Social Vulnerability," 5.

29. Ulrich Beck, *Risk Society: Towards a New Modernity*, translated by Mark Ritter (London: Sage Publications, 1992).

30. William I. Robinson, *Global Capitalism and the Crisis of Humanity* (Cambridge: Cambridge University Press, 2014), 1.

31. Dominic Boyer, "Energopower: An Introduction," *Anthropological Quarterly* 87:2 (2014): 325.

32. William R. Anderegg, Anna T. Trugman, Grayson Badgley, Christa M. Anderson, Ann Bartuska, Philippe Ciais, Danny Cullenward et al., "Climate-Driven Risks to the Climate Mitigation Potential of Forests," *Science* 368:6497 (2020): eaaz7005.

33. Henry A. Giroux, "Zombie Politics and Other Late Modern Monstrosities in the Age of Disposability," *Policy Futures in Education* 8:1 (2010): 1.

34. Dominic Boyer, "Anthropology Electric," *Cultural Anthropology* 30:4 (2015): 534.

35. Helen Nearing and Scott Nearing, *Living the Good Life: How to Live Sanely and Simply in a Troubled World* (New York: Schocken Books, 1973).

36. Douglas Stevenson, *Out to Change the World: The Evolution of the Farm Community* (Summertown, TN: Book Publishing Company, 2014).

37. Dan McKanan, *Eco-Alchemy: Anthroposophy and the History and Future of Environmentalism* (Berkeley: University of California Press, 2018).

38. Manuel Castells, *The Rise of the Network Society*, second edition (Malden, MA: Wiley-Blackwell, 2009); Manuel Castells, "The Network Society Revisited," *American Behavioral Scientist* (2022): 00027642221092803.

39. Claus Pias, *Cybernetics: The Macy Conference 1946–1953, The Complete Transactions*, revised edition (Zurich: Diaphanes, 2016), 1.

40. William I. Thomas, Jr., Carl Sauer, Marston Bates, and Lewis Mumford, editors, *Man's Role in Changing the Face of the Earth, Volumes 1 and 2* (Chicago: University of Chicago Press, 1956).

41. Paul Rabinow and Nikolas Rose, "Biopower Today," *Biosciences* 1:2 (2006): 195–217.

42. Thomas Lemke, *Biopolitics: An Advanced Introduction* (New York: New York University Press, 2011).

43. Michel Foucault, "Governmentality," translated by Rosi Braidotti and revised by Colin Gordon, in *The Foucault Effect: Studies in Governmentality*, edited by Graham Burchell, Colin Gordon, and Peter Miller (Chicago, IL: University of Chicago Press, 1991), 87–104.

44. Bruno Latour, *We Have Never Been Modern*, translated by Catherine Porter (Cambridge, MA: Harvard University Press, 1993).

45. Debra J. Borkovich and Philip D. Noah, "Big Data in the Information Age: Exploring the Intellectual Foundation of Communication Theory," *Information Systems Education Journal* 12:1 (2014): 15.

46. Johan Fredrikzon, "Review: *Cybernetics: The Macy Conferences*," *Sensorium Journal* 3 (2021): 57.

47. Alberto Melucci, "The Voice of the Roots: Ethno-National Mobilizations in a Global World," *Innovation: The European Journal of Social Science Research* 3:3 (1990): 359.

48. Langdon Winner, *Autonomous Technology; Technics-out-of-Control as a Theme in Political Thought* (Cambridge, MA: MIT Press, 1977).

49. Janna Anderson, Lee Rainie, and Alex Luchsinger, *Artificial Intelligence and the Future of Humans* (Washington, DC: Pew Research Center, 2018), 11.

50. William I. Robinson, "Accumulation Crisis and Global Police State," *Critical Sociology* 45:6 (2019): 8.

51. Robinson, "Accumulation Crisis and Global Police State," 10.

52. Monica Anderson and Jingjing Jiang, *Teens, Social Media & Technology 2018* (Washington, DC: Pew Research Center, 2018).

53. Monica Anderson, Emily A. Vogels, Andrew Perrin, and Lee Rainie, *Connection, Creativity and Drama: Teen Life on Social Media in 2022* (Washington, DC: Pew Research Center, 2022).

54. Herbert Marcuse, *One Dimensional Man: Studies in the Ideology of Advanced Industrial Society* (Boston: Beacon Press, 1964).

55. Albert Borgmann, *Technology and the Character of Contemporary Life: A Philosophical Inquiry* (Chicago: University of Chicago Press, 1984).

56. Michael Lerner, *Surplus Powerlessness: The Psychodynamics of Everyday Life and the Psychology of Individual and Social Transformation* (Lanham, MD: Humanities Press, 1998).

57. Braverman, *Labor and Monopoly Capital*, 34.

58. Stanley Aronowitz, *False Promises: The Shaping of American Working Class Consciousness* (Durham, NC: Duke University Press, 1992).

59. Neil Postman, *Amusing Ourselves to Death: Public Discourse in the Age of Show Business* (New York: Penguin Books, 1985).

60. Richard E. Sclove, *Democracy and Technology* (New York: Guilford Press, 1995), ix.

61. Hans Jonas, *The Imperative of Responsibility: In Search of an Ethics for the Technological Age* (Chicago: University of Chicago Press, 1984).

62. William Deresiewicz, *Excellent Sheep: The Miseducation of the American Elite and the Way to a Meaningful Life* (New York: Free Press, 2014).

63. Winner, *Autonomous Technology*, 8.

64. George Siemens, "Connectivism: A Learning Theory for the Digital Age," *International Journal of Instructional Technology and Digital Learning* 2:1 (2005): 5.

65. Siemens, "Connectivism," 6.

66. Siemens, "Connectivism," 7.

67. Paul Mason, *PostCapitalism: A Guide to Our Future* (London: Allen Lane, 2015), 215.

68. Taffel, "Scalar Entanglement in Digital Media Ecologies," 233–254.

69. Jonathan Crary, *Scorched Earth: Beyond the Digital Age to a Post-Capitalist World* (London: Verso, 2022), 18–19.

70. Mark Schorer, *William Blake: The Politics of Vision* (New York: Random House, 1959), 9–10.

71. Fred Turner, *From Counterculture to Cyberculture: Stewart Brand, the Whole Earth Network and the Rise of Digital Utopianism* (Chicago: University of Chicago Press, 2006).

72. Sy Taffel, "Escaping Attention: Digital Media Hardware, Materiality and Ecological Cost," *Culture Machine* 13 (2012): 19.

73. Charles Taylor, *Sources of the Self: The Making of Modern Identity* (Cambridge, MA: Harvard University Press, 1989).

74. Charles Taylor, Patricia Nanz, and Madeleine Beaubien Taylor, *Reconstructing Democracy: How Citizens Are Building from the Ground Up* (Cambridge, MA: Harvard University Press, 2020).

75. Charles Taylor, *A Secular Age* (Cambridge, MA: Harvard University Press, 2007), 473.

76. Taylor, *A Secular Age*, 475, 481.

77. Taylor, *A Secular Age*, 473–475.

78. Bruce R. Reichenbach, "Christianity, Science, and Three Phases of Being Human," *Zygon* 56:1 (2020): 96–117.

79. Reichenbach, "Christianity, Science, and Three Phases of Being Human," 108.

80. Reichenbach, "Christianity, Science, and Three Phases of Being Human," 109.

81. Ray Kurzweil, *The Age of Spiritual Machines* (New York, NY: Viking, 1999), 3.

82. Bernard Stiegler, *The Age of Disruption: Technology and Madness in Computational Capitalism*, translated by Daniel Ross (London: Polity Press, 2019).

83. Mark Featherstone, "Stiegler's Ecological Thought: The Politics of Knowledge in the Anthropocene," *Educational Philosophy and Theory* 52:4 (2020): 409–419.

84. Richard Niesche and Christina Gowlett, "Bernard Stiegler: Technics and Educational Leadership as a Form of *Psycho-Power*," *Social, Critical and Political Theories for Educational Leadership* (Singapore: Springer, 2019), 104.

85. IPCC, 2022: Summary for Policymakers. In *Climate Change 2022: Mitigation of Climate Change. Contribution of Working Group III to the Sixth Assessment Report of the Intergovernmental Panel on Climate Change*, edited by Priyadarshi R. Shukla, Jim Skea, Raphael Slade, Alaa Al Khourdajie, Renée van Diemen, David McCollum, Minal Pathak, Sheya Some, Purvi Vyas, Roger Fradera, Malek Belkacemi, Apoorva Hasija, Géninha Lisboa, Sigourney Luz, and Juliette Malley (Cambridge: Cambridge University Press, 2022), http://doi.org/10.1017/9781009157926.001.

86. Jake Bittle, *The Great Displacement: Climate Change and the Next American Migration* (New York: Simon & Schuster, 2023).

87. Karl Jaspers, *The Origin and Goal of History* (New Haven, CT: Yale University Press, 1953), 231.

88. Karl Jaspers, *Basic Philosophical Writings: Selections*, edited with translations by Leonard H. Ehrlich, Edith Ehrlich, and George B. Pepper (Athens, OH: Ohio University Press, 1986), 96.

89. Chris Thornhill and Ronny Miron, "Karl Jaspers," *The Stanford Encyclopedia of Philosophy*, edited by Edward N. Zalta (Spring 2022 edition), https://plato.stanford.edu/archives/spr2022/entries/jaspers/.

90. Viktor E. Frankl, *Man's Search for Meaning*, translated by Ilse Lasch (New York: Washington Square Press, 1959).

91. Geoffrey H. Hartman, *The Fateful Question of Culture* (New York: Columbia University Press, 1997), 103.

92. Hartman, *The Fateful Question of Culture*, 102–103.

93. Peter J Verovšek, "Integration after Totalitarianism: Arendt and Habermas on the Postwar Imperatives of Memory," *Journal of International Political Theory* 16:1 (2020): 2–24.

94. Hartman, *The Fateful Question of Culture*, 103.

95. George Steiner, *Grammars of Creation* (New Haven, CT: Yale University Press, 2001), 6.

96. George Steiner, *Grammars of Creation*, 6.

97. Jem Bendell, *Deep Adaptation: A Map for Navigating Climate Tragedy*, revised second edition, IFLAS Occasional Paper 2 (Carlisle: University of Cumbria Institute for Leadership and Sustainability, 2020), https://lifeworth.com/deepadaptation.pdf.

98. Gregory Wheeler, "Bounded Rationality," *The Stanford Encyclopedia of Philosophy*, edited by Edward N. Zalta (Fall 2020 edition), https://plato.stanford.edu/archives/fall2020/entries/bounded-rationality/.

99. Giorgio Agamben, *Homo Sacer: Sovereign Power and Bare Life*, translated by D. Heller-Roazen (Stanford, CA: Stanford University Press, 1998), 8.

100. Michael Hardt and Antonio Negri, *Empire* (Cambridge, MA: Harvard University Press, 2001), 366.

101. Hardt and Negri, *Commonwealth* (Cambridge, MA: Harvard University Press, 2009), 171.

102. Emanuele Leonardi, "Biopolitics of Climate Change: Carbon Commodities, Environmental Profanations, and the Lost Innocence of Use-Value," Electronic Thesis and Dissertation Repository, 959 (London: University of Western Ontario, 2012), https://ir.lib.uwo.ca/etd/959, iii.

103. Arendt, *The Human Condition*, 199.

104. Maurizio Passerin d'Entreves, "Hannah Arendt," *The Stanford Encyclopedia of Philosophy* (Fall 2022 edition), edited by Edward N. Zalta & Uri Nodelman (2022), https://plato.stanford.edu/archives/fall2022/entries/arendt/.

105. Ashley Dawson and A. Naomi Paik, "Germinations: An Introduction," *Radical History Review* 145 (2023): 1.

106. Dawson and Paik, "Germinations: An Introduction," 2.

107. Giroux, "Zombie Politics and Other Late Modern Monstrosities in the Age of Disposability," 2.

108. Henry A. Giroux, *America's Addiction to Terrorism* (New York: Monthly Review Press, 2016), 26.

109. William Barrett, *Time of Need: Forms of Imagination in the Twentieth Century* (New York: Harper & Row, 1972), 297.

110. Dipesh Chakrabarty, *The Climate of History in a Planetary Age* (Chicago: University of Chicago Press, 2021), 155.

111. Jene Bazzul and Sara Tolbert, "Reassembling the Natural and Social Commons," in *Educational Commons in Theory and Practice: Global Pedagogy and Politics*, edited by Alexander J. Means, Derek R. Ford, and Graham B. Slater (New York: Palgrave Macmillan, 2017), 5–73.

112. John R. Wagner, "Water and the Commons Imaginary," *Current Anthropology* 53:5 (2012): 620.

113. Wagner, "Water and the Commons Imaginary," 620.

114. Arjun Appadurai, "Disjunction and Difference in the Global Cultural Economy," *Theory, Culture & Society* 7 (1990): 296.

115. Sheila Jasanoff and Sang-Hyun Kim, "Containing the Atom: Sociotechnical Imaginaries and Nuclear Power in the United States and South Korea," *Minerva* 47 (2009): 122.

116. Joe Smith, Robert Butler, Rosie J. Day, Axel H. Goodbody, David H. Llewellyn, Mel Rohse, Bradon T. Smith, Renata A. Tyszczuk, Julia Udall, and Nicola M. Whyte, "Gathering Around Stories: Interdisciplinary Experiments in Support of Energy System Transitions," *Energy Research & Social Science* 31 (2017): 284–294.

117. Erling Holden, Kristin Linnerud, and Bente Johnsen Rygg, "A Review of Dominant Sustainable Energy Narratives," *Renewable and Sustainable Energy Reviews* 144 (2021): 1–2.

118. Laurence Delkina and Anthony Janetos, "Cosmopolitan, Dynamic, and Contested Energy Futures: Navigating the Pluralities and Polarities in the Energy Systems of Tomorrow," *Energy Research & Social Science* 35 (2018): 8.

119. Sheila Jasanoff, "Just Transitions: A Humble Approach to Global Energy Futures," *Energy Research & Social Science* 35 (2018): 13.

120. Margarita Balmaceda, Per Högselius, Corey Johnson, Heiko Pleines, Douglas Rogers, and Veli-Pekka Tynkkynen, "Energy Materiality: A Conceptual Review of Multi-Disciplinary Approaches," *Energy Research & Social Science* 56 (2019): 12.

121. Carina Wyborn, Jasper Montana, Nicole Kalas, F. Davila Cisneros, Sarah Clement, S. Izquierdo Tort, Natalie Knowles et al., *Research and Action Agenda for Sustaining Diverse and Just Futures for Life on Earth: Biodiversity Revisited* (Gland: Luc Hoffmann Institute, 2020).

122. Wyborn et al., *Research and Action Agenda for Sustaining Diverse and Just Futures for Life on Earth*, 11–12.

123. Wyborn et al., *Research and Action Agenda for Sustaining Diverse and Just Futures for Life on Earth*, 12.

124. Wyborn et al., *Research and Action Agenda for Sustaining Diverse and Just Futures for Life on Earth*, 13.

125. Wyborn et al., *Research and Action Agenda for Sustaining Diverse and Just Futures for Life on Earth*, 3.

126. Goffman, *Frame Analysis*, 21.

127. Robert D. Benford and David A. Snow, "Framing Processes and Social Movements: An Overview and Assessment," *Annual Review of Sociology* 26 (2000): 614.

128. Thomas Beery, Anton Stahl Olafsson, Sandra Gentin, Megan Maurer, Sanna Stålhammar, Christian Albert, Claudia Bieling, Arjen Buijs, Nora Fagerholm, Maria Garcia-Martin, Tobias Plieninger, and Christopher M. Raymond, "Disconnection from Nature: Expanding Our Understanding of Human–Nature Relations," *People and Nature* (2023): 482, https://doi.org/10.1002/pan3.10451.

129. Dimitris Stevis and Romain Felli, "Planetary Just Transition? How Inclusive and How Just?," *Earth System Governance* 6 (2020): 100065.

130. Arendt, *The Human Condition*, 57.

131. Anthony Giddens, *Modernity and Self-Identity: Self and Society in the Late Modern Age* (Stanford, CA: Stanford University Press, 1991), 221–223.

Bibliography

Aaron, Benjamin. "The Labor-Management Reporting and Disclosure Act of 1959." *Harvard Law Review* 73, no. 5 (1960): 851–907.

Abram, David. *The Spell of the Sensuous: Perception and Language in a More Than Human World.* New York: Pantheon, 1996.

Adams, Richard Newbold. *The Eighth Day: Social Evolution as the Self-Organization of Society.* Austin: University of Texas Press, 1988.

Adams, Sophie. "The Pragmatic Holism of Social-Ecological Systems Theory: Explaining Adaptive Capacity in a Changing Climate." *Progress in Human Geography* 45, no. 6 (2021): 1580–1600.

Addams, Jane. *The Spirit of Youth and the City Streets.* New York: Macmillan, 1909.

Agamben, Giorgio. *Homo Sacer: Sovereign Power and Bare Life*, translated by Daniel Heller- Roazen. Stanford: Stanford University Press, 1998.

Anderegg, William R., Anna T. Trugman, Grayson Badgley, Christa M. Anderson, Ann Bartuska, Philippe Ciais, Danny Cullenward et al. "Climate-Driven Risks to the Climate Mitigation Potential of Forests." *Science* 368, no. 6497 (2020): eaaz7005.

Anderson, Craig A. and Karen E. Dill. "Video Games and Aggressive Thoughts, Feelings, and Behavior in the Laboratory and in Life." *Journal of Personality and Social Psychology* 78, no. 4 (2000): 772–790.

Anderson, Gary L. "Critical Ethnography in Education: Origins, Current Status, and New Directions." *Review of Educational Research* 59, no. 1 (1989): 249–270.

Anderson, Gary L., Dipti Desai, Ana Inés Heras, and Carol Anne Spreen. *Creating Third Spaces of Learning for Post-Capitalism: Lessons from Educators, Artists, and Activists.* New York: Routledge, 2023.

Anderson, Janna, Lee Rainie, and Alex Luchsinger. *Artificial Intelligence and the Future of Humans.* Washington, DC: Pew Research Center 2018.

Anderson, Monica and Jingjing Jiang. *Teens, Social Media & Technology 2018.* Washington, DC: Pew Research Center, 2018.

Anderson, Monica, Emily A. Vogels, Andrew Perrin, and Lee Rainie. *Connection, Creativity and Drama: Teen Life on Social Media in 2022.* Washington, DC: Pew Research Center, 2022.

Appadurai, Arjun. "Disjunction and Difference in the Global Cultural Economy." *Theory, Culture & Society* 7 (1990): 295–310.

Apple, Michael W. "On Doing Critical Policy Analysis." *Educational Policy* 33, no. 1 (2019): 276–287.

Arendt, Hannah. *The Human Condition.* Chicago: University of Chicago Press, 1958.

Arendt, Hannah. *Between Past and Future: Eight Exercises in Political Thought.* New York: Viking, 1968.

Arendt, Hannah. "The Crisis in Education." In *Between Past and Future: Eight Exercises in Political Thought*, 173–196. New York: Viking, 1968.

Arendt, Hannah. *The Origins of Totalitarianism.* New York: Harcourt Brace Jovanovich, 1973.

Aronowitz, Stanley. *False Promises: The Shaping of American Working Class Consciousness.* Durham, NC: Duke University Press, 1992.

Aronowitz, Stanley. "Against Schooling: Education and Social Class." *Social Text* 22, no. 2 (2004): 13–36.

Avila, Eric. *Popular Culture in the Age of White Flight: Fear and Fantasy in Suburban Los Angeles.* Berkeley: University of California Press, 2006.

Avila, Eric. "All Freeways Lead to East Los Angeles: Rethinking the LA Freeway and its Meaning." In *Overdrive: L.A. Constructs the Future, 1940–1990*, edited by Wim de Wit and Christopher James Alexander, 35–47. Los Angeles: Getty Research Institute, 2013.

Balmaceda, Margarita, Per Högselius, Corey Johnson, Heiko Pleines, Douglas Rogers, and Veli-Pekka Tynkkynen. "Energy Materiality: A Conceptual Review of Multi-Disciplinary Approaches." *Energy Research & Social Science* 56 (2019): 101220.

Bandura, Albert. *The Social Foundations of Thought and Action: A Social-Cognitive Theory.* Englewood Cliffs, NJ: Prentice-Hall, 1986.

Banning, E. B. and Michael Chazan. "Structuring Interactions, Structuring Ideas: Domestication of Space in the Prehistoric Near East." In *Domesticating Space: Construction, Community, and Cosmology in the Late Prehistoric Near East*, edited by E. B. Banning and Michael Chazan, Studies in Early Near Eastern Production, Subsistence, and Environment, 5–14. Berlin: *ex oriente*, 2006.

Banta, Martha. *Imaging American Women: Idea and Ideals in Cultural History.* New York: Columbia University Press, 1987.

Barkin, David. "Shaping a Communitarian Ethos in an Era of Ecological Crisis." *Frontiers in Sustainability* 3 (2022): 1–14.

Barkun, Michael. *Disaster and the Millennium.* New Haven: Yale University Press, 1974.

Barrett, William. *Time of Need: Forms of Imagination in the Twentieth Century.* New York: Harper & Row, 1972.

Baym, Nina and Laurence B. Holland. "Introduction." In *The Norton Anthology of American Literature, Volume D: American Literature Between the Wars, 1914–1945*, edited by Nina Baym, Sixth Edition, 1071–1084. New York: Norton, 2003.

Bazzul, Jene and Sara Tolbert. "Reassembling the Natural and Social Commons." In *Educational Commons in Theory and Practice: Global Pedagogy and Politics*, edited by Alexander J. Means, Derek R. Ford, and Graham B. Slater, 5–73. New York: Palgrave Macmillan, 2017.

Beaton, Albert E., Ina V. S. Mullis, Michael O. Martin, Eugenio J. Gonzalez, Dana L. Kelly, and Teresa A. Smith. *The Mathematics Achievement in the Middle School Years: IEA's Third International Mathematics and Science Study (TIMSS).*Chestnut Hill, MA: Center for the Study of Testing, Evaluation, and Educational Policy, Boston College, 1996.

Beck, Ulrich. *Risk Society: Towards a New Modernity.* Translated by Mark Ritter. London: Sage Publications, 1992.

Beery, Thomas, Anton Stahl Olafsson, Sandra Gentin, Megan Maurer, Sanna Stålhammar, Christian Albert, Claudia Bieling, Arjen Buijs, Nora Fagerholm, Maria Garcia-Martin, Tobias Plieninger, and Christopher M. Raymond. "Disconnection from Nature: Expanding Our Understanding of Human–Nature Relations." *People and Nature* 5, no. 2 (2023): 470–488.

Behnke, Elizabeth A. "Edmund Husserl's Contribution to Phenomenology of the Body in Ideas II." *Phenomenology: Critical Concepts in Philosophy*, Volume 2. In *Phenomenology: Themes and Issues*, edited by Dermot Moran and Lester E. Embree with Tanja Staehler and Elizabeth A. Behnke, 235–264. London: Routledge, 2004.

Behnke, Elizabeth A. "Edmund Husserl: Phenomenology of Embodiment." *The Internet Encyclopedia of Philosophy*, ISSN 2161-0002, 2023. https://iep.utm.edu/husspemb/.

Behrend, Tara S., Michael R. Ford, Kathleen M. Ross, Edmund M. Han, Erin Peters Burton, and Nancy K. Spillane. Gary and Jerri-Ann Jacobs High Tech High: A Case Study of an Inclusive STEM-Focused High School in San Diego, California. (OSPrI Report 2014-03), George Washington University, Opportunity Structures for Preparation and Inspiration in STEM (2014). https://ospri.research.gwu.edu/sites/ospri.research.gwu.edu/files/downloads/OSPrI_Report_2014.

Bellack, Arno A. *Competing Ideologies in Research on Teaching*. Uppsala Reports on Education 1. Uppsala: Uppsala University Department of Education, 1978.

Bendell, Jem. *Deep Adaptation: A Map for Navigating Climate Tragedy*, revised second edition. IFLAS Occasional Paper 2. Carlisle: University of Cumbria Institute for Leadership and Sustainability, 2020. https://lifeworth.com/deepadaptation.pdf.

Bender, Thomas H. *The Unfinished City: New York and the Metropolitan Idea.* New York: The New Press, 2007.

Bendix, Reinhard. *Nation-Building and Citizenship: Studies of Our Changing Social Order.* Berkeley: University of California Press, 1977.

Benford, Robert D. and David A. Snow. "Framing Processes and Social Movements: An Overview and Assessment." *Annual Review of Sociology* 26 (2000): 611–639.

Benjamin, Walter. "The Work of Art in the Age of Mechanical Reproduction (1935)." Translated by Harry Zohn, *Illuminations*, edited by Hannah Arendt, 217–251. New York: Schocken Books, 1969.

Bennett, John W. *The Ecological Transition: Cultural Anthropology and Human Adaptation.* New York: Pergamon, 1976.

Bensman, Joseph and Robert Lilienfeld. *Craft and Consciousness: Occupational Techniques and the Development of World Images.* Hawthorne, NY: Aldine de Gruyter, 1991.

Berger, Peter L. *The Sacred Canopy: Elements of a Sociological Theory of Religion.* New York: Doubleday Anchor Books, 1969.

Berger, Peter L., Brigitte Berger, and Hansfried Kellner. *The Homeless Mind: Modernization and Consciousness.* New York: Random House, 1973.

Berger, Ron. *An Ethic of Excellence: Building a Culture of Craftsmanship with Students.* Portsmouth, NH: Heinemann, 2003.

Bernstein, Richard J. *The Pragmatic Turn.* Cambridge: Polity Press, 2010.

Berthoff, Warner. *A Literature without Qualities: American Writing since 1945.* Berkeley: University of California Press, 1979.

Bertolt, Brecht. "To Posterity." *Selected Poems*, translated by H. R. Hays, 172–177. New York: Grove Press, 1947.

Bettelheim, Bruno. *The Informed Heart: Autonomy in a Mass Age.* Glencoe, IL: Free Press, 1960.

Betts, Julian R. "The San Diego Blueprint for Student Success: A Retrospective Overview and Commentary." *Journal of Education for Students Placed at Risk* 14, no. 1 (2009): 120–129.

Betts, Julian R., Andrew C. Zau, and Kevin King. *From Blueprint to Reality: San Diego's Educational Reforms.* San Francisco: Public Policy Institute of California, 2005.

Beveridge, Charles E. and Paul Rocheleau. *Frederick Law Olmsted: Designing the American Landscape.* New York: Universe, 1998.

Bird, Kai and Martin J. Sherwin. *American Prometheus: The Triumph and Tragedy of J. Robert Oppenheimer.* New York: Knopf, 2005.

Bittle, Jake. *The Great Displacement: Climate Change and the Next American Migration.* New York: Simon & Schuster, 2023.

Blaustein, Richard. "The Encyclopedia of Life: Describing Species, Unifying Biology." *BioScience* 59, no. 7 (2009): 551–556.

Bledstein, Burton J. *The Culture of Professionalism: The Middle Class and the Development of Higher Education in America.* New York: Norton, 1976.

Bluestone, Bluestone and Bennett Harrison. *The Deindustrialization of America: Plant Closings, Community Abandonment and the Dismantling of Basic Industry.* New York: Basic Books, 1982.

Borgmann, Albert. *Technology and the Character of Contemporary Life: A Philosophical Inquiry.* Chicago: University of Chicago Press, 1984.

Boris, Eileen. *Art and Labor: Ruskin, Morris, and the Craftsman Ideal in America.* Philadelphia: Temple University Press, 1986.

Borkovich, Debra J. and Philip D. Noah. "Big Data in the Information Age: Exploring the Intellectual Foundation of Communication Theory." *Information Systems Education Journal* 12, no. 1 (2014): 15–26.

Boston University Center for Teaching & Learning. *Project-Based Learning Teaching Guide.* Boston: Boston University Center for Teaching & Learning, 2023. https://www.bu.edu/ctl/guides/project-based-learning/.

Bowen, James. *A History of Western Education, Volume 3: The Modern West, Europe and the New World.* New York: St. Martin's Press, 1981.

Bowles, Samuel and Herbert Gintis. *Schooling in Capitalist America: Educational Reform and the Contradictions of Economic Life.* New York: Basic Books, 1976.

Bowman, Leslie Greene. "The Arts and Crafts Movement in the Southland." In *The Arts and Crafts Movement in California: Living the Good Life*, edited by Kenneth R. Trapp, 163–201. New York: Abbeville Press, 1992.

Boyer, Dominic. "Energopower: An Introduction." *Anthropological Quarterly* 87, no. 2 (2014): 309–333.

Boyer, Dominic. "Anthropology Electric." *Cultural Anthropology* 30, no. 4 (2015): 531–539.

Boyer, Ernest L. *High School: A Report on Secondary Education in America.* New York: Harper & Row, 1983.

Bransford, John C., Brigid Barron, Roy D. Pea, Andrew Meltzoff, Pat Kuhl, Phil Bell, Reed Stevens, Daniel L. Schwartz, Nancy Vye, Byron Reeves, Jeremy Roschelle, and Nora H. Sabelli. "Foundations and Opportunities for an Interdisciplinary Science of Learning." In *The Cambridge Handbook of the Learning Sciences*, edited by R. Keith Sawyer, 39–77. New York: Cambridge University Press, 2006.

Braverman, Harry. *Labor and Monopoly Capital: The Degradation of Work in the Twentieth Century.* New York: Monthly Review Press, 1974.

Brenner, M. Harvey. *Mental Illness and the Economy.* Cambridge, MA: Harvard University Press, 1973.

Brenner, M. Harvey. *Estimating the Social Costs of Economic Policy: Implications for Mental and Physical Health and Criminal Aggression.* Report to the Congressional Research Service of the Library of Congress and Joint Economic Committee of Congress. Washington, DC: US Government Printing Office, 1976.

Brenner, M. Harvey. "Economic Change, Alcohol Consumption and Heart Disease Mortality in Nine Industrialized Countries." *Social Science and Medicine* 25, no. 2 (1987): 119–132.

Briggs, Charles L. and Daniel C. Hallin. "Biocommunicability: The Neoliberal Subject and Its Contradictions in News Coverage of Health Issues." *Social Text* 25, no. 4 (2007): 43–66.

Brignone, Emily, Daniel R. George, Lawrence Sinoway, Curren Katz, Charity Sauder, Andrea Murray, Robert Gladden, and Jennifer L. Kraschnewski. "Trends in the Diagnosis of Diseases of Despair in the United States, 2009–2018: A Retrospective Cohort Study." *BMJ Open* 10, no. 10 (2020): e037679.

Bronfenbrenner, Urie. *The Ecology of Human Development: Experiments by Nature and Design.* Cambridge, MA: Harvard University Press, 1979.

Brooks, H. Allen. "Chicago Architecture: Its Debt to the Arts and Crafts." *Journal of the Society of Architectural Historians* 30, no. 4 (1971): 312–317.

Brown, Richard D. *Modernization: The Transformation of American Life, 1600–1865.* New York: Hill and Wang, 1976.

Bruner, Jerome S. "The Act of Discovery." *Harvard Educational Review* 31 (1961): 21–32.

Bruner, Jerome S. "Vygotsky: A Historical and Conceptual Perspective." In *Culture, Communication, and Cognition: Vygotskian Perspectives*, edited by James V. Wertsch, 21–34. Cambridge: Cambridge University Press, 1985.

Bruner, Jerome S. "The Narrative Construction of Reality." *Critical Inquiry* 18, no. 1 (1991): 1–21.

Bruner, Jerome S. *The Culture of Education.* Cambridge, MA: Harvard University Press, 1996.

Bryk, Anthony S. *Improvement in Action: Advancing Quality in America's Schools.* Harvard Education Press, 2020.

Burawoy, Michael. *The Extended Case Method: Four Countries, Four Decades, Four Great Transformations, and One Theoretical Tradition.* Berkeley: University of California Press, 2009.

Burke, Kenneth. *The Philosophy of Literary Form: Studies in Symbolic Action.* Berkeley: University of California Press, 1973.

Burke, Kenneth. *Permanence and Change: An Anatomy of Purpose.* Berkeley: University of California Press, 1984.

Caillier, Stacey L. and Robert C. Riordan. "Teacher Education for the Schools We Need." *Journal of Teacher Education* 60, no. 5 (2009): 489–496.

Calhoun, Craig. "'New Social Movements' of the Early Nineteenth Century." *Social Science History* 17, no. 3 (1993): 385–427.

Camus, Albert. *The Rebel: An Essay of Man in Revolt.* Translated by Anthony Bower. New York: Knopf, 1956.

Cantwell, Robert. *When We Were Good: The Fold Revival.* Cambridge, MA: Harvard University Press, 1996.

Cappelli, Peter, Iwan Barankay, and David Lewin. "How the Great Recession Changed American Workers." *Wharton Business Daily* (September 10, 1918). https://knowledge.wharton.upenn.edu/podcast/knowledge-at-wharton-podcast/great-recession- american-dream/.

Caro, Robert A. *The Power Broker: Robert Moses and the Fall of New York.* New York: Knopf, 1974.

Carson, Rachel. *Silent Spring.* New York: Houghton Mifflin, 1962.

Case, Anne and Angus Deaton. "Rising Morbidity and Mortality in Midlife among White Non- Hispanic Americans in the 21st Century." *Proceedings of the National Academy of Sciences* 112, no. 49 (2015): 15078–15083.

Case, Anne and Angus Deaton. "Mortality and Morbidity in the 21st Century." *Brookings Papers on Economic Activity* (2017): 397–476.

Case, Anne and Angus A. Deaton. *Deaths of Despair and the Future of Capitalism.* Princeton: Princeton University Press, 2020.

Castagno, Angelina E. and Teresa McCarty, editors. *The Anthropology of Educational Policy: Ethnographic Inquiries into Policy as Sociocultural Process.* New York: Taylor and Francis, 2017.

Castells, Manuel. *The Rise of the Network Society.* Malden, MA: Wiley-Blackwell, 2009.

Castells, Manuel. "The Network Society Revisited." *American Behavioral Scientist* 67, no. 7 (2022): 940–946.

Cavallo, Dominick. *Muscles and Morals: Organized Playgrounds and Urban Reform, 1880–1920*. Philadelphia: University of Pennsylvania Press, 1981.

Centers for Disease Control and Prevention. "Suicide Trends among Youths and Young Adults Aged 10–24 Years—United States, 1990–2004." *Morbidity and Mortality Weekly Report* (MMWR) 56, no. 35 (2007): 905–908.

Chakrabarty, Dipesh. *The Climate of History in a Planetary Age*. Chicago: University of Chicago Press, 2021.

Chalmers, David. *And the Crooked Places Made Straight: The Struggle for Social Change in the 1960s*. Baltimore: Johns Hopkins University Press, 1996.

Chiasson, Phyllis. "Peirce's Design for Thinking: An Embedded Philosophy of Education." *Educational Philosophy and Theory* 37, no. 2 (2005): 207–226.

Christensen, Linda. "Stealing Home: Eminent Domain, Urban Renewal and the Loss of Community." *Rethinking Schools* 27, no. 4 (2013): 34–41.

Christian, David. *Maps of Time: An Introduction to Big History*. Berkeley: University of California Press, 2004.

Cockerham, William C. "Medical Sociology and Sociological Theory." In *The Blackwell Companion to Medical Sociology*, edited by William C. Cockerham, 22–44. Malden, MA: Blackwell, 2005.

Cohen, Lizabeth. *A Consumer's Republic: The Politics of Mass Consumption in Postwar America*. New York: Random House, 2003.

Cohen, Lizabeth. "The Lessons of the Great Depression." *The Atlantic* (May 17, 2020). https://www.theatlantic.com/ideas/archive/2020/05/how-rebuild-nation/611704/.

Cohen, Robert. *Freedom's Orator: Mario Savio and the Radical Legacy of the 1960s*. New York: Oxford University Press, 2009.

Cohen, Ronald D. *Rainbow Quest: The Folk Music Revival & American Society, 1940–1970*. Amherst: University of Massachusetts Press, 2002.

Cohn, Norman. *Europe's Inner Demons: The Demonization of Christians in Medieval Christendom*. Chicago: University of Chicago Press, 1993.

Cohn, Norman. *The Pursuit of the Millennium: Revolutionary Millenarians and Mystical Anarchists of the Middle Ages*. New York: Oxford University Press, 1970.

Cole, Michael. *Cultural Psychology: A Once and Future Discipline*. Cambridge, MA: Harvard University Press, 1996.

Comer-Schultz, Judith. "History and Historical Preservation in San Diego since 1945: Civic Identity in America's Finest City." PhD Dissertation. Arizona State University, 2011.

Cooper, Jr., John Milton. *Pivotal Decades: The United States, 1900–1920*. New York: Norton, 1990.

Cosgrove, Denis. "Contested Global Visions: One-World, Whole-Earth, and the Apollo Space Photographs." *Annals of the Association of American Geographers* 84, no. 2 (1994): 270–294.

Cotkin, George. *Existential America*. Baltimore: Johns Hopkins University Press, 2003.

Cowley, Malcolm. *Exile's Return: A Literary Odyssey of the 1920s*. New York: Viking, 1951.

Crary, Jonathan. *Scorched Earth: Beyond the Digital Age to a Post-Capitalist World*. London: Verso, 2022.

Crehan, Kate. *Gramsci, Culture and Anthropology*. Berkeley: University of California Press, 2002,

Cremin, Lawrence A. *The Transformation of the School: Progressivism in American Education, 1876–1957*. New York: Random House, 1961.

Cronon, William. "Foreword: Small Book, Big Story." In *Fire: A Brief History*, edited by Stephen J. Pyne, xi–xiv. Seattle: University of Washington Press, 2001.

Crowson, Robert L. and Ellen B. Goldring. "The New Localism: Re-examining Issues of Neighborhood and Community in Public Education." *Teachers College Record* 111, no. 13 (2009): 1–24.

Crutzen, Paul J. and Eugene F. Stoermer. "The Anthropocene." *Global Change Newsletter* 41 The Royal Swedish Academy of Sciences, Stockholm, Sweden (2000): 14–17.

Csikszentmihalyi, Mihaly. *Creativity: Flow and the Psychology of Discovery and Innovation*. New York: HarperCollins, 1996.

Csikszentmihalyi, Mihaly and Eugene Rochberg-Halton. *The Meaning of Things: Domestic Symbols of the Self*. Cambridge: Cambridge University Press, 1981.

Cuban, Larry. *How Teachers Taught: Constancy and Change in American Classrooms, 1890–1980*. New York: Longman, 1984.

Cullen, Dave. *Columbine*. New York: Twelve, 2009.

Dahrendorf, Ralf. *Life Chances*. Chicago: University of Chicago Press, 1979.

Dale, Roger and Susan Robertson. "Capitalism, Modernity and the Future of Education in the New Social Contract." *Teachers College Record* 111, no. 14 (2009): 111–129.

Davis, Joseph E. "Adolescents and the Pathologies of the Achieving Self." *Hedgehog Review* 11, no. 1 (2009): 36–49.

Davis, Mike. *Ecology of Fear: Los Angeles and the Imagination of Disaster*. New York: Metropolitan Books/Henry Holt and Company, 1998.

Davis, Mike. "Golden Ruins/Dark Ruptures: The Literary Destruction of Los Angeles." In *Critical Views: Essays on the Humanities and the Arts*, edited by Teresa Stojkov, 77–98. Berkeley: University of California Press, 2011.

Dawes, James. *The Language of War: Literature and Culture in the U.S. from the Civil War Through World War II*. Cambridge, MA: Harvard University Press, 2002.

Dawson, Ashley and A. Naomi Paik. "Germinations: An Introduction." *Radical History Review* 145 (2023): 1–11.

De Beauvoir, Simone. *The Second Sex*. Translated by Howard Madison Parshley. New York: Knopf, 1953.

de Sousa Santos, Boaventura. "Beyond Abyssal Thinking: From Global Lines to Ecologies of Knowledges." *Review* (*Fernand Braudel Center*) 30, no. 1 (2007): 45–89.

de Sousa Santos, Boaventura. "Encountering Other Cultural Universes on the Brink of Chaos." *Journal of World-Systems Research* 28, no. 2 (2022): 438–441.

Deane, Brianna. "UCLA School of Dentistry Creates Future Opportunities for Disadvantaged, Gifted High School Students." *UCLA Newsroom* (July 30, 2012).

Des Pres, Terrence. *The Survivor: An Anatomy of Life in the Death Camps*. New York: Oxford University Press, 1976.

DeVoe, Jill F., Katharin Peter, Phillip Kaufman, Sally A. Ruddy, Amanda K. Miller, Mike Planty, Thomas D. Snyder, Detis T. Duhart, and Michael R. Rand. *Indicators of School Crime and Safety: 2002*. NCES 2003–009/NCJ 196753. Washington, DC: US Departments of Education and Justice, 2002.

Delkina, Laurence and Anthony Janetos. "Cosmopolitan, Dynamic, and Contested Energy Futures: Navigating the Pluralities and Polarities in the Energy Systems of Tomorrow." *Energy Research & Social Science* 35 (2018): 1–10.

Demos, John and Virginia Demos. "Adolescence in Historical Perspective." *Journal of Marriage and Family* 31, no. 4 (1969): 632–638.

Deresiewicz, William. *Excellent Sheep: The Miseducation of the American Elite and the Way to a Meaningful Life*. New York: Free Press, 2014.

Dewey, John. *The School and Society*. Chicago: University of Chicago Press, 1900.

Dewey, John. *Democracy and Education*. New York: Macmillan, 1916.

Dewey, John. *Experience and Nature. (1925) John Dewey: The Later Works, 1925–1953, Volume 1: 1925*. Edited by Jo Ann Boydston. Carbondale: Southern Illinois University Press, 1981.

Dewey, John. *The Sources of a Science of Education*. New York: Horace Liveright, 1929.

Dewey, John. *How We Think: A Restatement of the Relation of Reflective Thinking to the Educative Process*. Boston: Heath and Company, 1933.

Dewey, John. "The Theory of the Chicago Experiment." In *John Dewey: The Later Works, Vol 11: 1935–1937*, edited by Jo Ann Boydston. Carbondale: Southern Illinois University Press, 1981.

Dewey, John. "Foreword." In *Primitive Man as Philosopher* by Paul Radin, xvii–xx. New York: Dover Publications, 1957.

Dewey, John. *Art as Experience*. New York: Putnam, 1958.

Diamond, Stanley. *In Search of the Primitive*. New Brunswick, NJ: Transaction Publishers, 1974.

Diamond, Stanley. "Paul Radin." In *Totems and Teachers: Key Figures in the History of Anthropology*, edited by Sydel Silverman, 51–64. Walnut Creek, CA: AltaMira Press, 2004.

Dickstein, Morris. *Gates of Eden: American Culture in the Sixties*. New York: Basic Books, 1977.

Donald, Merlin. *Origins of the Modern Mind*. Cambridge, MA: Harvard University Press, 1991.

Donald, Merlin. "Human Cognitive Evolution: What We Were, What We Are Becoming." *Social Research* 60, no. 1 (1993): 143–170.

Donnelly, Brian. *Digital Portfolios and Learning: The Students' Voices*. Doctoral Dissertation. University of California, Davis, 2010.

Douglas, Ann. *Terrible Honesty: Mongrel Manhattan in the 1920s*. New York: Farrar, Straus and Giroux, 1995.

Du Bois, W. E. B. *The Souls of Black Folk*. New York: Library of America, 1986.

Duckor, Brent and Daniel Perlstein. "Assessing Habits of Mind: Teaching to the Test at Central Park East Secondary School." *Teachers College Record* 116, no. 2 (2014): 1–33.

Dunlap, Peter T. "The Unifying Function of Affect: Founding a Theory of Psychocultural Development in the Epistemology of John Dewey and Carl Jung." *Educational Philosophy and Theory* 44, no. 1 (2012): 53–68.

Dylan, Bob. *Chronicles*. New York: Simon & Schuster, 2004.

Dylan, Bob. *The Lyrics 1961–2012*. New York: Simon & Schuster, 2016.

Eagleton, Terry. "Culture & Barbarism: Metaphysics in a Time of Terrorism." *Commonweal* 136, no. 6 (2009): 9–13.

Edwards, Rebecca. *New Spirits: Americans in the Gilded Age, 1865–1905*. New York: Oxford University Press, 2006.

Elementary & Secondary Education, Title I — Improving the Academic Achievement of the Disadvantaged, Sec. 101. *Improving the Academic Achievement of the Disadvantaged* (2002). Title I of the Act is amended to read as follows: Title I — Improving the Academic Achievement of the Disadvantaged Elementary and Secondary Education Act of 1965 (20 U.S.C. 6301 et seq.). Washington, DC: Public Law, 107–110—JAN. 8, 2002, 115 STAT. 1425.

Engelhardt, Tom. *The End of Victory Culture: Cold War America and the Disillusioning of a Generation*. New York: Basic Books, 1995.

Engeström, Yrjö. "Expansive Learning at Work: Toward an Activity Theoretical Reconceptualization." *Journal of Education and Work* 14, no. 1 (2001): 133–156.

Erikson, Erik H. *Childhood and Society*. New York: Norton, 1963.

Erikson, Erik H. *Identity: Youth and Crisis*. New York: Norton, 1968.

Erikson, Erik H. "Reflections on the Dissent of Contemporary Youth." *Daedalus* 99, no. 1 (1970): 154–176.

Erikson, Erik H. *Life History and the Historical Moment*. New York: Norton, 1975.

Evens, T. M. S. and Don Handelman. "Introduction: The Ethnographic Praxis of the Theory of Practice." In *The Manchester School: Practice and Ethnographic Praxis in Anthropology*, edited by T. M. S. Evens and Don Handelman, 1–11. New York: Berghahn Books, 2006.

Executive Order 9066, February 19, 1942; *General Records of the United States Government*, Record Group 11, National Archives.

Eyerman, Ron and Andrew Jamison. *Social Movements: A Cognitive Approach*. Cambridge: Polity, 1991.

Eyerman, Ron and Andrew Jamison. *Music and Social Movements: Mobilizing Traditions in the Twentieth Century*. Cambridge: Cambridge University Press, 1998.

Fantasia, Rick. *Cultures of Solidarity: Consciousness, Action, and Contemporary American Workers*. Berkeley: University of California Press, 1989.

Fass, Paula S. *The Damned and the Beautiful: American Youth in the 1920s*. New York: Oxford University Press, 1977.

Fassin, Didier and Richard Rechtman. *The Empire of Trauma: An Inquiry into the Condition of Victimhood*. Translated by Rachel Gomme. Princeton: Princeton University Press, 2009.

Faust, Drew Gilpin. *This Republic of Suffering: Death and the American Civil War.* New York: Knopf, 2008.

Faust, Drew Gilpin. "This Republic of Suffering: Death and the American Civil War." Address. Washington, DC: The National Archives, January 8, 2008. https://www.libraryofsocialscience.com/assets/pdf/Faust-This_Republic_of_Suffering.pdf.

Fear, Jeffrey and Paul Lerner. "Behind the Screens: Immigrants, Émigrés and Exiles in Mid- Twentieth-Century Los Angeles." *Jewish Culture and History* 17, no. 1–2 (2016): 1–21.

Featherstone, Mark. "Stiegler's Ecological Thought: The Politics of Knowledge in the Anthropocene." *Educational Philosophy and Theory* 52, no. 4 (2020): 409–419.

Fehrenbacher, Tom. "Logs from San Diego Bay." *UnBoxed* 13 (2015). http://www.hightechhigh.org/unboxed/issue13/logs_from_san_diego_bay/.

Fisher, Mark. *Capitalist Realism: Is There No Alternative?* London: John Hunt Publishing, 2009.

Fitzgerald, F. Scott. *The Great Gatsby.* New York: Charles Scribner's Sons, 1925.

Flavell, John H. "Cognitive Development: Children's Knowledge About the Mind." *Annual Review of Psychology* 50 (1999): 21–45.

Fligstein, Neil. "Social Skill and the Theory of Fields." *Sociological Theory* 19, no. 2 (2001): 106–124.

Flores, Adolfo. "Teens Get to Work with Lab Scientists at UCLA; Dental School Opens Lab to High Schoolers with Science Aptitude." *The Daily News of Los Angeles* (August 1, 2008); A4.

Flyn, Cal. *Islands of Abandonment: Nature Rebounding in the Post-Human Landscape.* New York: Viking, 2022.

Foner, Eric. *The Story of American Freedom.* New York: Norton, 1999.

Foote, Christopher L. and Richard W. Ryan. "Labor-Market Polarization over the Business Cycle." *NBER Macroeconomics Annual* 29, no. 1 (2015): 371–413.

Ford Foundation. "President's Message." *2000 Annual Report.* New York: Ford Foundation, 2001.

Foss, Daniel A. and Ralph Larkin. *Beyond Revolution: A Theory of Social Movements.* South Hadley, MA: Bergin & Garvey, 1986.

Foucault, Michel. *The History of Sexuality, Vol. 1: An Introduction.* New York: Vintage Books, 1990.

Foucault, Michel. "Governmentality." Translated by Rosi Braidotti and revised by Colin Gordon. In *The Foucault Effect: Studies in Governmentality*, edited by Graham Burchell, Colin Gordon, and Peter Miller, 87–104. Chicago: University of Chicago Press, 1991.

Foulkes, S. H. *Therapeutic Group Analysis.* London: Allen and Unwin, 1964.

Frank, Thomas. *The Conquest of Cool: Business Culture, Counterculture, and the Rise of Hip Consumerism.* Chicago: University of Chicago Press, 1997.

Frankl, Viktor E. *Man's Search for Meaning.* Translated by Ilse Lasch. New York: Washington Square Press, 1959.

Fredrikzon, Johan. "Review: *Cybernetics: The Macy Conferences.*" *Sensorium Journal* 3 (2021): 56–59.

Freire, Paulo. *The Politics of Education: Culture, Power, and Liberation.* South Hadley, MA: Bergin and Garvey, 1985.

Freire, Paulo. *Pedagogy of the Oppressed.* Translated by Myra Bergman Ramos. New York: Continuum, 2000.

Friedan, Betty. *The Feminine Mystique.* New York: Norton, 1963.

Fuchs, Thomas and Hanne De Jaegher. "Enactive Intersubjectivity: Participatory Sense-Making and Mutual Incorporation." *Phenomenology and the Cognitive Sciences* 8 (2009): 465–486.

Fuhrmann, Delia, Lisa J. Knoll, and Sarah-Jayne Blakemore. "Adolescence as a Sensitive Period of Brain Development." *Trends in Cognitive Sciences* 19, no. 10 (2015): 558–566.

Gamble, Clive. "The Anthropology of Deep History." *Journal of the Royal Anthropological Institute* 21, no. 1 (2015): 147–164.

George, Ann. *Kenneth Burke's Permanence and Change: A Critical Companion.* Columbia: University of South Carolina Press, 2018.

Giddens, Anthony. *The Consequences of Modernity.* Stanford, CA: Stanford University Press, 1990.

Giddens, Anthony. *Modernity and Self-Identity: Self and Society in the Late Modern Age.* Stanford, CA: Stanford University Press, 1991.

Gideon, Siegfried. *Mechanization Takes Command: A Contribution to Anonymous History.* Minneapolis: University of Minnesota Press, 2014.

Gilbert, Felix and David Clay Large. *The End of the European Era, 1890 to the Present.* New York: Norton, 2002.

Gimpel, James G., Frances E. Lee, and Rebecca U. Thorpe. "Geographic Distribution of the Federal Stimulus of 2009." *Political Science Quarterly* 127, no. 4 (2012): 567–595.

Giroux, Henry A. "Doing Cultural Studies: Youth and the Challenge of Pedagogy." *Harvard Educational Review* 64, no. 3 (1994): 278–309.

Giroux, Henry A. "Rethinking Cultural Politics and Radical Pedagogy in the Work of Antonio Gramsci." *Educational Theory* 49, no. 1 (1999): 1–19.

Giroux, Henry A. *The Abandoned Generation.* New York: Palgrave, 2003.

Giroux, Henry A. "Cultural Studies, Public Pedagogy, and the Responsibility of Intellectuals." *Communication and Critical/Cultural Studies* 1, no. 1 (2004): 59–79.

Giroux, Henry A. "Neoliberalism as Public Pedagogy." In *Handbook of Public Pedagogy: Education and Learning Beyond Schooling*, edited by Jennifer A. Sandlin, Brian D. Schultz, and Jake Burdick, 514–527. New York: Routledge, 2010.

Giroux, Henry A. "Zombie Politics and Other Late Modern Monstrosities in the Age of Disposability." *Policy Futures in Education* 8, no. 1 (2010): 1–7.

Giroux, Henry A. "Domestic Terrorism, Youth, and the Politics of Disposability." *Knowledge Cultures* 3, no. 5 (2015): 116–131.

Giroux, Henry A. *America's Addiction to Terrorism.* New York: Monthly Review Press, 2016.

Giroux, Henry A. *On Critical Pedagogy.* Bloomsbury Academic, 2020.

Giroux, Henry A. *Pedagogy of Resistance: Against Manufactured Ignorance.* London: Bloomsbury Academic, 2022.

Goffman, Erving. *Asylums: Essays on the Social Situation of Mental Patients and Other Inmates.* Garden City, NY: Doubleday, 1961.

Goffman, Erving. *Frame Analysis: An Essay on the Organization of Experience.* New York: Harper and Row, 1974.

Goldin, Claudia. "Egalitarianism and the Returns to Education during the Great Transformation of American Education." *Journal of Political Economy* 107, no. 6 (1999): S65–S94.

Goldin, Claudia. "The Human-Capital Century and American Leadership: Virtues of the Past." *The Journal of Economic History* 61, no. 2 (2001): 263–292.

Goldin, Claudia and Lawrence F. Katz. "The Origins of Technology-Skill Complementarity." *Quarterly Journal of Economics* 113, no. 3 (1998): 693–732.

Goldin, Claudia and Lawrence F. Katz. "Human Capital and Social Capital: The Rise of Secondary Schooling in America, 1910–1940." *Journal of Interdisciplinary History* 29, no. 4 (1999): 683–723.

Goldin, Claudia and Lawrence F. Katz. "Education and Income in the Early Twentieth Century: Evidence from the Prairies." *Journal of Economic History* 60, no. 3 (2000): 782–818.

Goldin, Claudia and Lawrence F. Katz. *The Race between Education and Technology: The Evolution of US Educational Wage Differentials, 1890–2005.* Working Paper 12984. Cambridge, MA: National Bureau of Economic Research, 2007.

Goldin, Claudia and Lawrence F. Katz. *The Race Between Education and Technology.* Cambridge, MA: Harvard University Press, 2008.

Goodlad, John I. *A Place Called School: Prospects for the Future.* New York: McGraw-Hill, 1984.

Goodman, Paul. *Growing Up Absurd: Problems of Youth in the Organized Society.* New York: Random House, 1960.

Gordon, Robert J. *Is U.S. Economic Growth Over? Faltering Innovation Confronts the Six Headwinds.* Working Paper 18135. Cambridge, MA: National Bureau of Economic Research, 2012.

Gordon, Robert J. "Perspectives on *The Rise and Fall of American Growth.*" *The American Economic Review* 106, no. 5 (2016): 72–76.

Goulet, Denis. "An Ethical Model for the Study of Values." *Harvard Educational Review* 41, no. 2 (1971): 205–227.

Graeber, David. "Enacting the Impossible: On Consensus Decision-Making." *Occupied Wall Street Journal* 3 (2011): 4.

Gramsci, Antonio. *Letters from Prison, Volume 2.* Edited by Frank Rosengarten. Translated by Ray Rosenthal. New York: Columbia University Press, 1994.

Gratton, Brian and Jon Moen. "Immigration, Culture, and Child Labor in the United States, 1880–1920." *Journal of Interdisciplinary History* 34, no. 3 (2004): 355–391.

Greenough, William T., James E. Black, and Christopher S. Wallace. "Experience and Brain Development." *Child Development* 58, no. 3 (1987): 539–559.

Greif, Mark. *The Age of the Crisis of Man: Thought and Fiction in America, 1933–1973.* Princeton: Princeton University Press, 2015.

Griswold del Castillo, Richard. "The Los Angeles 'Zoot Suit Riots' Revisited: Mexican and Latin American Perspectives." *Mexican Studies/Estudios Mexicanos* 16, no. 2 (2000): 367–391.

Gubrium, Aline and Krista Harper. *Participatory Visual and Digital Methods.* Walnut Creek, CA: Left Coast Press, 2013.

Gubrium, Aline, Krista Harper, and Marty Otanez, editors. *Participatory Visual and Digital Research in Action.* Walnut Creek, CA: Left Coast Press, 2015.

Gunderson, Lance H. "Ecological Resilience—in Theory and Application." *Annual Review of Ecology and Systematics* 31, no. 1 (2000): 425–439.

Gunderson, Lance H. and C. S. Holling, editors. *Panarchy: Understanding Transformations in Human and Natural Systems.* Washington, DC: Island Press, 2002.

Gunel, Zeynep and Lemi Baruh. "Social Networking Technologies and Social Movements." In *Social Networking: Redefining Communication in the Digital Age*, edited by Anastacia Kurylo and Tatyana Dumova, 131–144. Lanham, MD: Rowman and Littlefield, 2016.

Gutiérrez, Kris D. "Developing a Sociocritical Literacy in the Third Space." *Reading Research Quarterly* 43, no. 2 (2008): 148–164.

Gutiérrez, Kris D., Patricia Baquedano-López, and Carlos Tejeda. "Rethinking Diversity: Hybridity and Hybrid Language Practices in the Third Space." *Mind, Culture, and Activity* 6, no. 4 (1999): 286–303.

Haber, Samuel. *Efficiency and Uplift: Scientific Management in the Progressive Era, 1890–1920.* Chicago: University of Chicago Press, 1964.

Habermas, Jürgen. *The Philosophical Discourse on Modernity: Twelve Lectures.* Translated by Frederick G. Lawrence. Cambridge, MA: MIT Press, 1990.

Habermas, Jürgen. *The Structural Transformation of the Public Sphere: An Inquiry into a Category of Bourgeois Society.* Translated by Thomas Burger. Cambridge, MA: MIT Press, 1991.

Hales, Peter Bacon. *Outside the Gates of Eden: The Dream of America from Hiroshima to Now.* Chicago: University of Chicago Press, 2014.

Halpern, Robert. *The Means to Grow Up: Reinventing Apprenticeship as a Developmental Support in Adolescence.* New York: Routledge, 2009.

Halpern, Robert, Paul E. Heckman, and Reed W. Larson. *Realizing the Potential of Learning in Middle Adolescence.* West Hills, CA: Sally and Dick Roberts Coyote Foundation, 2013.

Hardt, Michael and Antonio Negri. *Empire.* Cambridge, MA: Harvard University Press, 2000.

Hardt, Michael and Antonio Negri. *Commonwealth.* Cambridge, MA: Harvard University Press, 2009.

Harrington, Michael. *The Other America: Poverty in the United States.* New York: Macmillan, 1962.

Harris, David R. "Introduction." In *The Archaeology of V. Gordon Childe: Contemporary Perspectives*, edited by David R. Harris, 1–7. Chicago: University of Chicago Press, 1994.

Hartman, Geoffrey H. *The Fateful Question of Culture.* New York: Columbia University Press, 1997.

Hawley, Ellis W. "Herbert Hoover, The Commerce Secretariat, and the Vision of an 'Associative State', 1921–1928." *The Journal of American History* 61, no. 1 (1974): 116–140.

Hawley, Ellis W. *The Great War and the Search for a Modern Order: A History of the American People and Their Institutions, 1917–1933.* New York: St. Martin's Press, 1979.

Hays, Samuel P. *Conservation and the Gospel of Efficiency: The Progressive Conservation Movement, 1890–1920.* Cambridge, MA: Harvard University Press, 1959.

Heckman, Paul E. and Viki L. Montera. "School Reform: The Flatworm in a Flat World: From Entropy to Renewal through Indigenous Invention." *Teachers College Record* 111, no. 5 (2009): 1328–1351.

Hegeman, Susan. *Patterns for Americans: Modernism and the Concept of Culture.* Princeton: Princeton University Press, 1999.

Heilbut, Anthony. *Exiled in Paradise: German Refugee Artists and Intellectuals in America from the 1930s to the Present.* Boston: Beacon Press, 1984.

Hemingway, Ernest. *A Farewell to Arms.* New York: Scribner, 1929.

Hemingway, Ernest. *A Moveable Feast.* New York: Scribners, 1964.

Henry, Holly and Amanda Taylor. "Re-Thinking Apollo: Envisioning Environmentalism in Space." *The Sociological Review* 57, no. 1, Supplement (2009): 190–203.

Hernández, Maciel M. and Mayra Y. Bámaca-Colbert. "A Behavioral Process Model of Familism." *Journal of Family Theory & Review* 8, no. 4 (2016): 463–483.

Hickman, Larry A. *John Dewey's Pragmatic Technology.* Bloomington: Indiana University Press, 1990.

Hiebert, James, Thomas P. Carpenter, Elizabeth Fennema, Karen Fuson, Piet Human, Hanlie Murray, Alwyn Olivier, and Diana Wearne. "Problem Solving as a Basis for Reform in Curriculum and Instruction: The Case of Mathematics." *Educational Researcher* 25, no. 4 (1996): 12–21.

Higham, John. *Strangers in the Land: Patterns of American Nativism, 1860–1925.* New Brunswick, NJ: Rutgers University Press.

Hightower, Ann M. "San Diego's Big Boom: Systemic Instructional Change in the Central Office and Schools." In *School Districts and Instructional Renewal*, edited by Amy M. Hightower, Michael S. Knapp, Julie A. Marsh, and Milbrey W. McLaughlin, 76–93. New York: Teachers College Press, 2002.

Hintikka, Jaakko. "Wittgenstein's Times (And Ours)." In *Time and History*, edited by Friedrich Stadler and Michael Stöltzner. Publications of the Austrian Ludwig Wittgenstein Society-New Series. Frankfurt: Ontos Verlag, 2006.

Hirst, William. "Cultural Trauma: Guest Editor's Introduction." *Social Research* 87, no. 3 (2020): xxxiii–xlix.

Hoetker, James and William P. Ahlbrand. "The Persistence of Recitation." *American Educational Research Journal* 6, no. 2 (1969): 145–167.

Holden, Erling, Kristin Linnerud, and Bente Johnsen Rygg. "A Review of Dominant Sustainable Energy Narratives." *Renewable and Sustainable Energy Reviews* 144 (2021): 110955.

Holling, C. S. "Foreword: The Backloop to Sustainability." In *Navigating Social-Ecological Systems*, edited by Fikret Berkes, Johan Colding, and Carl Folke. Cambridge: Cambridge University Press, 2003.

Holling, C. S. and Lance H. Gunderson. "Resilience and Adaptive Cycles." In *Panarchy: Understanding Transformations in Human and Natural Systems*, 25–62. Washington, DC: Island Press, 2002.

Hopper, Earl and Haim Weinberg, editors. *The Social Unconscious in Persons, Groups, and Societies. Vol. 3: The Foundation Matrix Extended and Re-Configured.* London: Karnac, 2017.

Howe, Paul. *Teen Spirit: How Adolescence Transformed the Adult World.* Ithaca, NY: Cornell University Press, 2020.

Hughes, Richard A. "Boston University School of Theology and the Civil Rights Movement." *Methodist History* 47, no. 3 (2009): 146–162.

Hynes, Samuel. *A War Imagined: The First World War and English Culture.* New York: Atheneum, 1990.

Ihde, Don. *Technology and the Lifeworld: From Garden to Earth.* Bloomington: Indiana University Press, 1990.

IPCC. 2022: Summary for Policymakers. In *Climate Change 2022: Mitigation of Climate Change. Contribution of Working Group III to the Sixth Assessment Report of the Intergovernmental Panel on Climate Change*, edited by Priyadarshi R. Shukla, Jim Skea, Raphael Slade, Alaa Al Khourdajie, Renée van Diemen, David McCollum, Minal Pathak, Sheya Some, Purvi Vyas, Roger Fradera, Malek Belkacemi, Apoorva Hasija, Géninha Lisboa, Sigourney Luz, and Juliette Malley. Cambridge: Cambridge University Press, 2022. https://doi.org/10.1017/9781009157926.001.

Isserman, Maurice. "50 Years Later: Poverty and The Other America." *Dissent* 59, no. 1 (2012): 83–89.

Iyengar, Shanto, Yphtach Lelkes, Matthew Levendusky, Neil Malhotra, and Sean J. Westwood. "The Origins and Consequences of Affective Polarization in the United States." *Annual Review of Political Science* 22 (2019): 129–146.

Jackman, Jarrell C. "German Émigrés in Southern California." In *The Muses Flee Hitler: Cultural Transfer and Adaptation*, edited by Jerrell C. Jackman and Carla M. Borden, 95–110. Washington, DC: Smithsonian Institution Press, 1983.

Jackson, Kenneth T. *Crabgrass Frontier: The Suburbanization of the United States.* New York: Oxford University Press, 1985.

Jackson, Shannon. *Lines of Activity: Performance, Historiography, Hull-House Domesticity.* Ann Arbor: University of Michigan Press, 2001.

James, William. "The Perception of Time." *The Journal of Speculative Philosophy* 20, no. 4 (1886): 374–407.

James, William. *The Principles of Psychology*, Volume 2. New York: Henry Holt & Company, 1890.

James, William. "Does Consciousness Exist?" (1904) *Essays in Radical Empiricism*, edited by Ralph Barton Perry, 3–22. New York: Dutton, 1971.

James, William. "The Thing and its Relations." (1905) *Essays in Radical Empiricism*, edited by Ralph Barton Perry, 49–64. New York: Dutton, 1971.

James, William. *Essays in Radical Empiricism*. (1912). Edited by Ralph Barton Perry. New York: Dutton, 1971.

Jamison, Andrew. "Climate Change Knowledge and Social Movement Theory." *Wiley Interdisciplinary Reviews: Climate Change* 1, no. 6 (2010): 811–823.

Jasanoff, Sheila. "Just Transitions: A Humble Approach to Global Energy Futures." *Energy Research & Social Science* 35 (2018): 11–14.

Jasanoff, Sheila and Sang-Hyun Kim. "Containing the Atom: Sociotechnical Imaginaries and Nuclear Power in the United States and South Korea." *Minerva* 47 (2009): 119–146.

Jaspers, Karl. *The Origin and Goal of History*. New Haven: Yale University Press, 1953.

Jaspers, Karl. *Basic Philosophical Writings: Selections*. Edited with translations by Leonard H. Ehrlich, Edith Ehrlich, and George B. Pepper. Athens, OH: Ohio University Press, 1986.

Jay, Martin. *Songs of Experience: Modern American and European Variations on a Universal Theme*. Berkeley: University of California Press, 2005.

Jessop, Bob. "Fordism and Post-Fordism: A Critical Reformulation." In *Pathways to Industrialization and Regional Development*, edited by Allen J. Scott and Michael J. Storper, 46–69. London: Routledge, 1992.

Johnson, Sandra. "The Neuroscience of the Mentor-Learner Relationship." *New Directions for Adult and Continuing Education* 110 (2006): 63–69.

Jonas, Hans. *The Imperative of Responsibility: In Search of an Ethics for the Technological Age*. Chicago: University of Chicago Press, 1984.

Juhn, Chinhui, Kevin M. Murphy, and Brooks Pierce. "Wage Inequality and the Rise in Returns to Skill." *The Journal of Political Economy* 101, no. 3 (1993): 410–432.

Kachur, S. Patrick, Gail M. Stennies, Kenneth E. Powell, William Modzeleski, Ronald Stephens, Rosemary Murphy, Marcie-jo Kresnow, David Sleet, and Richard Lowry. "School-Associated Violent Deaths in the United States, 1992 to 1994." *Journal of the American Medical Association* 275, no. 22 (1996): 1729–1733.

Kates, Robert W., Billie Lee Turner II, and William C. Clark. "The Great Transformation." In *The Earth As Transformed by Human Action*, edited by Billie Lee Turner II, William C. Clark, Robert W. Kates, John F. Richards, Jessica T. Mathews, and William B. Meyer, 1–17. Cambridge: Cambridge University Press, 1990.

Kendall, Philip C., editor. *Child and Adolescent Therapy: Cognitive-Behavioral Procedures*. New York: Guilford Press, 1991.

Kennedy, David M. *Over Here: The First World War and American Society*. New York: Oxford University Press, 1980.

Kerchner, Charles Taylor. *The Emperor's Clothes: Traditional and Avant Garde at High Tech High*. Claremont, CA: Claremont Graduate University, 2012.

Kett, Joseph F. "Reflections on the History of Adolescence in America." *The History of the Family* 8, no. 3 (2003): 355–373.

Kevles, Daniel J. *The Physicists: The History of a Scientific Community in Modern America*. Cambridge, MA: Harvard University Press, 1987.

Kilpatrick, William Heard. *Foundations of Method: Informal Talks on Teaching*. New York: Macmillan, 1925.

Kloppenberg, James T. *Uncertain Victory: Social Democracy and Progressivism in European and American Thought, 1870–1920.* New York: Oxford University Press, 1986.

Kluver, Jean and Larry Rosenstock. "Choice and Diversity: Irreconcilable Differences." *Principal Leadership* 3, no. 8 (2002): 12–18.

Kluver, Jean and Jeff Robin. *Changing the Subject. Twenty Years of Projects from High Tech High.* San Diego: High Tech High Graduate School of Education, 2021.

Koestler, Arthur. *The Ghost in the Machine.* New York: Macmillan, 1968.

Kohn, Alfie. *The Case Against Standardized Testing: Raising the Scores, Ruining the Schools.* Portsmouth, NH: Heinemann, 2000.

Kozaitis, Kathryn A. "The Rise of Anthropological Praxis." In *The Unity of Theory and Practice in Anthropology: Rebuilding a Fractured Synthesis*, edited by Carol E. Hill and Marietta L. Baba, 45–66. Washington: American Anthropological Association, 1997.

Kuk, Hye-Su and Rebecca Tarlau. "The Confluence of Popular Education and Social Movement Studies into Social Movement Learning: A Systematic Literature Review." *International Journal of Lifelong Education* 39, no. 5–6 (2020): 591–604.

Kurzweil, Ray. *The Age of Spiritual Machines.* New York, NY: Viking, 1999.

Landy, Marcia. "Culture and Politics in the Work of Antonio Gramsci." *boundary 2* 14, no. 3 (1986): 49–70.

Langer, Suzanne K. *Philosophy in a New Key.* Cambridge, MA: Harvard University Press, 1951.

Lasch, Christopher. *The Minimal Self: Psychic Survival in Troubled Times.* New York: Norton, 1984.

Latour, Bruno. *We Have Never Been Modern.* Translated by Catherine Porter. Cambridge, MA: Harvard University Press, 1993.

Laughton, Charles D., John McManus, and Eugene G. D'Aquili. *Brain, Symbol, and Experience: Towards a Neurophenomenology of Human Consciousness.* New York: Columbia University Press, 1993.

Lave, Jean. *Apprenticeship in Critical Ethnographic Practice.* Chicago: University of Chicago Press, 2011.

Lave, Jean. *Learning and Everyday Life: Access, Participation, and Changing Practice.* Cambridge University Press, 2019.

Lave, Jean and Etienne Wenger. *Situated Learning: Legitimate Peripheral Participation.* New York: Cambridge University Press, 1991.

Lears, Jackson. *No Place of Grace: Antimodernism and the Transformation of American Culture, 1880–1920.* New York: Pantheon, 1981.

Lears, Jackson. *Fables of Abundance: A Cultural History of Advertising in America.* New York: Basic Books, 1994.

Lears, Jackson. *Rebirth of a Nation: The Making of Modern America, 1877–1920.* New York: Harper Collins, 2009.

Lemann, Nicholas. *The Promised Land: The Great Black Migration and How It Changed America.* New York: Knopf, 1991.

Lemke, Thomas. *Biopolitics: An Advanced Introduction.* New York: New York University Press, 2011.

Lemon, James T. "Los Angeles, 1950: The Working Class Thriving on Military Largesse." In *Liberal Dreams and Nature's Limits: Great Cities of North American Since 1600*, 191–241. Toronto: Oxford University Press, 1996.

Leonardi, Emanuele. *Biopolitics of Climate Change: Carbon Commodities, Environmental Profanations, and the Lost Innocence of Use-Value.* London, ON: University of Western Ontario, 2012.

Lerner, Michael. *Surplus Powerlessness: The Psychodynamics of Everyday Life and the Psychology of Individual and Social Transformation.* Lanham, MD: Humanities Press, 1998.

Levin, Miriam R. "Dynamic Triad: City, Exposition, and Museum in Industrial Society." In *Urban Modernity: Cultural Innovation in the Second Industrial Revolution,* Levin et al, 1–12. Cambridge, MA: MIT Press, 2010.

Levin, Miriam R., Sophie Forgan, Martina Hessler, Robert H. Kargon, and Morris Low. *Urban Modernity: Cultural Innovation in the Second Industrial Revolution.* Cambridge, MA: MIT Press, 2010.

Lewin, Kurt. "Defining the 'Field at a Given Time.'" (1943) *Field Theory in Social Science: Selected Theoretical Papers*, edited by Dorwin Cartwright, 43–59. New York: Harper & Row, 1951.

Lewin, Kurt. "Psychological Ecology." (1943) *Field Theory in Social Science: Selected Theoretical Papers*, edited by Dorwin Cartwright, 170–187. New York: Harper & Row, 1951.

Lewin, Kurt. "Behavior and Development as a Function of the Total Situation." (1946) *Field Theory in Social Science: Selected Theoretical Papers*, edited by Dorwin Cartwright, 238–303. New York: Harper & Row, 1951.

Lidskog, Rolf. "Invented Communities and Social Vulnerability: The Local Post-Disaster Dynamics of Extreme Environmental Events." *Sustainability* 10, no. 12 (2018): 4457. https://doi.org/10.3390/su10124457.

Lindenmeyer, Kriste. *A Right to Childhood: The US Children's Bureau and Child Welfare, 1912–46.* Champaign: University of Illinois Press, 1997.

Linenthal, Edward T. *The Unfinished Bombing: Oklahoma City in American Memory.* New York: Oxford University Press, 2001.

Liroff, Richard A. *A National Policy for the Environment: NEPA and Its Aftermath.* Bloomington: Indiana University Press, 1976.

Livingston, James. *Pragmatism and the Political Economy of Cultural Revolution, 1850–1940.* Chapel Hill: University of North Carolina Press, 1994.

Loomer, Bernard M. "Tillich's Theory of Correlation." *The Journal of Religion* 36, no. 3 (1956): 150–156.

Lotchin, Roger W. "The City and the Sword through the Ages and the Era of the Cold War." In *Essays on Sunbelt Cities and Recent Urban America*, edited by Robert B. Fairbanks and Kathleen Underwood, 87–134. College Station: Texas A&M Press, 1990.

Lotchin, Roger W. "World War II and Urban California: City Planning and the Transformation Hypothesis." *Pacific Historical Review* 62, no. 2 (1993): 143–171.

Luciano, Eugenio and Elena Zanoni. "Antonio Stoppani's 'Anthropozoic'in the Context of the Anthropocene." *The British Journal for the History of Science* 56, no. 1 (2023): 103–114.

Luria, Alexander R. *The Working Brain: An Introduction to Neuropsychology*. Translated by Basil Haigh. New York: Basic Books, 1976.

Luria, Alexander R. *Higher Cortical Functions in Man.* Translated by Basil Haigh. New York: Basic Books, 1980.

Luria, Alexander R. "L.S. Vygotsky and the Problem of Functional Localization." *Journal of Russian and East European Psychology* 40, no. 1 (2002): 17–25.

Lynd, Robert S. and Helen Merrell Lynd. *Middletown: A Study in Modern American Culture.* New York: Harcourt, Brace and World, 1929.

Ma, Jennifer and Matea Pender. *Education Pays 2023: The Benefits of Higher Education for Individuals and Society.* New York: College Board, 2023.

Maida, Carl A. *Pathways through Crisis: Urban Risk and Public Culture.* Lanham, MD: Rowman & Littlefield, 2008.

Maida, Carl A. "Expert and Lay Knowledge in Pacoima: Public Anthropology and an Essential Tension in Community-Based Participatory Action Research." *Anthropology in Action* 16, no. 2 (2009): 14–26.

Maida, Carl A., Norma S. Gordon, and Norman L. Farberow. *The Crisis of Competence: Transitional Stress and the Displaced Worker.* New York: Routledge, 2015.

Manganaro, Marc. *Culture, 1922.* Princeton: Princeton University Press, 2002.

Mann, Eric. *Taking On General Motors: A Case Study of the UAW Campaign to Keep GM Van Nuys Open.* Los Angeles: UCLA Institute of Industrial Relations, 1987.

Marchand, Roland. *Advertising and the American Dream: Making Way for Modernity, 1920–1940.* Berkeley: University of California Press, 1985.

Marcus, George E. "Collaborative Options and Pedagogical Experiment in Anthropological Research on Experts and Policy Processes." *Anthropology in Action* 15, no. 2 (2008): 47–57.

Marcus, George E. and Michael M. J. Fischer. *Anthropology as Cultural Critique: An Experimental Moment in the Human Sciences.* Chicago: University of Chicago Press, 1999.

Marcus, Greil. "Who Put the Bomp in the Bomp De-Bomp De-Bomp?." In *Rock 'n' Roll Will Stand*, edited by Greil Marcus, 6–27. Boston: Beacon Press, 1969.

Marcus, Greil. *Mystery Train: Image of America in Rock 'N' Roll Music*, fourth edition. New York: Penguin Books, 1997.

Marcuse, Herbert. *One Dimensional Man: Studies in the Ideology of Advanced Industrial Society.* Boston: Beacon Press, 1964.

Mariscal, George, editor. *Aztlán and Viet Nam: Chicano and Chicana Experiences of the War.* Berkeley: University of California Press, 1999.

Marsh, George Perkins. *Man and Nature.* Edited by David Lowenthal. Cambridge, MA: Harvard University Press, 1965.

Mascolo, Armando. "L'évasion de l'être. Jean-Paul Sartre and the Phenomenology of Temporality." In *The Concept of Time in Early Twentieth Century Philosophy*, edited by Flavia Santoianni, 77–84. Cham: Springer, 2016.

Mason, Paul. *PostCapitalism: A Guide to Our Future.* London: Allen Lane, 2015.

Massey, Jonathan and Brett Snyder. "Occupying Wall Street: Places and Spaces of Political Action." *Places Journal* (2012). https://placesjournal.org/article/occupying-wall-street-places- and-spaces-of-political-action/.

Maxson, Cheryl L. and Malcolm W. Klein. "'Play Groups' No Longer: Urban Street Gangs in the Los Angeles Region." In *From Chicago to LA: Making Sense of Urban Theory*, edited by Michael J. Dear, 239–266. Thousand Oaks, CA: Sage Publications, 2002.

Maxwell, Richard. "Surveillance: Work, Myth, and Policy." *Social Text* 23, no. 2 (2005): 1–19.

May, Elaine Tyler. *Great Expectations: Marriage and Divorce in Post-Victorian America*. Chicago: University of Chicago Press, 1980.

May, Elaine Tyler. *Homeward Bound: American Families in the Cold War Era*. New York: Basic Books, 2017.

May, Larry. *Screening Out the Past: The Birth of Mass Culture and the Motion Picture Industry*. New York: Oxford University Press, 1980.

Mayhew, Katherine Camp and Anna Camp Edwards. *The Dewey School: The Laboratory School of the University of Chicago, 1896–1903*. Chicago: D. Appleton Century, 1936.

McAnany, Patricia A. and E. Christian Wells. "Toward a Theory of Ritual Economy." In *Dimensions of Ritual Economy*, edited by E. Christian Wells and Patricia A. McAnany, *Research in Economic Anthropology*, Volume 27, 1–16. Bingley: Emerald Group Publishing, 2008.

McCarthy, Christine L. "Knowing Truth: Pierce's Epistemology in an Educational Context." *Educational Philosophy and Theory* 37, no. 2 (2005): 157–176.

McDermott, John J. *The Culture of Experience: Philosophical Essays in the American Grain*. New York: New York University Press, 1976.

McDermott, John J. *Streams of Experience*. Amherst: University of Massachusetts Press, 1986.

McDermott, John J. "A Relational World: The Significance of the Thought of William James and John Dewey for Global Culture." In *Streams of Experience: Reflections on the History and Philosophy of American Culture*, 107–126. Amherst: University of Massachusetts Press, 1986.

McDermott, John J. *The Drama of Possibility: Experience as Philosophy of Culture*. Edited by Douglas R. Anderson. New York: Fordham University Press, 2007.

McDermott, John J. "The Aesthetic Drama of the Ordinary." In *The Drama of Possibility: Experience as Philosophy of Culture*, edited by Douglas R. Anderson, 390–402. New York: Fordham University Press, 2007.

McKanan, Dan. *Eco-Alchemy: Anthroposophy and the History and Future of Environmentalism*. Berkeley: University of California Press, 2018.

McKenna, Brian. "Paulo Freire's Blunt Challenge to Anthropology: Create a Pedagogy of the Oppressed for Your Times." *Critique of Anthropology* 33, no. 4 (2013): 447–475.

McLaren, Peter. "Afterword: Public Pedagogy and the Challenge of Historical Time." In *Handbook of Public Pedagogy: Education and Learning Beyond Schooling*, edited by Jennifer A. Sandlin, Brian D. Schultz, and Jake Burdick, 648–651. New York: Routledge, 2010.

McLuhan, Marshall and Quentin Fiore. *War and Peace in the Global Village.* New York: Bantam Books, 1968.

McNeill, John R. and Peter Engelke. *The Great Acceleration An Environmental History of the Anthropocene since 1945.* Cambridge, MA: Harvard University Press, 2014.

McShane, Clay. *Down the Asphalt Path: The Automobile and the American City.* New York: Columbia University Press, 1994.

Mehrotra, Ajay K. "American Economic Development, Managerial Corporate Capitalism, and the Institutional Foundations of the Modern Income Tax." *Law and Contemporary Problems* 73, no. 1 (2010): 25–61.

Meier, August. *Negro Thought in America, 1880–1915: Racial Ideologies in the Age of Booker T. Washington.* Ann Arbor: University of Michigan Press, 1963.

Meltzoff, Andrew N., Patricia K. Kuhl, Javier Movellan, and Terrence J. Sejnowsky. "Foundations for a New Science of Learning." *Science* 325, no. 5938 (2009): 284–288.

Melucci, Alberto. "An End to Social Movements? Introductory Paper to the Sessions on 'New Social Movements and Change in Organizational Forms'." *Social Science Information* 23, no. 4–5 (1984): 819–835.

Melucci, Alberto. "The Voice of the Roots: Ethno-National Mobilizations in a Global World." *Innovation: The European Journal of Social Science Research* 3, no. 3 (1990): 351–363.

Melucci, Alberto. "Liberation or Meaning? Social Movements, Culture and Democracy." *Development and Change* 23, no. 3 (1992): 43–77.

Melucci, Alberto. "The Global Planet and the Internal Planet: New Frontiers for Collective Action and Individual Transformation." In *Cultural Politics and Social Movements*, edited by Marcy Darnovsky, Barbara Epstein, and Richard Flacks, 287–198. Philadelphia: Temple University Press, 1995.

Melucci, Alberto. *Challenging Codes: Collective Action in the Information Age.* Cambridge: Cambridge University Press, 1996.

Melucci, Alberto. "Social Movements in Complex Societies: A European Perspective." *Arena Journal* 15 (2000): 81–99.

Menand, Louis. *The Metaphysical Club.* New York: Farrar, Straus, and Giroux, 2001.

Menand, Louis. *The Free World: Art and Thought of the Cold War.* New York: Farrar, Straus and Giroux, 2021.

Merleau-Ponty, Maurice. *The Phenomenology of Perception.* Translated by Donald A. Landes. New York: Routledge, 2012.

Mills, C. Wright. *White Collar: The American Middle Classes.* New York: Oxford University Press, 1951.

Mills, C. Wright. *The Power Elite.* New York: Oxford University Press, 1956.

Mills, C. Wright. *The Sociological Imagination.* New York: Oxford University Press, 1959.

Mills, C. Wright. "Letter to the New Left." *New Left Review* 5 (1960): 18–23.

Mills, C. Wright. *Letters and Autobiographical Writings*. Edited by Kathryn Mills with Pamela Mills. Berkeley: University of California Press, 2000.

Mills, C. Wright. *Sociology and Pragmatism: The Higher Learning in America*. Edited by Irving Louis Horowitz. New York: Oxford University Press, 1964.

Mintrop, Heinrich and Gail L. Sunderman. "Predictable Failure of Federal Sanctions-Driven Accountability for School Improvement–And Why We May Retain It Anyway." *Educational Researcher* 38, no. 5 (2009): 353–364.

Mintrop, Rick, Elizabeth Zumpe, Kara Jackson, Drew Nucci, and Jon Norman. *Designing for Deeper Learning as an Equity Approach: Schools and School Districts Serving Communities Disadvantaged by the Educational System Benefiting from Deeper Learning*. Stanford, CA: The Carnegie Foundation for the Advancement of Teaching, 2022.

Monahan, Torin. *Globalization, Technological Change, and Public Education*. New York: Routledge, 2005.

Morgan, Gareth. *Images of Organization*. Thousand Oaks, CA: Sage Publications, 2006.

Morrison, Philip, Phylis Morrison and The Office of Charles, and Ray Eames. *Powers of Ten*. New York: Scientific American Books, 1982.

Müller, Francis. *Design Ethnography: Epistemology and Methodology*. Translated by Anna Brailovsky. Cham: Springer, 2021.

Mullis, Ina V. S., Michael O. Martin, Pierre Foy, Dana L. Kelly, and Bethany Fishbein. *TIMSS 2019 International Results in Mathematics and Science*. Chestnut Hill, MA: TIMSS & PIRLS International Study Center, Lynch School of Education and Human Development, Boston College and International Association for the Evaluation of Educational Achievement, 2020.

Mumford, Lewis. "Botched Cities." *American Mercury* 28 (October 1929): 143–150.

Mumford, Lewis. *The City in History*. New York: Harcourt, Brace and World, 1961.

Mumford, Lewis, editor. *Roots of Contemporary American Architecture*. New York: Dover Books, 1972.

Nader, Ralph. *Unsafe at Any Speed: The Designed-In Dangers of the American Automobile*. New York: Grossman Publishers, 1965.

Nasaw, David. *Children of the City: At Work and Play*. Garden City, NY: Doubleday, 1985.

Nash, Gerald D. *The American West Transformed: The Impact of the Second World War*. Bloomington: Indiana University Press, 1985.

National Academies of Sciences, Engineering, and Medicine. *Ontologies in the Behavioral Sciences: Accelerating Research and the Spread of Knowledge*. Washington, DC: The National Academies Press, 2022. https://doi.org/10.17226/26464.

National Academies of Sciences, Engineering, and Medicine. *The Future of Education Research at IES: Advancing an Equity-Oriented Science*. Washington, DC: The National Academies Press, 2022.

National Center for Education Statistics. *The Nation's Report Card: Science 2000*. NCES 2003–453. Edited by Christine Y. O'Sullivan, Mary A. Lauko, Wendy S. Grigg, Jiahe Qian, and Jinming Zhang. Washington, DC: U.S. Department of Education, Institute of Educational Sciences, 2002.

National Commission on Excellence in Education. *A Nation At Risk: The Imperative for Educational Reform: A Report to the Nation and the Secretary of Education.* Washington, DC: United States Department of Education, 1983.

National Research Council and Institute of Medicine. *Deadly Lessons: Understanding Lethal School Violence, Case Studies of School Violence Committee*. Edited by Mark H. Moore, Carol V. Petrie, Anthony A. Braga, and Brenda L. McLaughlin. Division of Behavioral and Social Sciences and Education. Washington, DC: National Academies Press, 2003.

Nauert, Charles B. *Humanism and the Culture of Renaissance Europe*. Cambridge: Cambridge University Press, 2006.

Nearing, Helen and Scott Nearing. *Living the Good Life: How to Live Sanely and Simply in a Troubled World.* New York: Schocken Books, 1973.

Niebuhr, Reinhold. "The Nature and Destiny of Man." *Theology Today* 1, no. 2 (1944): 236–254.

Niesche, Richard and Christina Gowlett. "Bernard Stiegler: Technics and Educational Leadership as a Form of *Psycho-Power*." In *Social, Critical and Political Theories for Educational Leadership*, 85–109. Singapore: Springer, 2019.

Niesz, Tricia. "Social Movement Knowledge and Anthropology of Education." *Anthropology & Education Quarterly* 50, no. 2 (2019): 223–234.

Niesz, Tricia, Aaron M. Korora, Christy Burke Walkuski, and Rachel E. Foot. "Social Movements and Educational Research: Toward a United Field of Scholarship." *Teachers College Record* 120, no. 3 (2018): 1–41.

Nonini, Donald. "Praxis." *Dialectical Anthropology* 40, no. 3 (2016): 241–249.

Olney, Martha L. *Buy Now, Pay Later: Advertising, Credit, and Consumer Durables in the 1920s*. Chapel Hill: University of North Carolina Press, 1991.

Ottenheimer, Harriet Joseph. *The Anthropology of Language: An Introduction to Linguistic Anthropology*. Belmont, CA: Wadsworth, 2013.

Orr, David W. *Earth in Mind: On Education, Environment, and the Human Prospect*. Washington, DC: Island Press, 2004.

Orr, Julian E. *Talking About Machines: Ethnography of a Modern Job*. Ithaca, NY: Cornell University Press, 1996.

Ortner, Sherry B. *New Jersey Dreaming: Capital, Culture, and the Class of '58*. Raleigh, NC: Duke University Press, 2002,

Overy, Richard. "Heralds of Modernity: Cars and Planes from Invention to Necessity." In *Fin de Siècle and its Legacy*, edited by Mikulás Teich and Roy Porter, 54–79. Cambridge: Cambridge University Press, 1990.

Page, Reba. "The Turn Inward in Qualitative Research." *Harvard Educational Review* 70, no. 1 (2000): 23–38.

Pappas, Gregory. *The Magic City: Unemployment in a Working-Class Community*. Ithaca, NY: Cornell University Press, 1989.

Passerin d'Entreves, Maurizio. "Hannah Arendt." *The Stanford Encyclopedia of Philosophy* (Fall 2022 edition). Edited by Edward N. Zalta & Uri Nodelman (2022). https://plato.stanford.edu/archives/fall2022/entries/arendt/.

Patocka, Jan. *Heretical Essays in the History of Philosophy*. Translated by Erazim Kohak. Edited by James Dodd. Chicago: Open Court, 1996.

Pearlman, Bob. "Designing, and Making, the New American High School." *Technos-Bloomington* 11, no. 1 (2002): 12–19.

Peirce, Charles Sanders. *Collected Papers of Charles Sanders Peirce, Volumes V and VI, Pragmatism and Pragmatacism and Scientific Metaphysics*. Edited by Charles Hartshorne and Paul Weiss. Cambridge, MA: Harvard University Press, 1935.

Petrus, Stephen and Ronald D. Cohen. *Folk City: New York and the American Folk Music Revival*. New York: Oxford University Press, 2015.

Pianta, Robert C., Jay Belsky, Renate Houts, and Fred Morrison. "Opportunities to Learn in America's Elementary Classrooms." *Science* 315, no. 5820 (2007): 1795–1796.

Pias, Claus. *Cybernetics: The Macy Conference 1946–1953, The Complete Transactions*. Zurich: Diaphanes, 2016.

Piketty, Thomas. *Capital in the Twenty-First Century*. Translated by Arthur Goldhammer. Cambridge, MA: Harvard University Press, 2014.

Piketty, Thomas. "Thomas Piketty Looks Back at the Success of *Capital in the 21st Century*." *The Tocqueville Review/La revue Tocqueville* 37, no. 2 (2016): 57–76.

Piketty, Thomas. *Capital and Ideology*. Translated by Arthur Goldhammer. Cambridge, MA: Harvard University Press, 2020.

Pitt, Leonard and Dale Pitt. *Los Angeles A to Z: An Encyclopedia of City and County*. Berkeley, CA: University of California Press, 1997.

Postman, Neil. *Amusing Ourselves to Death: Public Discourse in the Age of Show Business*. New York: Penguin Books, 1985.

Postman, Neil. "The Humanism of Media Ecology." *Proceedings of the Media Ecology Association* 1, no. 1 (2000): 10–16.

Powell Jr., Lewis Franklin. "The Memo" (1971). *Powell Memorandum: Attack On American Free Enterprise System*. 1. https://scholarlycommons.law.wlu.edu/powellmemo/1.

Powell, Walter W. and Kaisa Snellman. "The Knowledge Economy." *Annual Review of Sociology* 30 (2004): 199–220.

Project GRAD. *Project GRAD: Working to Close the Academic Achievement Gap*. Houston, TX: Project GRAD, 2003.

Public Policy Institute of California. *The Success of San Diego School Reforms Could Serve as a Blueprint for the State*, Research Brief, Issue 104. San Francisco: Public Policy Institute of California, 2005.

Pulver, Simone, William R. Burnside, Kathryn J. Fiorella, Meghan L. Avolio, and Steven M. Alexander. "Introduction: Foundations of Socio-Environmental Research." In *Foundations of Socio-Environmental Research*, edited by William R. Burnside, Simone Pulver, Kathryn J. Fiorella, Meghan L. Avolio, and Steven M. Alexander. Cambridge, MA: Cambridge University Press, 2023.

Pyne, Stephen J. *Fire: A Brief History*. Seattle: University of Washington Press, 2001.

Pyne, Stephen J. "From Pleistocene to Pyrocene: Fire Replaces Ice." *Earth's Future* 8, no. 11 (2020): e2020EF001722. https://doi. org/10.1029/2020EF001722.

Rabinow, Paul. *The Accompaniment: Assembling the Contemporary*. Chicago: Unversity of Chicago Press, 2011.

Rabinow, Paul and Nikolas Rose. "Biopower Today." *Biosciences* 1, no. 2 (2006): 195–217.

Radin, Paul. *Primitive Man as Philosopher*. New York: Dover Publications, 1957.

Radin, Paul. *The Road of Life and Death: A Ritual Drama of the American Indians*. Vol. 5. Princeton: Princeton University Press, 1991.

Ramussen, Wayne D. *Taking the University to the People: Seventy-Five Years of Cooperative Extension*. West Lafayette, IN: Purdue University Press, 2002.

Rappaport, Roy A. *Pigs for the Ancestors: Ritual in the Ecology of a New Guinea People*. New Haven: Yale University Press, 1984.

Rappaport, Roy A. *Religion and Ritual in the Making of Humanity*. Cambridge: Cambridge University Press, 1999.

Raymer, Annalisa Lewis. *Pedagogy of Place Facilitation Guide: A Workshop for Cultivating and Promoting Place-Based Education*. Lexington, KY: University of Kentucky Appalachian Center and the Appalachian Rural Education Network (AREN), 2001.

Reichenbach, Bruce, R. "Christianity, Science, and Three Phases of Being Human." *Zygon* 56, no. 1 (2020): 96–117.

Ritchhart, Ron. *Intellectual Character: What It Is, Why It Matters, and How to Get It*. San Francisco: Jossey-Bass, 2002.

Ritchhart, Ron. *Creating Cultures of Thinking: The 8 Forces We Must Master to Truly Transform Our Schools*. San Francisco: Jossey-Bass, 2015.

Riordan, Robert C. "Change the Subject: Making the Case for Project-Based Learning." *Edutopia*, January 17, 2013. https://www.edutopia.org/blog/21st-centuryskills- changing-subjects-larry-rosenstock-rob-riordan.

Riordan, Robert C. and Stacey Callier. "Schools as Equitable Communities of Inquiry." In *Sustainability, Human Well-Being, and the Future of Education*, edited by Justin W. Cook, 121–160. Cham: Palgrave Macmillan, 2019.

Riordan, Robert C. and Larry Rosenstock. "Changing the Subject: Making the Case for Project- Based Learning." *Edutopia: Education Trends* (2013). https://www.edutopia.org/blog/21st- century-skills-changing-subjects-larry-rosenstock-rob-riordan.

Rivers, W. H. R. "War Neurosis and Military Training." *Instinct and the Unconscious*, 208. Originally a report to the Medical Research Committee, London, and published in *Mental Hygiene* 2 (1918): 531–533.

Robinson, William I. *Global Capitalism and the Crisis of Humanity*. Cambridge: Cambridge University Press, 2014.

Robinson, William I. "Accumulation Crisis and Global Police State." *Critical Sociology* 45, no. 6 (2019): 845–858.

Rockoff, Hugh. *Until It's Over, Over There: The U.S. Economy in World War I*. Working Paper 10580. Cambridge, MA: National Bureau of Economic Research, 2004.

Rodger, Daniel T. *Atlantic Crossings: Social Politics in a Progressive Age*. Cambridge, MA: Harvard University Press, 1998.

Rogoff, Barbara. *The Cultural Nature of Human Development*. New York: Oxford University Press, 2002.

Rolfe, Brett. "Building an Electronic Repertoire of Contention." *Social Movement Studies* 4, no. 1 (2005): 65–74.

Roller, Bill. "Book Review: *The Social Unconscious in Persons, Groups, and Societies, Volume 3: The Foundation Matrix Extended and Re-Configured.*" *International Journal of Group Psychotherapy* 69, no. 3 (2019): 559–563.

Rome, Adam. *The Genius of Earth Day: How a 1970 Teach-In Unexpectedly Made the First Green Generation*. New York: Macmillan, 2013.

Roosevelt, Franklin D. Radio Address From Albany, New York: "The 'Forgotten Man' Speech." Online by Gerhard Peters and John T. Woolley. *The American Presidency Project*. https://www.presidency.ucsb.edu/node/288092.

Rosenstock, Larry. "Technology at High Tech High." 2004. http://www.bobpearlman.org/BestPractices/TechnologyatHighTechHigh.pdf.

Rosenstock, Larry. "High Tech High Design Principles." 2017. https://www.hightechhigh.org/about-us/.

Rothenberg, Jerome and Diane Rothenberg. *Symposium of the Whole: A Range of Discourse Toward an Ethnopoetics*. Berkeley: University of California Press, 1983,

Rubin, Lillian B. *Worlds of Pain: Life in the Working Class Family*. New York: Basic Books, 1976.

Ryan, Morgan, Gaël McGill, and Edward O. Wilson. *E. O. Wilson's Life on Earth*. Durham, NC: E. O. Wilson Biodiversity Foundation, 2014.

Rykwert, Joseph. "House and Home." *Social Research* 58, no. 1 (1991): 51–62.

Sable, Michael. "The Impact of the Biotechnology Industry on Local Economic Development in the Boston and San Diego Metropolitan Areas." *Technological Forecasting and Social Change* 74, no. 1 (2007): 36–60.

Sandlin, Jennifer A., Brian D. Schultz, and Jake Burdick. "Understanding and Exploring the Terrain of Public Pedagogy." In *Handbook of Public Pedagogy: Education and Learning Beyond Schooling*, edited by Jennifer A. Sandlin, Brian D. Schultz, and Jake Burdick, 1–6. New York: Routledge, 2010.

Sarason, Seymour B. *Revisiting "The Culture of the School and the Problem of Change."* New York: Teachers College Press, 1996.

Sartre, Jean-Paul. *Existentialism is a Humanism*. Translated by Carol Macomber. Edited by John Kulka. New Haven: Yale University Press, 2007.

Sartre, Jean-Paul. *Being and Nothingness*. Translated by Sarah Richmond. London: Routledge, 2018.

Schnaubelt, Christopher M. "Making BRAC Politically Palatable." *The RAND Blog* (March 16, 2017). https://www.rand.org/blog/2017/03/making-brac- politically palatable.html.

Schön, Donald A. *The Reflective Practitioner: How Professionals Think in Action*. New York: Basic Books, 1981.

Schorer, Mark. *William Blake: The Politics of Vision*. New York: Random House, 1959.

Schragger, Richard C. "The Limits of Localism." *Michigan Law Review* 100, no. 2 (2001): 371– 472.

Schulte, Joachim. "Wittgenstein on Time (1929–1933)." In *Time and History,* edited by Friedrich Stadler and Michael Stöltzner. Publications of the Austrian Ludwig Wittgenstein Society-New Series. Frankfurt: Ontos Verlag, 2006.

Sclove, Richard E. *Democracy and Technology*. New York: Guilford Press, 1995.

Seigfried, Charlene Haddock. "Socializing Democracy: Jane Addams and John Dewey." *Philosophy of the Social Sciences* 29, no. 2 (1999): 207–230.

Sernett, Milton C. *Bound for the Promised Land: African American Religion and the Great Migration*. Raleigh: Duke University Press, 1997.

Shah, Archit. "Emergency Economic Stabilization Act of 2008." *Harvard Journal on Legislation* 46 (2009): 569–584.

Shapiro, Michael J. "Every Move You Make: Bodies, Surveillance, and Media." *Social Text* 23, no. 2 (2005): 22–34.

Sheppard, Eric, Vinay Gidwani, Michael Goldman, Helga Leitner, Ananya Roy, and Anant Maringanti. "Introduction: Urban Revolutions in the Age of Global Urbanism." *Urban Studies* 52, no. 11 (2015): 1947–1961.

Shiller, Robert J. *The New Financial Order: Risk in the 21st Century*. Princeton: Princeton University Press, 2004.

Shor, Ira. "What is Critical Literacy?" *Journal of Pedagogy, Pluralism & Practice* 4, no. 1 (1999). Cambridge, MA: Lesley College. http://www.lesley.edu/journals/jppp/4/index.html.

Shultz, Benjamin. *Land of Abundance: A History of Settler Colonialism in Southern California*. Electronic Theses, Projects, and Dissertations. 985. https://scholarworks.lib.csusb.edu/etd/985 (2020).

Siemens, George. "Connectivism: A Learning Theory for the Digital Age." *International Journal of Instructional Technology and Digital Learning* 2, no. 1 (2005): 3–10.

Sifton, Elisabeth. *The Serenity Prayer: Faith and Politics in Times of Peace and War*. New York: Norton, 2003.

Silvert, Kalman H. *Expectant Peoples: Nationalism and Development*. New York: Random House, 1963.

Simmel, Georg. *Simmel on Culture: Selected Writings*. Edited by David Frisby and Mike Featherstone. London: Sage Publications, 1997.

Sizer, Theodore R. *Horace's Compromise: The Dilemma of the American High School: The First Report from a Study of High Schools, Co-sponsored by the National Association of Secondary School Principals and the Commission on Educational Issues of the National Association of Independent Schools*. Boston: Houghton Mifflin, 1984.

Sizer, Theodore R. *Horace's Hope: What Works for the American High School*. Boston: Houghton Mifflin, 1996.

Skelchy, Russell P. and Jeremy E. Taylor. *Sonic Histories of Occupation: Experiencing Sound and Empire in a Global Context*. Edited by Russell P. Skelchy and Jeremy E. Taylor. London: Bloomsbury Academic, 2022.

Sklar, Martin J. *The United States as a Developing Country: Studies in US History in the Progressive Era and the 1920s*. Cambridge: Cambridge University Press, 1992.

Smith, Joe, Robert Butler, Rosie J. Day, Axel H. Goodbody, David H. Llewellyn, Mel Rohse, Bradon T. Smith, Renata A. Tyszczuk, Julia Udall, and Nicola M. Whyte. "Gathering Around Stories: Interdisciplinary Experiments in Support of Energy System Transitions." *Energy Research & Social Science* 31 (2017): 284–294.

Smith, Kennedy, Kate Joncas, and Bill Parrish. *Revitalizing Downtown: The Professional's Guide to the Main Street Approach.* Washington, DC: National Main Street Center, National Trust for Historic Preservation, 1996.

Smith-Rosenberg, Carroll. *Disorderly Conduct: Visions of Gender in Victorian America.* New York: Knopf, 1985.

Snyder, Gary. *Earth House Hold.* New York: New Directions, 1969.

Snyder, Gary. "The Politics of Ethnopoetics." *Alcheringa: Ethnopoetics: A First International Symposium* 2 (1976): 13–22.

Sorrentino, Monica. "Space in Education." In *The Concept of Time in Early Twentieth Century Philosophy*, edited by Flavia Santoianni, 15–25. Cham: Springer, 2016.

Spear, Linda P. *The Behavioral Neuroscience of Adolescence.* New York: Norton, 2010.

Springsteen, Bruce. *Born to Run.* New York: Simon & Schuster, 2016.

Stanford University. The Martin Luther King Jr. Research and Education Institute. 2023. https://kinginstitute.stanford.edu/king-papers/documents/reinhold-niebuhr-1.

Stanford University. The Martin Luther King Jr. Research and Education Institute. 2023. https://kinginstitute.stanford.edu/encyclopedia/tillich-paul.

Stansell, Christine. *American Moderns: Bohemian New York and the Creation of a New Century.* New York: Metropolitan Books, 2000.

Starr, Kevin. *Material Dreams: Southern California through the 1920s.* New York: Oxford University Press, 1991.

Starr, Kevin. *The Dream Endures: California Enters the 1940s.* New York: Oxford University Press, 2002.

Stavrianakis, Anthony, Paul Rabinow, and Trine Mygind Korsby. "In the Workshop: Anthropology in a Collaborative Zone of Inquiry." In *The Composition of Anthropology: How Anthropological Texts are Written*, edited by Morten Nielsen and Nigel Rapport, 169–192. London: Routledge, 2018.

Steffen, Will, Katherine Richardson, Johan Rockström, Sarah E. Cornell, Ingo Fetzer, Elena M. Bennett, Reinette Biggs, Stephen R. Carpenter, Wim de Vries, Cynthia A. de Wit, Carl Folke, Dieter Gerten, Jens Heinke, Georgina M. Mace, Linn M. Persson, Veerabhadran Ramanathan, Belinda Reyers, and Sverker Sörlin. "Planetary Boundaries: Guiding Human Development on a Changing Planet." *Science* 347 (2015): 6223.

Steinbeck, John. *The Log from the Sea of Cortez.* New York: The Viking Press, 1951.

Steiner, George. *Grammars of Creation.* New Haven: Yale University Press, 2002.

Stemhagen, Kurt and David Waddington. "Beyond the 'Pragmatic Acquiescence' Controversy: Reconciling the Educational Thought of Lewis Mumford and John Dewey." *Educational Studies*, 47, no. 5 (2011): 469–489.

Stephen, David and Eve Goldberg. *Profile: High Tech High Network—Student-Centered Learning in Action.* Quincy, MA: Nellie Mae Education Foundation, 2013. https://www.nmefoundation.org/getmedia/aa1d06c3-dd1d-460b-b99f-5b283d03e008/ PROFILE- HighTechHighNetwork-NMEF.

Stevenson, Douglas. *Out to Change the World: The Evolution of the Farm Community.* Summertown, TN: Book Publishing Company, 2014.

Stevis, Dimitris and Romain Felli. "Planetary Just Transition? How Inclusive and How Just?" *Earth System Governance* 6 (2020): 100065.

Stewart, Julie Pearson and Bárbara M. Brizuela. "Habits of Thought and Work—The Disciplines and Qualitative Research: An Introduction." *Harvard Educational Review* 70, no. 1 (2000): 22.

Stiegler, Bernard. *The Age of Disruption: Technology and Madness in Computational Capitalism.* Translated by Daniel Ross. London: Polity, 2019.

Stocking, Jr., George W. "Franz Boas and the Culture Concept in Historical Perspective." In *Race, Culture, and Evolution: Essays in the History of Anthropology*, 195–233. Chicago: University of Chicago Press, 1982.

Stroup, Herbert. *Social Welfare Pioneers.* Chicago: Nelson-Hall Publishers, 1986.

Students of the Gary and Jerri-Ann Jacobs High Tech High. *The Two Sides of the Boat Channel: A Field Guide.* San Diego: BuenFeVA Press, 2004.

Students of the Gary and Jerri-Ann Jacobs High Tech High. *Perspectives of San Diego Bay: A Field Guide.* Providence: Next Generation Press, 2005.

Students of the Gary and Jerri-Ann Jacobs High Tech High. *San Diego Bay: A Story of Exploitation and Restoration.* El Cajon, CA: Sunbelt Publications, 2007.

Students of the Gary and Jerri-Ann Jacobs High Tech High. *San Diego Bay: A Call for Conservation.* El Cajon, CA: Sunbelt Publications, 2009.

Students of the Gary and Jerri-Ann Jacobs High Tech High. *Biomimicry: Respecting Nature through Design.* San Diego: High Tech High, 2014.

Susman, Warren I. "History and the American Intellectual: Uses of a Usable Past." *American Quarterly* 16, no. 2 Part 2: Supplement (1964): 243–263.

Taffel, Sy. "Escaping Attention: Digital Media Hardware, Materiality and Ecological Cost." *Culture Machine* 13 (2012): 1–28.

Taffel, Sy. "Scalar Entanglement in Digital Media Ecologies." *European Journal of Media Studies* 2, no. 1 (2013): 233–254.

Taipale, Joona. *Phenomenology and Embodiment: Husserl and the Constitution of Subjectivity.* Evanston: Northwestern University Press, 2014.

Taylor, Charles. *Sources of the Self: The Making of Modern Identity.* Cambridge, MA: Harvard University Press, 1989.

Taylor, Charles. *A Secular Age.* Cambridge, MA: Harvard University Press, 2007.

Taylor, Charles, Patricia Nanz, and Madeleine Beaubien Taylor. *Reconstructing Democracy: How Citizens Are Building from the Ground Up.* Cambridge, MA: Harvard University Press, 2020.

Teich, Mikulás and Roy Porter, editors. *Fin de Siècle and Its Legacy.* Cambridge: Cambridge University Press, 1990.

Thayer-Bacon, Barbara J. "An Exploration of Myles Horton's Democratic Praxis: Highlander Folk School." *Educational Foundations* 18, no. 2 (2004): 5–23.

Thomas, Christopher D. *Reclaiming Democratic Education: Student and Teacher Activism and the Future of Education.* New York: Teachers College Press, 2022.

Thomas, Jr., William I., Carl Sauer, Marston Bates, and Lewis Mumford, editors. *Man's Role in Changing the Face of the Earth, Volumes 1 and 2.* Chicago: University of Chicago Press, 1956.

Thompson, Evan. *Mind in Life: Biology, Phenomenology, and the Sciences of Mind.* Cambridge, MA: Harvard University Press, 2007.

Thoreau, Henry David. *Walden.* Edited by J. Lyndon Shanley. Princeton, NJ: Princeton University Press, 1971.

Thornhill, Chris and Ronny Miron. "Karl Jaspers." *The Stanford Encyclopedia of Philosophy*, edited by Edward N. Zalta (Spring 2022 edition), 2022. https://plato.stanford.edu/archives/spr2022/entries/jaspers/.

Tillich, Paul. *The Courage To Be.* New Haven: Yale University Press, 1952.

Tilly, Charles. *Big Structures Large Processes Huge Comparisons.* New York: Russell Sage Foundation, 1984.

Tilly, Charles. "The Trouble with Stories." In *The Social Worlds of Higher Education: Handbook for Teaching in a New Century*, edited by Ronald Aminzade and Bernice Pescosolido, 256–270. Thousand Oaks, CA: Pine Forge Press, 1999.

Tilly, Charles. "Spaces of Contention." *Mobilization: An International Journal* 5, no. 2 (2000): 135–159.

Tilly, Charles and Sidney Tarrow. *Contentious Politics.* New York: Oxford University Press, 2015.

Touraine, Alain. *Return of the Actor.* Minneapolis: University of Minnesota Press, 1988.

Touraine, Alain. "History, Modernity and Global Identities." *Glocalism: Journal of Culture, Politics and Innovation* 2 (2020): 1–9.

Trachtenberg, Alan. *The Incorporation of America: Culture and Society in the Gilded Age.* New York: Hill and Wang, 1982.

Trotter Jr., William, editor. *The Great Migration in Historical Perspective: New Dimensions of Race, Class, and Gender.* Bloomington: Indiana University Press, 1991.

Turner, Fred. *From Counterculture to Cyberculture: Stewart Brand, the Whole Earth Network and the Rise of Digital Utopianism.* Chicago: University of Chicago Press, 2006.

Turner, Victor W. *The Ritual Process.* Chicago: Aldine, 1969.

Turner, Victor W. "Dewey, Dilthey, and Drama: An Essay in the Anthropology of Experience." In *The Anthropology of Experience*, edited by Victor W. Turner and Edward M. Bruner, 33–44. Urbana: University of Illinois Press, 1986.

Turner, Billie Lee, William C. Clark, Robert W. Kates, John F. Richards, Jessica T. Mathews, and William B. Meyer, editors. *The Earth as Transformed by Human Action.* New York: Cambridge University Press, 1993.

Twain, Mark and Charles Dudley Warner. *The Gilded Age.* New York: New American Library, 1969.

Tyack, David and Larry Cuban. *Tinkering Toward Utopia: A Century of Public School Reform.* Cambridge, MA: Harvard University Press, 1995.

United States Environmental Protection Agency. "Superfund Site Information: San Diego Naval Training Center." https://cumulis.epa.gov/supercpad/cursites/csitinfo.cfm?id=0902730.

United Way of Greater Los Angeles. *Los Angeles County Ten Years Later: A Tale of Two Cities – One Future.* Los Angeles, CA: United Way of Greater Los Angeles

2010. https://www.yumpu.com/en/document/view/45985294/la-county-10-years-later-a-tale-of-two- cities-one-future.

US Census Bureau. *Statistical Abstract of the United States: 2000.* Washington, DC: US Census Bureau, 2000.

Varela, Francisco J., Evan Thompson, and Eleanor Rosch. *The Embodied Mind: Cognitive Science and Human Experience.* Cambridge, MA: MIT Press, 2016.

Vavra, Jay. "The San Diego Bay Study: Community-Based Conservation." In *Proceedings of the Pacific Division, American Association for the Advancement of Science,* Volume 28, Part. 72. Ashland, OR: American Association for the Advancement of Science, Pacific Division, 2009.

Verbeek, Peter-Paul. "Don Ihde: The Technological Lifeworld." In *American Philosophy of Technology: The Empirical Turn,* edited by Herman John Achterhuis, 119–146. Bloomington: Indiana University Press, 2001.

Verovšek, Peter J. "Integration after Totalitarianism: Arendt and Habermas on the Postwar Imperatives of Memory." *Journal of International Political Theory* 16, no. 1 (2020): 2–24.

Vidich, Arthur J. and Stanford M. Lyman. *American Sociology: Worldly Rejections of Religion and Their Directions.* New Haven: Yale University Press, 1985.

Visser-Maessen, Laura. *Robert Parris Moses: A Life in Civil Rights and Leadership at the Grassroots.* Chapel Hill: University of North Carolina Press, 2016.

Vygotsky, Lev S. "The Development of Concept Formation in Adolescence." In *The Vygotsky Reader*, edited by René Van der Veer and Jaan Valsiner, 2660287. Oxford: Blackwell, 1994.

Vygotsky, Lev S. *The History of the Development of Higher Mental Functions, The Collected Works of L. S. Vygotsky: Vol. 4.* Translated by Marie J. Hall. Edited by Robert W. Rieber and Aaron S. Carton. New York: Plenum Press, 1997.

Wagner, John R. "Water and the Commons Imaginary." *Current Anthropology* 53, no. 5 (2012): 617–641.

Wakefield, Stephanie. *Anthropocene Back Loop: Experimentation in Unsafe Operating Space.* London: Open Humanities Press, 2020.

Wallace, Anthony F. C. "Revitalization Movements." *American Anthropologist* 58, no. 2 (1956): 264–281.

Wallace, Anthony F. C. "Mazeway Resynthesis: A Biocultural Theory of Religious Inspiration." *Transactions of the New York Academy of Sciences* 18, no. 7 Series II (1956): 626–638.

Wallace, Anthony F. C. "Mazeway Disintegration: The Individual's Perception of Socio- Cultural Disorganization." *Human Organization* 16, no. 2 (1957): 23–27.

Wallace, Anthony F. C. *The Death and Rebirth of the Seneca.* New York: Knopf, 1970.

Wallace, Anthony F. C. "Paradigmatic Processes in Culture Change." *American Anthropologist* 74, no. 3 (1972): 467–478.

Wallace, Anthony F. C. *Rockdale: The Growth of an American Village in the Early Industrial Revolution.* New York: Knopf, 1978.

Wallerstein, Immanuel. "A World-System Perspective on the Social Sciences." *The British Journal of Sociology* 27, no. 3 (1976): 343–352.

Ward, David and Olivier Zunz, editors. *The Landscape of Modernity: New York City, 1900–1940.* Baltimore: Johns Hopkins University Press, 1992.

Watkins, Trevor. "Architecture and 'Theatres of Memory' in the Neolithic of Southwest Asia." In *Rethinking Materiality: The Engagement of Mind with the Material World*, edited by Elizabeth DeMarrais, Chris Gosden, and Colin Renfrew, 97–106. Cambridge: McDonald Institute of Archaeological Research, 2004.

Watkins, Trevor. "Architecture and the Symbolic Construction of New Worlds." In *Domesticating Space: Construction, Community, and Cosmology in the Late Prehistoric Near East*, edited by E. B. Banning and Michael Chazan. Studies in Early Near Eastern Production, Subsistence, and Environment, 6, 15–23. Berlin: *ex oriente,* 2006.

Wax, Murray L., Stanley Diamond, and Fred O. Gearing. *Anthropological Perspectives on Education.* New York: Basic Books, 1971.

Weber, Max. *The Protestant Ethic and the Spirit of Capitalism.* Translated by Talcott Parsons. New York: Scribner, 1958.

Weber, Max. *Economy and Society: An Outline of Interpretive Sociology.* Edited by Guenther Roth and Claus Wittich. Berkeley: University of California Press, 1978.

Wenger, Etienne. "Communities of Practice and Social Learning Systems: The Career of a Concept." In *Social Learning Systems and Communities of Practice*, edited by Chris Blackmore. London: Springer, 2010, 180.

Wenger, Etienne and William M. Snyder. "Communities of Practice: The Organizational Frontier." *Harvard Business Review* 78, no. 1 (2000): 139–146.

Westbrook, Robert B. *John Dewey and American Democracy.* Ithaca: Cornell University Press, 1991.

Wheeler, Gregory. "Bounded Rationality." In *The Stanford Encyclopedia of Philosophy.* Edited by Edward N. Zalta. (Fall 2020 edition) 2020. https://plato.stanford.edu/archives/fall2020/entries/bounded-rationality/.

Wiebe, Robert H. *The Search For Order, 1877–1920.* New York: Hill and Wang, 1967.

Wilkinson, Richard and Kate Pickett. *The Inner Level: How More Equal Societies Reduce Stress, Restore Sanity and Improve Everyone's Well-Being.* New York: Penguin, 2019.

Williams, Raymond. "Brecht and Beyond." In *Politics and Letters: Interviews with New Left Review*, 214–234. London: Verso, 2015.

Willis, Paul. *Learning to Labor: How Working Class Kids Get Working Class Jobs.* New York: Columbia University Press, 1977.

Willis, Paul. "Foot Soldiers of Modernity: The Dialectics of Cultural Consumption and the 21st Century School." *Harvard Educational Review* 73, no. 3 (2003): 390–415.

Wilson, Edward O. "The Encyclopedia of Life." *Trends in Ecology & Evolution* 18, no. 2 (2003): 77–80.

Wilson, Richard Guy. "America and the Machine Age." In *The Machine Age in America, 1918–1941*, edited by Richard Guy Wilson, Dianne H. Pilgrim, and Dickran Tashjian, 23–41. New York: Abrams, 1996.

Winner, Langdon. *Autonomous Technology: Technics-Out-of-Control as a Theme in Political Thought.* Cambridge, MA: MIT Press, 1978.

Wittgenstein, Ludwig. *Tractatus Logico-Philosophicus.* Translated by David F. Pears and Brian F. McGuinness. London: Routledge & Kegan Paul, 1961.

Wohl, Robert. *The Generation of 1914.* Cambridge, MA: Harvard University Press, 1979.

Wolf, Eric R. *Anthropology.* Englewood Cliffs, NJ: Prentice-Hall, 1964.

Wolf, Eric R. *Europe and the People Without History.* Berkeley: University of California Press, 1982.

Wolfe, Tom. "The New Life Out There." *New York Herald Tribune* (November 21, 1965). Reprinted in *McLuhan: Hot and Cool.* Edited by Gerald Emanuel Stearn, 30–48. New York: The New American Library, 1969.

Wolff, Geoffrey. *Black Sun: The Brief Transit and Violent Eclipse of Harry Crosby.* New York: Random House, 1976.

Wright, Frank Lloyd. "The Art and Craft of the Machine." In *Roots of Contemporary American Architecture*, edited by Lewis Mumford, 169–185. New York: Dover Books, 1972.

Wyborn, Carina, Jasper Montana, Nicole Kalas, F. Davila Cisneros, Sarah Clement, S. Izquierdo Tort, Natalie Knowles et al. *Research and Action Agenda for Sustaining Diverse and Just Futures for Life on Earth: Biodiversity Revisited.* Gland: Luc Hoffmann Institute, 2020.

Young, Michael P. "A Revolution of the Soul: Transformative Experiences and Immediate Abolition." In *Passionate Politics: Emotions and Social Movements*, edited by Jeff Goodwin, James M. Jasper, and Francesca Polletta, 199–214. Chicago: University of Chicago Press, 2001.

Young, Michael P. "Confessional Protest: The Religious Birth of U.S. National Social Movements." *American Sociological Review* 67, no. 5 (2002): 660–688.

Young, Michael P. *Bearing Witness Against Sin: The Evangelical Birth of the American Social Movement.* Chicago: University of Chicago Press, 2006.

Zilberman, David B. and Susan Layton. "Culture-Historical Reconstruction and Mythology in the Anthropology of Paul Radin." *Dialectical Anthropology* 6, no. 4 (1982): 275–290.

Žižek, Slavoj. "Occupy Wall Street: What is to Be Done Next?" *The Guardian* (April 24, 2012). https://www.theguardian.com/commentisfree/cifamerica/2012/apr/24/occupy- wall-street-what-is-to-be-done-next.

Zucker, Lynne G., Michael R. Darby, and Marilynn B. Brewer. "Intellectual Human Capital and the Birth of U.S. Biotechnology Enterprises." *American Economic Review* 88, no. 1 (1998): 290–306.

Zucker, Robert A., Raul Gonzalez, Sarah W. Feldstein Ewing, Martin P. Paulus, Judith Arroyo, Andrew Fuligni, Amanda Sheffield Morris, Mariana Sanchez, and Thomas Wills. "Assessment of Culture and Environment in the Adolescent Brain and Cognitive Development Study: Rationale, Description of Measures, and Early Data." *Developmental Cognitive Neuroscience* 32 (2018): 107–120.

Index

About the Author

Carl A. Maida is a professor emeritus of public and population health at the University of California, Los Angeles (UCLA), where he currently teaches seminars on the anthropology of biomedicine, global health, and scientific research ethics in the UCLA Graduate Division. He holds a joint faculty appointment at the Institute of the Environment and Sustainability in the UCLA College of Letters and Science, where he teaches courses on action research methods in sustainability. He has carried out school- and community-based studies of intergenerational participatory action research and project-based learning, and also participatory evaluation research on treatment of complex trauma effects in multiply traumatized, socially marginalized adolescents. He currently studies ethnic cultural disparities in health care, focusing on child and adolescent health in clinical and school-based settings. He directed the UCLA Pre-College Science Education Program, a two-year high-school apprenticeship program for gifted African American, and Latino, African, East Asian, and Pacific Island second-generation immigrant youth. His recent books, *Sustainability and Communities of Place* (2007) and *Common Worlds: Paths Toward Sustainable Urbanism* (2019), explore sustainable development as a local practice worldwide. He serves as a member of the UCLA Campus Sustainability Committee and is also a member of the Commission on Anthropology and Education of the International Union of Anthropological and Ethnological Sciences. He is a Fellow of the American Association for the Advancement of Science, the American Anthropological Association, and the Society for Applied Anthropology.